Raising Farm Animals: Livestock Management

Raising Farm Animals: Livestock Management

Edited by Roger Greer

SYRAWOOD
PUBLISHING HOUSE
New York

Published by Syrawood Publishing House,
750 Third Avenue, 9th Floor,
New York, NY 10017, USA
www.syrawoodpublishinghouse.com

Raising Farm Animals: Livestock Management
Edited by Roger Greer

International Standard Book Number: 978-1-68286-663-4 (Hardback)

Cataloging-in-Publication Data

Raising farm animals : livestock management / edited by Roger Greer.
p. cm.
Includes bibliographical references and index.
ISBN 978-1-68286-663-4
1. Livestock. 2. Livestock--Breeding. 3. Animal culture. I. Greer, Roger.
SF65.2 .R35 2019
636--dc23

TABLE OF CONTENTS

PREFACE

Livestock management is an integral part of animal husbandry. Livestock are the domesticated animals raised on a farm or an agricultural setting to obtain products like fur, meat, milk, eggs, leather, etc. or for the purpose of farm labour. Livestock management encompasses the complete well-being and health of all farm animals. This book presents researches and studies performed by experts across the globe. The chapters compiled herein also shed light on adoption of new methods and practices as a result of scientific advancement over the years. Thus, from theories to research to practical applications, case studies related to all contemporary topics of relevance to this field have been included in this book. It will prove to be immensely beneficial to students and researchers in this field.

Various studies have approached the subject by analyzing it with a single perspective, but the present book provides diverse methodologies and techniques to address this field. This book contains theories and applications needed for understanding the subject from different perspectives. The aim is to keep the readers informed about the progresses in the field; therefore, the contributions were carefully examined to compile novel researches by specialists from across the globe.

Indeed, the job of the editor is the most crucial and challenging in compiling all chapters into a single book. In the end, I would extend my sincere thanks to the chapter authors for their profound work. I am also thankful for the support provided by my family and colleagues during the compilation of this book.

Editor

1

Whip Rule Breaches in a Major Australian Racing Jurisdiction: Welfare and Regulatory Implications

Jennifer Hood [1,*], Carolyn McDonald [1], Bethany Wilson [1], Phil McManus [2] and Paul McGreevy [1]

[1] Faculty of Veterinary Science, University of Sydney, Room 206, R.M.C. Gunn Building, Sydney 2006, New South Wales, Australia; carolyn131283@gmail.com (C.M.); bethany.wilson@sydney.edu.au (B.W.); paul.mcgreevy@sydney.edu.au (P.M.)

[2] School of Geosciences, University of Sydney, Room 435, F09, Madsen Building, Sydney 2006, New South Wales, Australia; phil.mcmanus@sydney.edu.au

* Correspondence: jenih@iinet.net.au

Academic Editor: Clive J. C. Phillips

Simple Summary: An evidence-based analysis of whip rule breaches in horse racing is needed to address community expectations that racehorses are treated humanely. The study provides the first peer-reviewed characterisation of whip rule breaches and their regulatory outcomes in horseracing, and considers the relationship between rules affecting racing integrity and the welfare of racehorses in a major Australian racing jurisdiction.

Abstract: Whip use in horseracing is increasingly being questioned on ethical, animal welfare, social sustainability, and legal grounds. Despite this, there is weak evidence for whip use and its regulation by Stewards in Australia. To help address this, we characterised whip rule breaches recorded by Stewards using Stewards Reports and Race Diaries from 2013 and 2016 in New South Wales (NSW) and the Australian Capital Territory (ACT). There were more recorded breaches at Metropolitan (M) than Country (C) or Provincial (P) locations, and by riders of horses that finished first, second, or third than by riders of horses that finished in other positions. The most commonly recorded breaches were forehand whip use on more than five occasions before the 100-metre (m) mark (44%), and whip use that raises the jockey's arm above shoulder height (24%). It is recommended that racing compliance data be analysed annually to inform the evidence-base for policy, education, and regulatory change, and ensure the welfare of racehorses and racing integrity.

Keywords: animal welfare; horseracing; racing integrity; whip rule breaches; whip use

---

1. Introduction

The use of whips in horse racing is increasingly being challenged on ethical, welfare [1–11], social sustainability [10], and legal grounds [8]. Industry proponents argue "acceptable use of the whip ... means that the whip is used for safety (of both jockey and horse) or to encourage the horse to perform to its best when in contention" [12] (p. 5). The ability of the whip to achieve these goals remains unproven [8], while there is evidence that striking a horse with a padded racing whip would be at least aversive and at worst, possibly painful [4,13,14]. An Australian study by McGreevy et al. [4] found 83% of whip strikes caused indentations of the skin of the horses whipped, and comparative studies in mice and humans showed such deformation is likely to be detected by cutaneous nociceptors [13], as did a recent study in horses [14].

Increasingly, the community is concerned with the humane treatment of animals and, indeed, there are growing calls for the whip to be banned [15,16]. There has been strong criticism of whip use by some experienced industry observers [17], and RSPCA Australia "is opposed to the use of whips on racehorses for the purpose of enhancing performance as they inflict pain and distress" [18].

Self-regulation of the Australian horseracing industry means Stewards enforce the whip rules, including interpreting and determining penalties. Since 2009, there have been a number of controversial reforms made to whip rules in Australia. While these have sought to restrict the number and type of whip strikes allowed, like the whip rules, they are not evidence-based. Indeed, there is a paucity of published data on whip use in Australia (and worldwide), including regulatory outcomes.

In "Responsible Regulation: A Review of the use of the whip in Horseracing" (UK Review) [19], a statistical analysis of whip offences was provided. In their critique of this report, Jones et al. [8] argued the UK Review was not "a "scientific", peer-reviewed paper" and concluded "further independent scientific review is needed to reach definitive conclusions about whip use on racehorse welfare". In its defence, the UK Review at least provided some useful statistics on whip use in the UK that, up until the current study, have not been readily available for racing in Australia.

Our study examined data from Racing NSW Stewards Reports (Stewards Reports) [20] and Racing NSW Race Diary (Race Diary) [21] for 2013 (and for 4 months in 2016) in New South Wales (NSW) and the Australian Capital Territory (ACT), which is a major horseracing jurisdiction in Australia. To the authors' knowledge, this is the first peer-reviewed study to characterise whip use from a regulatory and welfare perspective.

2. Materials and Methods

2.1. Data Sources

Stewards Reports [20] 1 January 2013–31 December 2013, and 1 January 2016–30 April 2016, were examined to identify numbers and types of breaches of the Australian Rules of Racing (AR) pertaining to whip use and their regulatory outcomes, in each race at every Racing NSW race meeting. Other information, including Racing NSW Breeder Owner Bonus Scheme (BOBS) [22], prize money details, and lengths of races, was obtained from the Race Diary [21].

2.2. Other Terms Used

BOBS refers to the BOBS in NSW and includes BOBS Extra incentive scheme [22];

breach refers to a breach of the AR whip rules recorded in the Stewards Reports, as distinct from a breach that may have occurred but was not recorded;

breach code refers to the code we used to denote each whip rule (see Table 1);

first breach refers to the first breach recorded by the Stewards in a start (see below);

location refers to whether the race track was at a Country (C), Metropolitan (M), or Provincial (P) location as defined in the NSW Local Rules of Racing (LR); C includes races in the ACT (these were excluded from the BOBS analysis as the BOBS status of races in the ACT is not routinely provided in the Race Diary);

prize money won refers to the total prize money won by an individual horse in a race;

riders refers to both jockeys and apprentices;

repeat offender refers to a rider who has more than one whip rule breach recorded in 2013, and includes those with second breaches (see below);

Rules of Racing refers to Racing NSW's Rules of Racing and includes the AR and the LR [23];

second breach refers to a breach following a first breach in a start;

start refers to a start by an individual horse in a race;

the 100-metre (m) mark refers to the start of the final 100 m before the finishing post;

total prize money on offer refers to the total prize money offered for a race (including all place getters);

whip rules refers to AR 137A. (1)–(9) current in 2013 (see Table 1 for breach codes used);

$ refers to Australian dollars.

Table 1. Whip rules breached in 348 first breaches.

Breach Code	Whip Rules Breached	Whip Rule Description	Number (out of 348)	Percentage
1	AR 137A. (1)(a)	Only padded whips of a design and specifications approved by a panel appointed by the ARB may be carried in races or official trials;	0	0
2	(1)(b)	Every such whip must be in satisfactory condition and not modified in any way;	0	0
3	(1)(c)	Stewards may confiscate any whip not in a satisfactory condition or modified;	0	0
4	(1)(d)	Any rider found guilty of above may be penalised. Provided the master and/or other person in charge of an apprentice jockey at the relevant time may also be penalised unless they satisfy Stewards all proper care to ensure the apprentice complied with the rule was taken;	0	0
5	AR 137A. (2)	Only whips of a design and specifications approved by ARB may be carried in track work;	0	0
6	AR 137A. (3)	Excessive/unnecessary/improper whip use;	10	2.87
7	AR 137A. (4)(a)	Whip use forward of horse's shoulder/vicinity of head;	6	1.72
8	(4)(b)	Whip use that raises arm above jockey's shoulder height;	87	25.00
9	(4)(c)	Whip use when horse is out of contention;	44	12.64
10	(4)(d)	Whip use when horse is showing no response;	0	0
11	(4)(e)	Whip use after passing winning post;	0	0
12	(4)(f)	Whip use causing injury to horse;	0	0
13	(4)(g)	Whip use when horse is clearly winning;	1	0.29
14	(4)(h)	Whip use when horse has no reasonable prospect of improving/losing position;	1	0.29
15	(4)(i)	Whip use in such manner that the seam of the flap is the point of contact with horse, unless rider satisfies Stewards this was neither deliberate nor reckless;	0	0
16	AR 137A. (5)(a)(i)	Forehand whip use * in consecutive strides prior to 100 m mark;	33	9.48
17	(5)(a)(ii)	Forehand whip use * on more than 5 occasions prior to 100 m mark;	157	45.11
18	(5)(a)(iii)	Rider may at his discretion use whip with slapping motion down horse's shoulder, with whip hand remaining on reins, or alternatively in a backhand manner;	0	0
19	(5)(b)	In the final 100 m a rider may use whip at his discretion;	0	0
20	AR 137A. (7)(a)	Any trainer, owner or authorised agent must not give instructions to a rider regarding the use of the whip, which, if carried out, might result in a breach of this rule;	0	0
21	(7)(b)	No person may offer inducements to a rider to use the whip in such a way that, if carried out, might result in a breach of this rule;	0	0
22	AR 137A. (8)	Any person who fails to comply with any provisions of this rule is guilty of an offence;	0	0
23	AR 137A. (9)	An owner or authorised representative, trainer, rider or Steward may lodge an objection against the placing of a horse where the rider contravenes AR 137A. (3) or (5);	0	0
24	Unspecified whip use	Whip Rule breach not specified by Stewards.	9	2.59

The wording of the whip rules has been simplified for the purposes of this article (for exact wording, see Rules of Racing [23]; * Whip rules were amended 1 December 2015 to include backhand whip strikes as well as existing forehand restrictions; ARB—Australian Racing Board (now Racing Australia).

2.3. Data Analysis

Data were analysed using R Core Team statistical and computing software (R Foundation for Statistical Computing, Vienna, Austria) [24]. Associations between categorical variables were explored using χ^2 tests for independence with a p value of less than 0.05 as the cut-off for rejecting the null hypothesis. For some analyses, following the rejection of the null hypothesis, additional pairwise χ^2 tests were employed using a more stringent p value (0.05/m, where m is the number of post-hoc hypotheses being tested) to ensure that the familywise type 1 error rate remained acceptable. χ^2 tests were undertaken using the "chisq.test" function of *r*, which includes a Yates Continuity Correction for 2×2 tables but not for other $i \times j$ tables. When one or more cell in the table of expected frequencies was <5, a Fisher's exact test was performed instead, using the function "fisher.test". Where pertinent, adjusted standardised residuals were calculated manually.

In addition to categorical variables, three continuous variables (race length, prize money won, and total prize money on offer) were examined for an association with breach code of the first breach among starts with breaches. Race length was examined by ANOVA using the r functions "anova" and "lm". After detecting an interaction between the breach code and whether the start was C, M, or P, the dataset was subset into C, M and P datasets and separate ANOVAs were performed, using a p value cut-off of $p < 0.017$. The null hypothesis was rejected for M tracks, leading to post-hoc Tukey testing on this dataset using *r*'s "TukeyHSD" function with a 98.3333% family-wise confidence level. The distribution of prize money won and total prize money on offer were positively skewed with outliers, so a non-parametric Kruskal-Wallis rank sum test was performed using the "kruskal.test" function in r. Because the C, M, and P datasets were analysed separately, a p value cut-off of $p < 0.017$ was used.

Descriptive data comparing fines and suspensions imposed in the first four months of 2016 with those in the similar 2013 period were analysed manually.

3. Results

3.1. Overall Frequency of Breaches in 2013

Between 1 January and 31 December 2013, there were 56,456 starts in 5604 races at 785 race meetings on 122 different tracks at Country (C), Metropolitan (M) and Provincial (P) locations in NSW and the ACT. A breach of the whip rules was reported in 348 starts (0.62%) and, of these, 37 included a second breach (corresponding to 10.63% of starts with breaches and 0.07% of all starts). This equates to a breach or breaches being reported in 332 of the 5604 races (5.92%).

Of the 348 starts with breaches, 317 (91.09%) represented the only horses in their races with a breach or breaches; 28 (8.05%) occurred in races in which one other horse had a breach reported (14 races); and three (0.86%) occurred in one race in which two other horses had a breach reported. Further details are provided in Supplementary Materials Table S1.

3.2. Locations of Breaches in 2013

The 348 starts resulting in a breach or breaches occurred at 241 of 785 race meetings (30.70%) and at 64 of the 122 racetracks (52.46%). The 37 starts with second breaches occurred at 37 race meetings at 20 different tracks. Despite hosting 67.34% of starts, C tracks recorded only 57.47% of starts with a breach or breaches. Conversely, M tracks hosted 14.53% of starts and yet recorded 22.13% of starts with a breach or breaches. Further details are provided in Supplementary Materials Table S1.

Using a chi-square test, we revealed an association between the frequency of breaches and whether a track was at a C, M or P location ($\chi^2 = 19.9703$, $p < 0.01$). Post hoc chi-square testing showed starts at C locations were significantly less likely to result in a breach than starts at P and M locations ($\chi^2 = 15.0727$, $p < 0.01$) and starts at M locations were significantly more likely to result in a breach than starts at C and P locations ($\chi^2 = 15.6525$, $p < 0.01$).

3.3. *Whip Rules Breached in 2013*

3.3.1. First Breaches

Table 1 shows the whip rule most commonly breached as a first breach was AR 137A. (5)(a)(ii) (Code 17), which accounted for 45.11% of first breaches. Breaches of AR 137A. (4)(b) (Code 8) accounted for 25.00% of first breaches, while breaches of AR 137A. (4)(c) (Code 9) and AR 137A. (5)(a)(i) (Code 16) accounted for 12.64% and 9.48% of first breaches, respectively. This means 54.59% of first breaches were attributable to whip use prior to the 100 m mark (Codes 16 and 17). No breaches were recorded for several whip rules (see Codes 1–5, 10–12, 15, 18–23).

(i) Location of First Breaches

Table 2 shows breach codes for the first breach of a start classified by C, M, P location of the track at which the start occurred.

Table 2. Whip rules breached in 348 first breaches classified by Country, Metropolitan, and Provincial location.

Breach Code	Country Number out of 200 (%)		Metropolitan Number out of 77 (%)		Provincial Number out of 71 (%)	
6	7	(3.50%)	1	(1.30%)	2	(2.82%)
7	3	(1.50%)	2	(2.60%)	1	(1.41%)
8	32	(16.00%)	40	(51.95%)	15	(21.13%)
9	30	(15.00%)	4	(5.19%)	10	(14.08%)
13	1	(0.50%)	0	(0.00%)	0	(0.00%)
14	0	(0.00%)	0	(0.00%)	1	(1.41%)
16	11	(5.50%)	10	(12.99%)	12	(16.90%)
17	107	(53.50%)	20	(25.97%)	30	(42.25%)
24	9	(4.50%)	0	(0.00%)	0	(0.00%)

See Table 1 for breach code key.

(a) Location of Code 8 First Breaches (Whip Use That Raises Arm above Jockey's Shoulder Height)

The probability of a Code 8 first breach was associated with whether a start occurred at an M, C or P location (χ^2 = 39.0306, $p \leq 0.01$). Post-hoc testing revealed starts at M locations were significantly more likely to result in the recording of a Code 8 first breach than starts at a C track location (χ^2 = 38.5024, $p \leq 0.01$).

(b) Location of Code 9 First Breaches (Whip Use When Horse Is Out of Contention)

This less common rule breach accounted for approximately 1 in 8 first breaches overall. There was no association between Code 9 breaches and C, M, P locations (χ^2 = 5.007, p = 0.0818).

(c) Location of Code 16 First Breaches (Forehand Whip Use in Consecutive Strides prior to 100 m Mark)

The probability of a Code 16 breach was associated with whether a start occurred at a C, M, or P location (χ^2 = 9.35, $p \leq 0.01$). Post-hoc testing revealed starts at C locations were significantly less likely to result in a Code 16 breach than starts at other locations (χ^2 = 7.6338, $p \leq 0.01$).

(d) Location of Code 17 First Breaches (Forehand Whip Use on More Than Five Occasions prior to 100 m Mark)

The probability of a Code 17 breach was associated with whether a start occurred at a C, M, or P location (χ^2 = 17.3068, $p \leq 0.01$). Post hoc testing revealed starts at C locations were significantly more likely to result in a Code 17 first breach than starts at an M location (χ^2 = 15.8759, $p \leq 0.01$).

(ii) First Breaches by Rider Gender

Of 348 starts where a code was recorded as breached, 296 (85.06%) had male riders and 52 (14.94%) had female riders. A Fisher's exact test, examining whether the relative proportions of breach codes (16, 17, 9, 8, or other) were independent of rider gender, was performed using r. The overall p value was <0.0001. This suggests recorded breaches depend on whether the rider was male or female. The observed frequencies of breaches of Code 16 (Forehand whip use in consecutive strides prior to 100 m mark) and Code 8 (Whip use that raises arm above jockey's shoulder height) are similar to expected under independence. Comparing adjusted standardised residuals to the Z distribution, female riders appeared slightly, but not significantly, overrepresented for codes with breaches other than Code 8, Code 9 (Whip use when horse is out of contention), Code 16, and Code 17 (Forehand whip use on more than 5 occasions prior to 100 m mark) ($p = 0.095$); significantly underrepresented for Code 17 ($p < 0.001$); and significantly overrepresented for Code 9 ($p < 0.001$).

3.3.2. Second Breaches

Of 348 starts with first breaches, 37 (10.63%) had second breaches. The majority (21% or 56.76%) were breaches of Code 16 (Forehand whip use in consecutive strides prior to 100 m mark), while another 11 (29.73%) were breaches of Code 17 (Forehand whip use on more than five occasions prior to 100 m mark). The five remaining second breaches (13.51%) were breaches of Code 8 (Whip use that raises arm above jockey's shoulder height). This means 86.49% of second breaches occurred prior to the 100 m mark.

(i) Location of Second Breaches

While second breaches occurred in fewer than 10% of first breaches at C and M locations, at P locations this number was significantly higher at 22.54% ($\chi^2 = 11.7733$, $p < 0.01$). Using a Fisher's exact test, we found a significant association ($p = 0.02294$) between the C, M, P location of the track and the breach code of a second breach. This association most likely relates to the apparent underrepresentation of breaches of Code 17 (Forehand whip use on more than five occasions prior to 100 m mark) and overrepresentation of breaches of Code 16 (Forehand whip use in consecutive strides prior to 100 m mark) at C locations, especially when compared to P locations. Further details are provided in Supplementary Materials Table S2.

(ii) Second Breaches by Rider Gender

Of 37 starts with second breaches, 34 (91.89%) were by male and three (8.11%) by female riders. Once a first breach was recorded, we found no association between rider gender and a second breach being recorded ($\chi^2 = 0.9793$; $p = 0.3224$). This suggests that once a first breach is recorded, the frequency of a second breach is independent of gender.

3.3.3. Whip Rules Breached in First and Second Breaches

Code 17 (Forehand whip use on more than five occasions prior to 100 m mark) was the most common whip rule breached in first breaches and Code 16 (Forehand whip use in consecutive strides prior to 100 m mark) the most common second breach. Of 37 starts with two breaches recorded, 19 (51.35%) followed this pattern. The next most common pattern, seen in 10 horses (27.03%), was the reverse of this. This means where two breaches were recorded in one start, over 75% (29 out of 37) were breaches of these two whip rules and as such occurred prior to the 100 m mark. When first and second breaches are considered together, breaches of Code 17 were the most common (44%), followed by breaches of Code 8 (Whip use that raises arm above jockey's shoulder height) (24%).

3.4. Breach Outcomes in 2013

Our examination of the Stewards Reports showed the format varied between racetracks. While all contained the major findings, these were presented differently and in varying detail. For example, some Stewards Reports listed Conviction Recorded next to some whip rule breaches, even though a recorded breach of any whip rule constitutes a Conviction Recorded [25].

The most common outcome for first and second breaches of a whip rule was a reprimand (48.28% and 64.86%, respectively). Where a fine was the outcome, as in 104 cases (29.88%) of first breaches, the most common value (mode) and the median value was $200. There was only one fine ($200) awarded for a second breach (see Table 3).

Table 3. Outcomes of first and second breaches of whip rules.

Outcome	First Breach (n)	Second Breach (n)
$100	12	0
$200	61	1
$300	16	0
$400	8	0
$500	4	0
$600	1	0
$800	2	0
Not Known	8	0
Caution	56	2
No Outcome	1	0
Reprimand	170	25
Suspension	9	4
Conviction Recorded	0	5
Total	348	37

3.4.1. First Breach Outcomes

All Code 6 first breaches (Excessive/unnecessary/improper whip use) resulted in cautions (n = 6) or reprimands (n = 4). Breaches of Code 17 (Forehand whip use on more than five occasions prior to 100 m mark), Code 8 (Whip use that raises arm above jockey's shoulder height), Code 9 (Whip use when horse is out of contention), and Code 16 (Forehand whip use in consecutive strides prior to 100 m mark), were the most common first breaches (n = 321), and had variable outcomes. There were only nine suspensions in total and these resulted from breaches of Codes 16 and 17. The median suspension imposed was eight days (range 7–10). Further details are provided in Table 4.

Table 4. Outcomes of first breaches classified by whip rule breached.

First Breach Code	C	R	100	200	300	400	500	600	800	S	NO	NK
6	6	4	0	0	0	0	0	0	0	0	0	0
7	4	0	0	2	0	0	0	0	0	0	0	0
8	28	32	6	9	1	6	2	1	2	0	0	0
9	16	21	5	1	1	0	0	0	0	0	0	0
13	1	0	0	0	0	0	0	0	0	0	0	0
14	0	1	0	0	0	0	0	0	0	0	0	0
16	0	21	0	7	3	0	0	0	0	2	0	0
17	1	91	1	42	11	2	2	0	0	7	0	0
24	0	0	0	0	0	0	0	0	0	0	1	8

C—Caution; R—Reprimand; 100–800—$ Fine; S—Suspension; NO—No Outcome; NK—Not Known; Code 6—Excessive/unnecessary/improper whip use; Code 7—Whip use forward of horse's shoulder/vicinity of head; Code 8—Whip use that raises arm above jockey's shoulder height; Code 9—Whip use when horse is out of contention; Code 13—Whip use when horse is clearly winning; Code 14—Whip use when horse has no reasonable prospect of improving/losing position; Code 16—Forehand whip use * in consecutive strides prior to 100 m mark; Code 17—Forehand whip use * on more than 5 occasions prior to 100 m mark; Code 24—Whip Rule breach not specified by Stewards; * Whip rules were amended 1 December 2015 to include backhand whip strikes as well as existing forehand restrictions.

For first breaches, 104 fines were imposed resulting in a total of $25,600. Further details are provided in Supplementary Materials Table S3. When second breaches are included (see Section 3.4.2 below) the total is $25,800.

3.4.2. Second Breach Outcomes

Second breaches ($n = 37$) were recorded for breaches of Code 16 (Forehand whip use in consecutive strides prior to 100 m mark) ($n = 21$), Code 17 (Forehand whip use on more than five occasions prior to 100 m mark) ($n = 11$), and Code 8 (Whip use that raises arm above jockey's shoulder height) ($n = 5$). The four suspensions imposed were for Code 16 and 17 breaches. The median suspension imposed was 7.5 days (range 5–14). The only fine imposed for a second breach was $200 (1× Code 17). No cautions were issued for Code 16 or 17 second breaches (this is similar to first breaches in which there was only one caution issued for these breaches). Further details are provided in Supplementary Materials Table S4.

3.5. Prize Money in 2013

3.5.1. Prize Money Won by Horses with Breaches

Using the Kruskal-Wallis test, we found horses whose riders had Code 9 breaches (Whip use when horse is out of contention) won significantly less money than horses whose riders breached other codes in races at C, M, and P locations ($p < 0.0001$; $p = 0.0086$; $p = 0.0002$, respectively).

The total prize money won by horses in starts where a first breach resulted in a fine was $1,108,425, and for the single start where a second breach had a fine imposed ($200), the total prize money won was $17,500, making the total prize money won by horses in starts resulting in fines $1,125,925. The median prize money won where there was one breach reported was $1000 (range $0–$105,000); for two breaches it was $2100 (range $0–$400,000), and for all starts with breaches it was $1325 (range $0–$400,000).

Overall, fines represented about 2.29% of the total prize money won by these horses. The median percentage of a fine compared to the prize money won in an individual case was 13.9%. The number ranges very widely from 0.19% (for a horse that won $105,000 and the rider received a $200 fine) to eight cases in which the fine exceeded the prize money won.

3.5.2. Total Prize Money on Offer

(i) Total Prize Money on Offer in Races with Breaches

Breaches occurred in 332 races in which the total prize money on offer was $12,636,450 ($2,860,200 was from 194 C races (*mean = $14,743.30 per race*); $2,014,500 was from 65 P races (*mean = $30,992.31 per race*), and $7,761,750 was from 73 M races (*mean = $106,325.34 per race*)). Overall, fines represented about 0.20% of the total prize money on offer in races with breaches.

The median total prize money on offer in races in which there was one whip rule breach was $15,000 (range $2500–$250,000), for two breaches it was $22,000 (range $6000–$2,250,000), and for all starts with breaches it was $15,000 (range $2500–$2,250,000).

(ii) Total Prize Money on Offer in Races with Breaches by Location

There was no significant (Kruskal-Wallis $\chi^2 = 10.3853$) association between total prize money on offer for races and the frequency of code breaches at M ($p = 0.0344$), C ($p = 0.5327$) or P locations ($p = 0.1771$). Further details are provided in Supplementary Materials Table S5.

3.6. Race Length and Breaches in 2013

Breaches occurred in races ranging from 900 to 2400 m (median 1400 m). Using ANOVA with post-hoc Tukey tests, we found the length of races where there were breaches of Code 8 (Whip use that raises arm above jockey's shoulder height) or Code 9 (Whip use when horse is out of contention) was significantly lower ($p < 0.01$; $p = 0.01$, respectively) than the length of races where there were Code 17 breaches (Forehand whip use on more than five occasions prior to 100 m mark) at M tracks (but not at C or P tracks). Further details are provided in Supplementary Materials Table S6.

Length of Races with Breaches by Type of Race and Whip Rule Breached

Table 5 treats first (n = 348) and second breaches (n = 37) equally (Total 385), and shows 56.88% of all breaches occurred in sprint races. It also shows that overall Code 17 breaches were the most common.

Table 5. Length of races with breaches by type of race and whip rule breached.

Breach Code	Sprint ≤1400 m (n)	Mile 1406–1750 m (n)	Middle 1800–2400 m (n)	Total (n)
8	55	24	13	92 (23. 90%)
9	29	11	4	44 (11.43%)
16	33	10	11	54 (14.03%)
17	84	42	42	168 (43.64%)
Other	18	6	3	27 (7.01%)
Total	219 (56.88%)	93 (24.16%)	73 (18.96%)	385

First and second breaches are included for a total number of 385 (348 first breaches + 37 second breaches); Code 8—Whip use that raises arm above jockey's shoulder height; Code 9—Whip use when horse is out of contention; Code 16—Forehand whip use * in consecutive strides prior to 100 m mark; Code 17—Forehand whip use * on more than 5 occasions prior to 100 m mark; Other—see Table 1 for other breach codes; * Whip rules were amended 1 December 2015 to include backhand whip strikes as well as existing forehand restrictions.

3.7. *Riders with Breaches in 2013*

3.7.1. Frequency of Riders with Breaches

One hundred and thirty nine riders were responsible for the 348 starts with breaches, and of these, 51.08% were repeat offenders (riders who had more than one whip rule breach recorded in 2013, including those with second breaches in a start). Further details are provided in Supplementary Materials Table S7.

(i) Riders with the Highest Number of Breaches

Of the 15 riders with the highest numbers of breaches, only one was an apprentice and one was female. Further details are provided in Supplementary Materials Table S8.

(ii) Riders Breaching Whip Rules in Two Races at the Same Race Meeting

There were 13 race meetings (3 M, 4 P, 6 C) where the same riders had breaches in two races, with one rider having breaches in two races at two different race meetings. There were no instances of a rider breaching in more than two races at the same race meeting. Four of the riders were apprentices (two males, two females). Seven of these 13 riders were jockeys with the highest numbers of breaches overall and were male. Further details are provided in Supplementary Materials Table S9.

3.7.2. Fines and Prize Money of Riders with Breaches

The $ value of fines imposed on riders for whip rule breaches as a percentage of the prize money won by the horse and as a percentage of the rider's 4.95% share of the prize money, varied widely (see Table 6).

Table 6. Fines imposed and prize money won by riders (n = 5) with the most breaches in 2013.

Rider Code	Number of Starts with Breaches	Number of Starts with Breaches When Fine Was Imposed	Whip Rule Fines $	Prize Money Won by Horse $	4.95% Rider's Share of Prize Money Won $	Fine as % Prize Money Won	Fine as % Rider's 4.95% Share of Prize Money Won
A	13	13	4300	238,335	11,797.58	1.80	36.45
B	12	8	2100	66,230	3278.39	3.17	64.06
C	12	4	1000	178,425	8832.00	0.56	11.3
D	10	2	600	2975	147.26	20.17	407.44
E	9	7	1700	9580	474.21	17.7	358.49

Rider code—each letter represents the name of a different rider and is the same as designated in Supplementary Materials Tables S8 and S9.

3.7.3. Whip Rule Breaches by Apprentices and Jockeys

We were unable to consider the overall percentage of riders in 2013 that were apprentices as opposed to jockeys. However, we did investigate whether, among riders with breaches, apprentices were associated with different breach codes than jockeys. A χ^2 analysis of rider status and breaches of Code 16 (Forehand whip use in consecutive strides prior to 100 m mark); Code 17 (Forehand whip use on more than five occasions prior to 100 m mark); Code 8 (Whip use that raises arm above jockey's shoulder height); and Code 9 (Whip use when horse is out of contention), plus a pool of the remaining breach codes, suggested that, among riders with breaches, the whip rule breached depends on jockey status ($\chi^2 = 29.4424$, $p < 0.0001$). Comparing adjusted standardised residuals to the Z distribution, apprentices appear slightly, but not significantly, underrepresented among breaches of Code 8 ($p = 0.09653$) and Code 17 ($p = 0.0452$, which is insufficient after correction for family error rate), and significantly overrepresented among breaches of Code 9 (<0.0001). As there were different breach code patterns depending on C, M, or P location, and because there might be differences in the percentage of jockeys versus apprentices at these locations, we treated this as a potential confounding factor and conducted separate post-hoc Fisher's exact tests for independence of breach code and rider status on breaches at C ($p = 0.0003$), P ($p = 0.0008$), and M ($p = 0.4705$) locations separately. The apparent overrepresentation of apprentices among Code 9 breaches remained evident at C and P locations in the post-hoc testing.

3.8. Race Finishing Positions and Percentage of Whip Rule Breaches in 2013

Figure 1 shows the highest percentage of whip rule breaches occurred in horses that ran second, closely followed by horses that ran first. Horses in first place had riders with the highest number of first breaches, but when second breaches were included, second place finishing horses had the highest number of breaches, although the difference between first and second horses was not significant in this sample. The next highest percentage was in horses that ran third. Overall, horses finishing first, second, or third had significantly more breaches than horses finishing in other positions (Chi-squared = 69.4457, *p*-value < 0.0001, if we assume a single first, second, and third finisher in each race of the sample). The next highest percentage was seen in horses that ran last.

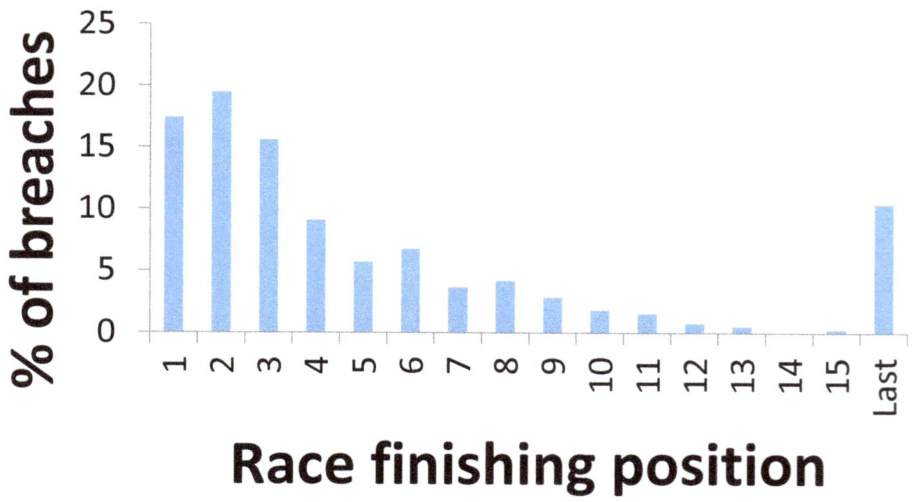

Figure 1. Race finishing positions and percentage of whip rule breaches.

When we examined breach types in horses that ran last, we found 26 (65.00%) of the total 40 (39 first breaches and one second breach) were Code 9 breaches.

3.9. Breeder Owner Bonus Scheme (BOBS) in 2013

As explained in Materials and Methods, these analyses exclude ACT data.

3.9.1. Frequency of BOBS and Non-BOBS Races by Location

Non-BOBS races were more likely to be held at M locations (over 7 out of 10 Non-BOBS races occurred at M tracks, despite about 15.5% of races being held at these tracks). Further details are provided in Supplementary Materials Table S10. A chi-square test for independence of BOBS/Non-BOBS races and location gives a chi-square statistic of 527.1027 and a p value well <0.01.

3.9.2. Frequency of Whip Rule Breaches in BOBS Races

At C locations, 187 (97.91%) of the 191 starts resulting in a breach or breaches were in BOBS races, which comprised over 99% of races. At M tracks, 62 (80.52%) of 77 starts resulting in a breach or breaches occurred in BOBS races, which comprised 82% of races. At P tracks, 67 (94.37%) of 71 starts resulting in a breach or breaches occurred in BOBS races, which comprised 97.39% of races. These data do not suggest BOBS races are more likely to result in breaches. Further details are provided in Supplementary Materials (see Tables S10 and S11, and Section S3.9.2.).

3.9.3. Percentage of Breaches in BOBS Horses

Of 54,380 starts in NSW, approximately two of every nine horses with a breach or breaches were BOBS horses. Our data do not suggest breaches are more likely in BOBS horses in BOBS races than in other horses. For further details see Supplementary Materials Table S12.

3.9.4. Whip Rule First Breaches in BOBS and Non-BOBS Races

Prima facie, there is not a large difference between the frequency of whip rule breaches for BOBS and Non-BOBS horses in BOBS races, compared with horses in Non-BOBS Races. Further details are provided in Supplementary Materials (see Table S13, and Section S3.9.4. (i) and (ii)).

3.10. Fines and Suspensions—2013 Compared to 2016

We compared the number and $ value of fines and the number and length (days) of suspensions imposed in the first four months of 2016 with those in the similar period of 2013 in NSW/ACT. We found there were 15 suspensions (range 4–21 days) and 115 fines (Total $42,650) in the designated 2016 period compared to five suspensions (range 7–10 days) and 36 fines (Total $9400) in the equivalent 2013 period [20].

4. Discussion

We identified a need for a peer-reviewed study that characterised whip rule breaches in horse racing from a welfare and regulatory perspective. To achieve this, we used Stewards Reports [20] and Race Diaries [21] from 2013 (and 2016) in a major Australian racing jurisdiction (NSW/ACT). Given the extensive data generated and the number of significant findings, we discuss these under the headings below.

Firstly, however, we provide evidence regarding the nociceptive potential of whip use. Whipping racehorses and its regulation would be irrelevant as animal welfare concerns, if whipping was not potentially painful. McGreevy et al. (2012) reported that 83% of whips strikes viewed in slow motion made a visible indentation [4]. After investigating the mechanical nociceptive thresholds on the dorsal metatarsus in horses, Taylor et al. (2016) stated that "When a noxious mechanical stimulus is applied, pressure distorts the nociceptive nerve endings, activating the nociceptive pathway to the spinal cord" [14]. It was clear these authors were referring to nociception in the horse in general when they explained that the mechanical threshold (MT) is affected by "Numerous additional factors ... including the operator, the environment, the anatomical site, the rate of stimulus application and the characteristics of the tissue". The dorsal aspect of the cannon bone is favoured when measuring MTs as there are minimal anatomical variations between horses, not least because there is very little soft tissue between the skin and periosteum [26].

We argue, therefore, that while the MT for nociception would be expected to vary according to anatomic location, no external part of a normal horse would be insensate. Indeed, the whip is used exactly because the horse feels it. Nociceptive threshold testing in animals is still most commonly used to assess the effects of analgesic drugs in a research setting [14] and studies of potential tissue injury and pain from whipping are scant. Besides the study by McGreevy et al. [4], there appears to be only one article, a conference proceeding from Australia in 1996, which investigated "the potential of whips to injure horses". This found that "Two factors determine the biophysics of impacts to the body: (1) displacement—when skin and internal structures are stretched/crushed; (2) rate of displacement—reflecting the mechanical "strain-rate sensitive" properties of tissues as pressure waves form during fast impacts" [27].

While the report said "An unequivocal statement on the biomechanics of injury by whips in general is not possible because of the many physiological and physical factors determining tissue damage", it followed this with "It is, however, possible to develop general principles about a whip's potential to injure which can be used in a more objective assessment of abuse, thereby allowing the Rules of Racing to be modified to exclude whips with a *greater* potential to injure" (our emphasis). It is disappointing that the authors' suggestion that "Non-invasive techniques, including thermography and ultrasound ... could be useful ... to assess whip damage potential and to quantify whip injury after suspected whip abuse" has not, to our knowledge, yet been undertaken.

Notwithstanding this knowledge gap, Taylor et al. (2016) said that while "In animals ... the precise sensation experienced is unknown ... it is assumed that if a stimulus that produces pain when applied to humans elicits an appropriate response in an animal, this stimulus represents the threshold for pain (Le Bars et al. 2001)" [14]. A 2004 review of the known physiological properties of specialised mechanoreceptors in humans and experimental animals stated "Pain is very often but not always associated with a mechanical stimulus. The mechanical stimulus that causes pain might be an intense one such as pinching of the skin, traumatic injury, or under neuropathic conditions brush or light pressure" [13].

Further, in Australia there is a legal obligation under jurisdictional animal welfare acts to comply with the Australian code for the care and use of animals for scientific purposes 8th edition 2013 (the Code) [28] when research using animals, including horses, is conducted. Section 1.10 states "Animals have a capacity to experience pain and distress, even though they may perceive and respond to circumstances differently from humans. Pain and distress may be difficult to evaluate in animals. Unless there is evidence to the contrary, it must be assumed that procedures and conditions that would cause pain and distress in humans cause pain and distress in animals. Decisions regarding the possible impact of procedures or conditions on an animal's wellbeing must be made in consideration of an animal's capacity to experience pain and distress". Accordingly, we argue that whip use that indents a horse's skin [4] must be assumed to be painful, unless there is evidence to the contrary, which there is not, and that restrictions on whipping can generally be seen as a step toward the more humane treatment of horses in racing.

4.1. Frequency and Location of Whip Rule Breaches

We found whip rule breaches were recorded by Stewards in NSW/ACT in 2013 in 1% of starts; 6% of races; 31% of race meetings, and at 53% of racetracks. Only a small number of starts with breaches had second breaches in the same race (11%), and over 90% of horses were the only horses in their races with a breach or breaches.

In a study by McGreevy et al. [4] using side-on high speed footage (2000 frames per sec) of 15 race finishes at 2 P NSW race meetings in 2011, 28 whip rule breaches in 9 horses were identified that were not recorded by Stewards. Using the Race Diary [21] to obtain the number of starts in these races, we calculated 6% of these starts had at least one whip rule breach. For the same starts, Stewards recorded a breach frequency of 1% that corresponded to one whip rule breach in a horse that ran tenth and hence was not included in the McGreevy study. Given this study used footage obtained from only one side of the track, and at race finishes, it provides strong evidence that the Stewards under-reported whip rule breaches in the 15 races studied in 2011. As the same surveillance techniques were also used by Stewards in 2013, it is possible that the 1% frequency of whip rule breaches recorded by Stewards for that year, may also underestimate the actual number of whip rule breaches. Of course, it is possible that the breach frequency observed in the small number of races looked at by McGreevy et al. was an anomaly, but it may also indicate the need for faster footage than that currently examined by Stewards (25 frames per sec) [4], which views races head-on. It may also be possible that Stewards focus on the worst offending rider in a race, perhaps because of time pressures, and as such may overlook other riders who are also breaching whip rules. Current surveillance methods should be investigated to ensure these are not failing to detect breaches.

The UK Review found a breach percentage of 0.74 for flat turf racing between 2004 and 2011, which appears slightly higher than the raw 2013 NSW/ACT percentage. However, it needs to be considered that the UK whip rules set a maximum of 15 whip strokes per race [12] (p. 14), whereas that was not the case in 2013 in Australia, nor is it the case now where "In the final 100 m of a race ... a rider may use his whip at his discretion" (AR. 137A. (5)(b)).

Following this Review, whip use was further restricted in the UK and is currently capped at 7 strikes per flat race, but penalties are imposed at the discretion of the Stewards [29], as also occurs in Australia. In 2015, the number of UK whip offences, expressed as a percentage of total rides, was 0.61% compared with 1.12% in 2010 [29]. As Racing Australia (RA) does not publish these data, it is not currently possible to say whether the number of whip breaches in NSW/ACT is increasing, decreasing, or static. This shortcoming should be addressed by an industry commitment to reporting, or by mandatory reporting, as part of incremental improvement in racehorse welfare and racing integrity.

Our study found starts at C locations were significantly less likely to result in a breach being recorded than starts at P and M locations, and starts at M locations were significantly more likely to result in a breach being recorded than starts at C and P locations. We were advised there was no difference between the level of surveillance used by Stewards to detect whip breaches at these locations [30] and that Stewards are not attached to individual tracks. Instead, a Stewards panel based in the metropolitan area services M and P race meetings and, in regional centres, Stewards are based in five locations [31]. It is possible, therefore, that reported differences in breach frequencies at C, M and P locations may reflect true differences in the numbers of breaches occurring, or they may reflect behavioural differences in reporting among individual Stewards or between the two pools of Stewards. Further studies are needed to confirm this.

We thought total prize money on offer might incentivise whipping, but our results did not demonstrate this. As Racing NSW describes the BOBS as "the most popular racing incentive scheme in Australia" [32] (p. 30), we also investigated whether it might incentivise whip breaches, but again this was not the case. However, it is possible that the desire to win might have been an incentive, as we demonstrated riders of horses finishing first, second, or third had significantly more whip rule breaches recorded against them than riders of horses finishing in other positions. While our findings support those of Evans and McGreevy that "whippings were associated with superior performance and that there was an association between final placing and the number of whip strikes in the final 200 m section" [3], it should be remembered that in that study, only whip-rule-compliant whip use was included and race footage was not examined by the authors. In this instance, we would want to exclude the possibility that Stewards may tend to focus on the horses running first, second or third, and, as such, possibly miss whip rule breaches by riders of other horses.

4.2. Types and Location of Whip Rule Breaches

Our findings also characterise the types and locations of whip rule breaches recorded by Stewards in NSW/ACT in 2013. We now know that the whip rules most commonly recorded as first breaches were forehand whip use on more than five occasions prior to the 100 m mark (Code 17), and using an action that raises the jockey's arm above shoulder height (Code 8), which accounted for over 45% and 25% of first breaches, respectively. The finding that 70% of first breaches consisted of breaches of only two whip rules deserves further attention, as does the finding that 55% of first breaches consisted of breaches of Code 16 (Forehand whip use in consecutive strides prior to the 100 m mark) and Code 17, which also applies before the 100 m mark.

Possible reasons for these findings include that these breaches may be more easily detected than some others, as Stewards view the race footage head-on [4]; Stewards prioritise these breaches (this does not appear to be the case from the Racing NSW Rider Penalty Guidelines for Whip Rule Breaches (see Appendix A) [33] that were used in 2013, but discretionary biases might occur in practice), or simply these breaches are what might be expected if whip strike frequency and force are perceived to be needed for a horse to win: that is, there is an increased number of whip strikes, and an increased number of more potentially powerful whip strikes with the whip wielded above the jockey's shoulder. However, McGreevy et al. [4] stated "It is not clear how much, if any, amplification in force that whipping from this height creates. From the perspective of force and therefore pain, the recoil of the whip may be more important than the height from which it descends during its trajectory". Given we found over a quarter of all first breaches involved whip use when the jockey's arm is raised above his shoulder height, this question regarding the force of these whip strikes requires clarification.

Importantly, an increase in the number of whip strikes prior to the 100 m mark does not tell us about the actual number of whip strikes used in the final 100 m (except it seems unlikely riders would decrease whip use in this crucial race segment). This contention is supported by Evans and McGreevy [3] who found 98% of horses were whipped in the final 200 m of five sprint races at an M location in NSW, when they were slowing. Evans and McGreevy concluded their data "make whipping tired horses in the name of sport very difficult to justify". Our data also show whip breaches occurring when horses are likely to be fatigued.

While Whip Rule AR 137A. (3) (Code 6) prohibits excessive, unnecessary, or improper whip use throughout the race, only 3% of first breaches fell into this category. This might suggest such whip breaches do not occur frequently in the critical last 100 m of the race, but this is unlikely given even a cursory examination of race finish footage in the media. One of Australia's leading sports journalists, Patrick Smith, wrote: "as the rules are now, jockeys will whack their horses non-stop down the straight from the 100 m post ... " [17]. The low number of breaches recorded for this whip rule might also reflect problems in detection (as suggested by McGreevy et al. [4]), or simply the Stewards' discretion.

It is also unclear what constitutes excessive, unnecessary, or improper whip use (AR 137A. (3)) (Code 6), given AR 137A. (4) prohibits whip use (a) forward of horse's shoulder/vicinity of head (Code 7); (b) that raises arm above jockey's shoulder height (Code 8); (c) when horse is out of contention (Code 9); (d) when horse is showing no response (Code 10); (e) after passing winning post (Code 11); (f) causing injury to the horse (Code 12); (g) when horse is clearly winning (Code 13); (h) when horse has no reasonable prospect of improving/losing position (Code 14); and (i) in such manner that the seam of the flap is the point of contact with horse, unless rider satisfies Stewards this was neither deliberate nor reckless (Code 15).

Considering that, in 2013, AR 137A. (5)(a) prohibited (i) forehand whip use in consecutive strides prior to the 100 m mark (Code 16) and (ii) forehand whip use on more than five occasions prior to the 100 m mark (Code 17), it appears by default that Code 6 breaches involving whip use in consecutive strides and excessive numbers of whip strikes applied only to the final 100 m of the race. The very low frequency of Code 6 breaches (excessive, unnecessary, or improper whip use) recorded in 2013 provides an impetus to investigate at what stages of races these breaches are reported in, what constitutes these breaches, and in general what percentage of all breaches occur and are recorded in the final 100 m.

Additionally, Code 6 breaches raise the important issue of what Stewards consider to constitute "unnecessary" whipping. This notion logically relies on the concept of "necessary" whipping and there remains a paucity of peer-reviewed evidence to support this, either for safety, or to improve performance [8]. Further, there is a legal argument to consider in relation to what constitutes "unnecessary" under the NSW *Prevention of Cruelty to Animals Act 1979* (POCTA) [34]. Section 5 prohibits cruelty to animals, including racehorses, and Section 4(2) explains " ... a reference to an act of cruelty committed upon an animal includes a reference to any act or omission as a consequence of which the animal is unreasonably, unnecessarily or unjustifiably:

(a) beaten, kicked, killed, wounded, pinioned, mutilated, maimed, abused, tormented, tortured, terrified or infuriated,
(b) over-loaded, over-worked, over-driven, over-ridden or over-used ...

or

(d) inflicted with pain", where "pain" includes suffering and distress (Section 4(1)).

It appears critical this question of whether whipping racehorses could constitute cruelty under the POCTA is considered. Jones et al. [8] discuss "necessary" in relation to animal welfare legislation by using the leading case on this question—Ford v. Wiley, an English High Court decision from 1889 that developed a test to determine when suffering caused to an animal could be deemed to be unnecessary in a particular set of circumstances. It would appear overdue that this be applied to whip use in horse racing in Australia, and particularly in relation to the regulatory outcomes for Code 6 breaches (excessive, unnecessary, or improper whip use).

We also found significant differences between the likelihood of certain whip rule breaches being recorded at the three race locations. Starts at M locations were significantly more likely to result in Code 8 first breaches (Using an action that raises the jockey's arm above shoulder height) than starts at C locations, and starts at C locations were significantly less likely to result in Code 16 breaches (Forehand whip use in consecutive strides prior to the 100 m mark) than starts at other locations, and significantly more likely to result in Code 17 first breaches (Forehand whip use on more than five occasions prior to the 100 m mark) than starts at M locations.

As discussed in relation to the significant differences we found in the frequency of whip breaches recorded between locations, the significant differences in the types of breaches being recorded at the different locations may reflect true differences in the types of breaches occurring, or perhaps differences in detection and recording amongst individual Stewards or between the two pools of Stewards. Further studies should investigate this.

We found no differences between locations for the recording of Code 9 breaches (Horse whipped when out of contention), which comprised 13% of first breaches. This may reflect our earlier discussion that the desire to win, or possibly not to come last, may lead riders into this whip use, and that these incentives apply at all locations. It is also possible Stewards may be more likely to record a Code 9 breach for a horse that runs last at any location, as it may be more obvious the horse was out of contention than for other non-winning/non-place getter horses. It is worth noting that being out of contention is an holistic evaluation that many observers could arrive at, despite the difficulties in providing a quantitative approach to this rule.

When we examined the types of breaches recorded in horses that ran last, we found most were Code 9, and conversely most Code 9 breaches were recorded in horses that came last. Not surprisingly, horses with these breaches recorded won significantly less money than horses with other breaches in races at all locations.

McGreevy et al. [9] provided insights into Code 9 breaches when they noted in relation to UK and Australian horseracing " ... no definition of being "out of contention" is offered by either set of rules and therefore is open to considerable interpretation". Evans and McGreevy [3] further highlighted the conundrum facing a jockey to both "ride his horse out (i.e., ensures the horse gave of its best) to the end ... " in accordance with AR 137. (b). It appears that in 2013 these issues had not yet been sufficiently clarified, and this remains the case currently.

In interpreting our data on second breaches, it must be remembered that while the level of surveillance may be the same at all locations, Stewards may use their discretion to order first and second breaches differently at the three locations, or there may be no difference in the importance of what breach is recorded first. Notwithstanding this, we are able to say that where two breaches were recorded in one start, over 75% involved the same two whip rules (Codes 16 and 17) and hence occurred prior to the 100 m mark. Attempts to clarify what determined the order of two breaches recorded in a start and whether this affected what penalties were imposed, were answered simply as "both penalised" [30]. This highlights the need for a transparent prosecution policy or penalty guidelines in Australia, as now provided in the UK by the British Horseracing Authority (BHA) [35]. Likewise, our examination of Stewards Reports showed a need for greater consistency in how information is recorded between racetracks.

Several whip rules had no breaches recorded. In some cases this may reflect a lack of clarity of the rules, or difficulties in detection, rather than indicating a 100% compliance rate (though that is probably the case with AR 137A. (1) and (2), which pertain to the correct, padded whip being used).

For example, in the same study that revealed indentations in the skin of horses in over 80% of whip strikes in 15 race finishes at two P NSW race meetings in 2011, McGreevy et al. [4] raised concerns "the seam rule" (Code 15) "is virtually impossible to police, even using significantly more detailed footage than is usually reviewed by racing Stewards, and its inclusion is therefore futile". The authors noted "the number of times breaches of the seam contact rule have ever been prosecuted since the current rules were established is negligible". Our finding that there were no breaches of this whip rule recorded by Stewards in 2013 supports McGreevy et al.'s contention, and further questions the adequacy of monitoring of this rule by Stewards, and the validity of the rule itself.

Similarly, it appears likely that current methodology employed by Stewards is not sufficiently sensitive to detect tissue injury caused to horses from whipping, except in extreme cases where a weal (or welt) is visible to the naked eye. This may explain why there were no recorded breaches of the "injury" whip rule (Code 12) in 2013, despite evidence that whip use in racing frequently indents the skin of horses and as such may injure it [4,13,14,26,27].

Currently, horses that have raced are examined by industry veterinarians before they are removed from the racecourse. We argue this may fail to expose inflammatory processes and, hence, ensure whip rules prohibiting injury to the horse are complied with. Where feasible, the examination of horses the day after racing with advanced tools, such as thermography, would help clarify the extent of whip-related injuries.

In their 2012 study, McGreevy et al. also showed that the unpadded section of the whip made contact with the skin of the horse in 64% of impacts, and challenged the notion padded whips (as prescribed in AR 137A. (1) and (2)) prevent injury to the horse. The reason proponed was the padding is held onto the shaft of the whip with unpadded binding that strikes the horse whenever the padded section does [4].

Our study considered whip use that was recorded by Stewards as breaching the whip rules and, hence, in all likelihood analysed only a small proportion of whip strikes overall. This can be easily appreciated when you consider AR 137A. (5)(b) allows the rider to use "his whip at his discretion" in the final 100 m of every race, where the rules restricting the number of strikes and whip use in consecutive strides do not apply. The Stewards reports show therefore that the minimum number of horses whipped in NSW/ACT in 2013 was 384 (including 37 horses in which second breaches were recorded). In light of McGreevy's findings [4], our contention is that these horses would also have been struck with the unpadded part of the whip in many of these impacts. Indeed, we argue that this would also have been the case in many of the whip strikes that did not breach the whip rules and hence were not the focus of our study.

Likewise, another of McGreevy et al.'s findings [4] was that the horse's flank (abdomen) is "twice as likely to be struck with the entirety of the whip than is the hindleg". There is no evidence to suggest this unintended whip action was not still occurring in 2013. As McGreevy et al. highlighted, although this area of the horse is regarded as being "particularly sensitive to tactile stimulation", and UK whip rules restrict whip use to the quarters of the horse, no similar safeguard is provided in Australia. This is surprising given Australia is one of 49 countries that signed the International Agreement on Breeding, Racing and Wagering [36], which prohibits the use of the whip on the flank. It is unclear why flank strikes are not reported under Whip Rule AR 137A. (3) (Code 6), which prohibits improper whip use. It may be bilateral race footage is needed before any assurance about proper whip use can be provided.

Similarly, the International Agreement's prohibition on the use of the whip with excessive force cannot be effectively policed in the absence of whips that detect force. One such electronic whip, which counts whip strikes and their force, is currently being advertised [37]. While this type of device might work to limit the force of whip impacts, it would be unlikely to prevent possible injury resulting from repeated strikes to the same area, or take into account differences in pain thresholds and susceptibilities to injury in individual horses.

Further, as the Stewards did not record any breaches in 2013 involving whip use when the horse was showing no response (Code 10), this implies firstly that the Stewards are able to detect such responses, even when horses are slowing; and secondly, that all whipped horses responded to the Stewards' satisfaction. The training and evidence-base required to produce and assess this skill should be critically examined.

These problems in the detection and enforcement of whip rule breaches raise important issues about Australian whip rules. The need for laws (or rules) to be clear is the first Rule of Law Principle identified by the Law Council of Australia [38] (p. 1). This is not only so those who are regulated understand what they cannot do, but also to prevent arbitrary enforcement and erosion of respect for the law. Sophocles warned "What you cannot enforce, do not command", and this is relevant to the integrity issues facing racing today. Given the industry is currently permitted to self-regulate, it is even more important that the AR are clear, enforceable, and enforced fairly and consistently. If breaches of rules are hard to quantify, then the rules need to be reworded to allow detection without quantification so that a binary (yes/no) outcome can be recorded.

4.3. Race Length, Location and Whip Rule Breached

We found the length of races at M locations varied significantly with some whip rule breaches. It may be at these tracks, riders wield the whip above the shoulder more (Code 8) and also whip when out of contention (Code 9) more in shorter races, while in longer races there is more opportunity to use additional whip strokes (hence, more Code 17 breaches). So, in shorter races we might be seeing "Fast and Furious" whip use, in which the rider uses whip strokes that arguably have greater force, while also having less time to decide if the horse is out of contention or, indeed, responding.

When we examined race length according to race type, and considered first and second breaches equally, we found most breaches were recorded in sprint races (which may support the "Fast and Furious" hypothesis), with less than half as many breaches recorded in mile and middle distance races. Not surprisingly, Code 16 and 17 breaches accounted for the majority of breaches recorded in all race types, and again reflects the strong recording of these whip breaches that occur prior to the 100 m mark.

4.4. Rider Gender, Type and Whip Breaches

It was outside the scope of this study to collate the raw data for the total number of jockeys and apprentices, and total number of male versus female riders, in all starts in 2013. Hence, we are unable to say which gender/type of rider is more likely to have whip rule breaches recorded. However, we noticed a different pattern of breaches between these groups.

Nearly six times as many male riders had starts with breaches recorded compared to female riders. However, as only the gender of riders with breaches recorded was known, the pertinence of this difference is unclear. Certainly, once one breach was recorded, there was no association between gender and the recording of a second breach. However, we did find that of riders with breaches recorded, female riders were significantly underrepresented for Code 17 breaches (Forehand whip use on more than five occasions prior to 100 m mark), and overrepresented for Code 9 breaches (Whip use when the horse is out of contention). These are interesting findings regarding possible gender effects, but again require further investigation.

Aligning with a previous study of Australian jockeys and apprentices [39], our results also showed that the type of whip rule recorded as breached depended on jockey status. Among riders with breaches, apprentices were significantly more likely than jockeys to have breaches of Code 9 (Whip use when the horse is out of contention) recorded at C and P locations. Given we also found no differences between locations for Code 9 breaches recorded overall, it is possible that, at M locations, jockeys may also be breaching this whip rule at the same frequency as apprentices. This may reflect a greater desire by all riders at the high profile M races to win, or not to run last, or again reflect differences relating to individual Stewards or between the two pools of Stewards.

While we acknowledge that only 1% of starts had whip rule breaches recorded in NSW/ACT in 2013, it is still concerning that just over 50% of riders responsible for breaches were repeat offenders (that is, they had more than one whip rule breach recorded in 2013, including second breaches in a start). Nearly twice as many male riders than female riders were in this category. Of these, there were nearly twice as many jockeys (of which just over half were repeat offenders) than apprentices (of which just under half were repeat offenders). The number of repeat offenders in all categories is troubling, especially apprentices as they are still in training and should be "on probation". It would appear, in 2013 at least, there were insufficient deterrents to repeat offending, and possibly a lack of effective education regarding whip rules for apprentices. In a pilot study, McGreevy et al. found 15 of 24 jockeys said trainers and owners were sources of pressure to use the whip [40]. This and other possible reasons should be further investigated in relation to gender, type of rider, type of whip rule breach, and location.

4.5. Breach Outcomes

Overall, breach outcomes or penalties were low in our view. Reprimands were the most common outcome for first and second breaches, with less than a third of first breaches resulting in fines, and even then the total was a modest $25,600 (median value $200). Further, all Code 6 first breaches (Excessive, unnecessary, or improper whip use) resulted in cautions or reprimands only.

The four most commonly recorded first breaches had variable outcomes. For example, Code 8 first breaches (Whip use that raises arm above jockey's shoulder height) had the highest fines imposed overall (2 × $800), while fines were rare for Code 9 breaches (Whip use when the horse is out of contention), with the highest fine imposed being $300.

In 2013, Stewards imposed only 13 suspensions for whip rule breaches, which ranged from 5 to 14 days, and all resulted from breaches of Codes 16 and 17. No cautions were issued for Code 16 breaches, and only one was issued for a Code 17 breach, which might imply Stewards regard these more seriously than other whip breaches.

We sought the Stewards' "prosecution policy or similar" that helped determine the penalties imposed for whip rule breaches in 2013, as this was not available on Racing NSW's website. The resultant advice was there "is only a guideline" (see Appendix A) [33], and this related to Code 16 and 17 whip breaches only. Although not stated in the Guidelines, suspensions appear to be a serious outcome and are not suggested as outcomes for breaches of Code 16 (Forehand whip use in consecutive strides prior to 100 m mark) or Code 17 (Forehand whip use on more than five occasions prior to the 100 m mark), until the sixth offence for both 4–5 whip strikes in consecutive strides (Code 16) and 4–5 whip strikes in excess of the five allowed under Code 17; for the second offence for six or more whip strikes in consecutive strides (Code 16), and for six or more whip strikes in excess of the five allowed under Code 17.

While Guidelines were provided to us for only these two whip rules, it may be possible Stewards use additional guidelines. Penalties are also always at the Stewards' discretion. Nonetheless, it is telling that in 2013 suspensions were imposed only for Code 16 and 17 breaches, and thus were for whip breaches that took place prior to the 100 m mark.

Although Stewards can impose loss of winnings and earnings on riders breaching whip rules (AR 196. (2)), this did not occur [41]. Overall, the penalties imposed for whip rule breaches in 2013 appear low, with the majority not involving fines or suspensions, and when these were imposed they appeared insufficient to act as serious deterrents (as seen in, what the current authors consider to be, the high number of repeat offenders).

4.6. Riders with the Highest Number of Whip Breaches

Our aim here was to provide a preliminary insight into the type of rider whose whip use attracted the Stewards' attention in 2013. Of 15 riders with the highest numbers of breaches, most were jockeys and male, the number of starts with one or more breaches ranged from six to 13, and nearly 75% of these riders had starts with second breaches.

There were also 13 race meetings where the same riders had breaches in two races, with one rider having breaches in two races at two different race meetings. Four of the riders were apprentices (two males, two females) and seven were jockeys among the 15 riders with the highest number of breaches overall and were male.

When we considered the five riders with the four highest numbers of starts with breaches, their fines as a percentage of their 4.95% share of the total prize money won ranged from 11% to 407%. So, for some riders, the fines for whip breaches appear to have considerable impact on their share of the prize money, but we do not know whether this is compensated for in other races in which they do not breach the whip rules, or are not recorded as doing so. It is interesting that the jockey with the highest number of starts with breaches and the highest fines overall ($4300), rode horses with winnings of $238,335, of which his share was $11,797. For this jockey, 36% of his winnings were lost as fines. It may be that what riders lose in individual races in $ terms from whip rule breaches, is out-weighed by improvements to the riders' winning statistics and, hence, translate to more and higher quality race work.

4.7. Possible Whip Breach Deterrents

While over a million dollars in total prize money was won by horses in starts with breaches, the median percentage of a fine compared to the prize money won in an individual case was only 14%, though this ranged widely from almost zero % (for a horse that won $105,000 but the rider was fined $200) to eight cases in which the fine exceeded the prize money won.

Overall, fines for whip rule breaches represented only 2% of the total prize money won by these horses in 2013. Similarly, we found fines represented almost zero % of total prize money on offer in races with breaches, and like prize money won, was unlikely to act as a meaningful deterrent.

4.8. The 2015 Whip Reforms and the Current Situation

In 2013, backhand whip use, where the whip is held like a ski pole, as opposed to forehand whip use, where the whip is gripped like a tennis racquet [9], was exempted from AR 137A. (5)(i) and (ii) (Codes 16 and 17). McGreevy et al. opined in 2012 [4] that it was possible "that the rules have inadvertently encouraged jockeys to use the backhand rather than the forehand actions to avoid being penalised". The veracity of this concern was reinforced by whip rule reforms enacted on 1 December 2015 by RA. In a letter to jockeys, John Messara AM (Chairman, RA) said:

"Since the introduction of the 2009 reforms, riders have become increasingly proficient at whipping in the backhand manner. Many backhand strikes can be equated in force with forehand strikes.

Australia's current whip rules are not best international practice when benchmarked against other major racing jurisdictions. Racing is accountable for the highest standards of animal welfare, in line with community standards. Community standards require a new regime of whip usage which is tailored towards principles of horsemanship rather than punishment. Australia has an international reputation for leading the world on animal welfare issues and we want to maintain it". [42]

An RA Media Release on 23 October 2015 further explained:

"In respect of Australian Rule of Racing AR. 137A., the Board has decided:

1. To remove the distinction between forehand and backhand whip strikes so that there is a limit of five forehand or backhand whip strikes prior to the 100 m.
2. To introduce stronger penalties for whip offences including greater emphasis on suspensions for serious breaches and for breaches in Group and Listed races.

The rule changes are largely an extension of the whip reforms of 2009. The changes to the whip rules in 2009 introduced limits on the number and manner of whip strikes which in conjunction with a padded whip has ensured the welfare of the horse.

However, too great a reliance on the backhand application of the whip has developed in response to the limits imposed on the forehand application. After careful consideration, the Board decided that backhand strikes should be treated in the same way as forehand strikes so as to leave no room for misinterpretation of the rules against excessive use". [43]

Unfortunately, the 2015 reforms do little to assist interpretation of what constitutes excessive/unnecessary/improper whip use (Code 6-AR 137A. (3)), including in the final 100 m. While whip rule reforms were much needed, it is also unfortunate the reform process was not evidence-based, more transparent, and inclusive to stakeholders including scientists, welfare groups, and the community. What the reforms tell us, though, is that our 2013 data very much underestimated the true extent of whip use prior to the 100 m mark, because backhand strikes were excluded. This is concerning, especially as McGreevy et al. [9] found that while the forehand versus backhand action does not influence the force on impact when using the non-dominant hand, when using the dominant hand, the jockeys in the study (using a model horse), struck with more force in the backhand.

We briefly reviewed the fines and suspensions imposed in the first four months of 2016 with those in the similar 2013 period in NSW/ACT, given RA's 2015 pledge to introduce stronger penalties for whip offences. We found there were four suspensions (range 7–10 days) and 35 fines totalling $9400 in 2013, and in the same period in 2016, there were 15 suspensions (range 4–21 days) and 115 fines totalling $42,650 [20]. It is beyond the scope of the current study to consider this in detail, but given the increase in the number and $ value of fines, and the number and length of suspensions, these data should be examined further. While RA has been very vocal in justifying the higher penalties it imposed at this year's Autumn Racing Carnival [44], industry reporter Adrian Dunn wrote that for this period "NSW—which has held more meetings than any other state—is not facing the same issue as Victoria and Queensland. While NSW has issued more whip-infringement suspensions (12) than Victoria, it has delivered only 99 fines—less than half that of Victoria" [45].

On 24 June 2016, RA announced a review of the new whip rules since their introduction last December [46]. Hopefully, this will examine the reasons for the cited jurisdictional differences above, as well as evaluate whether the reforms have achieved the intended outcomes, which presumably include rectifying the assessment that Australia's whip rules, prior to the reforms, were "not best international practice" [42].

Based on our findings from 2013, when the whip rules were the same as they were immediately prior to the December 2015 reforms, there is a case to argue that a review of all the whip rules is warranted, and not just those pertaining to whip use prior to the 100 m mark. Any review should be science-based and include independent comparison with the UK where, unlike Australia, the total number of whip strikes is restricted. Nonetheless, the increased number of fines and suspensions in the 2016 period may indicate a step toward better regulation of whip use and welfare, even if it is just that backhand strikes are now being penalised.

Knight and Hamilton [47] stated in relation to the UK Review that "The conclusion that the whip contributes to rider safety has important implications under Australian Work Health and Safety legislation and therefore affects the ability of racing authorities to change the rules relating to whip use". We argue this makes it all the more important that studies that direct policy and regulatory change are based on peer-reviewed studies using race data, much of which are already being collected by industry. Some whip use information is not currently available, but must be if properly informed decisions are to be made about the use of whips in racing. In 2012, McGreevy and Ralston raised the case "for publication of the numbers of official whip strikes per horse per race, a move that is likely to increase transparency in the activity of stewards" [39].

Further, in her paper "Sustainability, thoroughbred racing and the need for change", Bergmann wrote: "Improving transparency and regulation is important and can improve welfare outcomes, if transparency and regulation go beyond the aim of protecting the integrity of the race and shift the focus on protecting the horse" [10]. Similarly, in their paper "Science alone is not always enough; the importance of ethical assessment for a more comprehensive view of equine welfare" [11], Heleski and Anthony quote Grandin's saying that it is easy for "bad to become normal" [48].

5. Conclusions

This multi-disciplinary study provides the first peer-reviewed characterisation of whip rule breaches recorded by Stewards and their regulatory outcomes in horseracing. This was achieved using Stewards Reports and Race Diaries, and improves understanding of whip use and compliance in a major Australian racing jurisdiction. Importantly, it is a step toward rectifying the current knowledge gap in this increasingly contentious area.

Whip use in horse racing is an important animal welfare issue. While there is a lack of peer-reviewed evidence that specifically demonstrates that whipping horses causes pain, we argue there is sufficient evidence now and a moral and legal imperative to assume that, in the absence of evidence to the contrary, whipping is potentially painful. As such, restrictions on whip use have the real potential to improve the welfare of horses in racing.

Not only is there growing community concern about the welfare of horses being whipped in races, whipping, and its regulation, is also a racing integrity issue. Our finding that there was a significantly higher frequency of recorded breaches by riders of horses that finished first, second, or third than by riders of horses that finished in other positions, suggests a desire to win may incentivise whip rule breaches and potentially affect race and betting outcomes. Likewise our finding that there was a significantly higher frequency of recorded breaches at M than at C or P locations may reflect the desire to win may be particularly high at the more prestigious 'city' race tracks. This preliminary study raises a number of questions that need to be answered including whether the number of breaches recorded by Stewards aligns with the actual number of whip rule breaches at all locations, and what Steward and surveillance-related factors might affect whether a breach is recorded or not.

It is also concerning from a regulatory point of view that a small number of types of whip breach predominated in 2013, with no breaches recorded for several whip rules, despite evidence to suggest some of these may be breached regularly. Our study showed that over half of all first breaches recorded occurred prior to the 100 m mark and involved two whip rules. Unfortunately, with the exception of these, the Stewards reports do not specify at which race stage whip breaches occur. Hence it is not clear how many breaches are recorded in the crucial final 100 m of races. Regulatory outcomes or penalties were generally low and, in our view, insufficient to act as deterrents. Our findings strongly suggest that the whip rules, their surveillance and recording, and the penalties imposed warrant urgent and independent review.

Likewise, the associations our study reveals between recorded whip rule breaches, racing data variables, and outcomes provide a strong case for such analyses to be undertaken annually to inform the evidence-base for policy, education, and regulatory change, including breaches and penalties.

Indeed, this study raises several issues pertinent to horse welfare and racing integrity that warrant further investigation. We hope it will also lead to more evidence-based whip rules that meet growing community expectations that horses in racing will be treated humanely.

Supplementary Materials: The following are available online at www.mdpi.com/2076-2615/7/1/4/s1. Table S1: Details of tracks, race meetings, races, races with breaches, starts, and starts with breaches for Country, Metropolitan, and Provincial locations, Table S2: Whip rules breached in second breaches classified by Country, Metropolitan, and Provincial location, Table S3: Fines resulting from first breaches, Table S4: Outcomes of second breaches classified by whip rule breached, Table S5: Total prize money on offer in races with breaches by location, Table S6: Length of races with breaches, Table S7: Riders with breaches (including repeat offenders), Table S8: Riders with the highest numbers of breaches, Table S9: Riders breaching whip rules in two races at the same race meeting, Table S10: BOBS and Non-BOBS races classified by Country, Metropolitan, or Provincial location, Table S11: Percentage of BOBS horses in starts classified by Country, Metropolitan or Provincial location, Table S12: BOBS status of 339 horses with breaches classified by Country, Metropolitan or Provincial location, Table S13: Whip rules breached in first breaches in BOBS and Non-BOBS races.

Acknowledgments: This work was a collaboration between the University of Sydney (Phil McManus and Paul McGreevy) and independent researchers (Jennifer Hood, Carolyn McDonald and Bethany Wilson). This study and the costs of publishing in open access were funded by an Australian Research Council grant (Caring for Thoroughbreds) held by Phil McManus and Paul McGreevy. The authors thank the stewards at Racing NSW for their advice on a number of regulatory matters. The authors also thank the independent reviewers of this paper for their invaluable assistance.

Author Contributions: Paul McGreevy and Jennifer Hood conceived and designed the experiments; Jennifer Hood and Carolyn McDonald collected the data; Jennifer Hood and Bethany Wilson analysed the data; Jennifer Hood, Bethany Wilson, Carolyn McDonald, Paul McGreevy and Phil McManus wrote the paper and approved the article's content.

Conflicts of Interest: The authors declare no conflict of interest.

Abbreviations

The following abbreviations are used in this manuscript:

ACT	Australian Capital Territory
AR	Australian Rules of Racing
ARB	Australian Racing Board
BHA	British Horseracing Authority
BOBS	Breeder Owner Bonus Scheme
C	Country
LR	Local Rules of Racing
M	Metropolitan
NSW	New South Wales
P	Provincial
RA	Racing Australia
$	AUS dollar

Appendix

Racing NSW

Rider Penalty Guidelines for Whip Rule breaches

(To be used as a guide only with all circumstances of the breach to be considered, penalties to remain at the discretion of the Stewards)

Table A1. 400–100 metres Successive strokes within the 5 permitted strokes AR 137A (5)(a)(i)

Offence	Successive (2)	Successive (3)	Successive (4–5)	Successive (6+)
1st	Reprimand	Reprimand	$200	$500
2nd	Reprimand	Reprimand	$300	Suspension up to 1 week
3rd	Reprimand	$200	$400	Suspension up to 1 week
4th	$200	$200	$500	Suspension up to 2 weeks
5th	$200	$200	$600	Suspension up to 2 weeks
6th	$200	$300	Suspension up to 1 week	Suspensions up to 3 weeks
7th	$300	$300	Suspension up to 1 week	Suspensions up to 3 weeks
8th	$300	$300	Suspension up to 2 weeks	Suspension up to 4 weeks
9th	$300	$400	Suspension up to 2 weeks	Suspension up to 4 weeks
10th	$400	$400	Fine & Suspension 2 weeks +	Suspension 4 weeks +

Please note number of successive strokes includes total strokes ie 3 = 3 in successive.

Table A2. 400–100 Metres Additional Strokes in excess of 5 strokes AR 137(A) (5)(a)(ii)

Offence	1 Additional	2 Additional	3 Additional	4–5 Additional	6 + Additional
1	Reprimand	Reprimand	Reprimand	$200	Suspension up to 1 week
2	Reprimand	Reprimand	Reprimand	$300	Suspension 1 week +
3	Reprimand	Reprimand	$200	$400	Suspension up to 2 weeks
4	Reprimand	$200	$200	$500	Suspension 2 weeks +
5	$200	$200	$200	$600	Suspension up to 3 weeks
6	$200	$200	$300	Suspension up to 1 week	Suspension 3 weeks +
7	$200	$300	$300	Suspension up to 1 week	Suspension up to 4 weeks
8	$200	$300	$300	Suspension up to 2 weeks	Suspension 4 weeks +
9	$300	$300	$400	Suspension up to 2 weeks	Suspension 4 weeks +
10	$300	$400	$400	Fine & Suspension 2 weeks +	Suspension 4 weeks +

For repeated offences involving 6 or more additional or successive strides the penalty may include a fine or the equivalent to the riders percentage of prize money; As a deterrent breaches in Group and Feature races may attract a heavier penalty; If a rider commits a minor breach of both rules, only one, penalty shall apply, that being the greater penalty provided for in the two respective tables i.e., a 7th time offender using one additional stroke of the whip, including one successive application will be eligible to be fined $300 in total; A riders record refreshes after 400 rides or 12 calendar months; Percentage of Prize money means the riders percentage is expressed as a maximum figure for penalty; A Riders breach of the whip rules prior to the 26th of September 2009 are not to be taken into account when assessing penalty.

References

1. Jones, B.; McGreevy, P.D. Ethical equitation: Applying a cost-benefit approach. *J. Vet. Behav.* **2010**, *5*, 196–202. [CrossRef]
2. McGreevy, P.D.; Oddie, C. Holding the whip hand—A note on the distribution of jockeys' whip hand preferences in Australian thoroughbred racing. *J. Vet. Behav.* **2011**, *6*, 287–289. [CrossRef]
3. Evans, D.; McGreevy, P. An investigation of racing performance and whip use by jockeys in thoroughbred races. *PLoS ONE* **2011**, *6*, e15622. [CrossRef] [PubMed]
4. McGreevy, P.D.; Corken, R.A.; Salvin, H.; Black, C.M. Whip use by jockeys in a sample of Australian thoroughbred races—An observational study. *PLoS ONE* **2012**, *7*, e33398. [CrossRef] [PubMed]
5. Graham, R.; McManus, P. Changing human-animal relationships in sport: An analysis of the UK and Australian horse racing whips debates. *Animals* **2016**, *6*, 32. [CrossRef] [PubMed]
6. Horsetalk.co.nz. Ban on Whip Use Would Be Positive for Racing, Suggests John Francome. Available online: http://horsetalk.co.nz/2015/11/12/ban-whip-use-positive-racing-francome/#axzz40UAtJE5b (accessed on 23 June 2016).
7. McLean, A.N.; McGreevy, P.D. Ethical equitation: Capping the price horses pay for human glory. *J. Vet. Behav.* **2010**, *5*, 203–209. [CrossRef]
8. Jones, B.; Goodfellow, J.; Yeates, J.; McGreevy, P. A critical analysis of the British Horseracing Authority's review of the use of the whip in horseracing. *Animals* **2015**, *5*, 138–150. [CrossRef] [PubMed]
9. McGreevy, P.D.; Hawson, L.A.; Salvin, H.; McLean, A.N. A note on the force of whip impacts delivered by jockeys using forehand and backhand strikes. *J. Vet. Behav.* **2013**, *8*, 395–399. [CrossRef]
10. Bergmann, I. Sustainability, thoroughbred racing and the need for change. *Pferdeheilkunde* **2015**, *31*, 490–498.

11. Heleski, C.R.; Anthony, R. Science alone is not always enough: The importance of ethical assessment for a more comprehensive view of equine welfare. *J. Vet. Behav.* **2012**, *7*, 169–178. [CrossRef]
12. British Horseracing Authority. Responsible Regulation: A Review of the Use of the Whip in Horseracing. 2011. Available online: http://www.britishhorseracing.com/wp-content/uploads/2014/03/WhipReview.pdf (accessed on 23 June 2016).
13. Lewin, G.R.; Moshourab, R. Mechanosensation and pain. *J. Neurobiol.* **2004**, *61*, 30–44. [CrossRef] [PubMed]
14. Taylor, P.M.; Crosignani, N.; Lopes, C.; Rosa, A.C.; Luna, S.P.L.; Puoli Filho, J.N.P. Mechanical nociceptive thresholds using four probe configurations in horses. *Vet. Anaesth. Analg.* **2016**, *43*, 99–108. [CrossRef] [PubMed]
15. SBS News. Greens back RSPCA Call to Ban Horse Whips. 26 February 2015. Available online: http://www.sbs.com.au/news/article/2012/03/21/greens-back-rspca-call-ban-horse-whips (accessed on 8 July 2016).
16. The Coalition for the Protection of Racehorses. Proposal for the Phasing out of the Whip in Australian Thoroughbred Racing March 2015. Available online: https://www.horseracingkills.com/campaigns/the-whip/proposal-for-phasing-out-the-whip/ (accessed on 30 November 2016).
17. The Australian. As Public Turns against Racing Codes, Whip Goes Under Review. Available online: http://www.theaustralian.com.au/sport/opinion/patrick-smith/as-public-turns-against-racing-codes-whip-goes-under-review/news-story/f3cc57b7ae1402a3a8874623a20d233c (accessed on 23 June 2016).
18. RSPCA Policy C05 Horse Racing. Available online: http://kb.rspca.org.au/RSPCA-Policy-C05-Horse-Racing_642.html (accessed on 23 June 2016).
19. British Horseracing Authority. Responsible Regulation: A Review of the Use of the Whip in Horseracing. Annex A. 2011. Available online: http://www.britishhorseracing.com/wp-content/uploads/2014/03/WhipReview.pdf (accessed on 15 July 2016).
20. Racing NSW Stewards Reports. Available online: http://new.racingnsw.com.au/default.aspx?s=race-diary-stewards&filter=Stewards (accessed on 23 June 2016).
21. Racing NSW Race Diary. Available online: http://racing.racingnsw.com.au/FreeFields/Calendar_Meetings.aspx?State=NSW (accessed on 23 June 2016).
22. Racing NSW BOBS General Information. Available online: http://www.racingnsw.com.au/default.aspx?s=general-information-bobs (accessed on 23 June 2016).
23. Racing NSW. Rules of Racing of Racing NSW. Available online: http://www.racingnsw.com.au/site/_content/document/00000401-source.pdf (accessed on 23 June 2016).
24. R Core Team. *R: A Language and Environment for Statistical Computing*; R Foundation for Statistical Computing: Vienna, Austria, 2016; Available online: http://www.R-project.org/ (accessed on 23 June 2016).
25. Murrihy, R. Chief Steward NSW Racing, Sydney, NSW, Australia. 12 May 2016.
26. Love, E.J.; Murrell, J.; Whay, H.R. Thermal and mechanical nociceptive threshold testing in horses: A review. *Vet. Anaesth. Analg.* **2011**, *38*, 3–14. [CrossRef] [PubMed]
27. Mills, P.C.; Higgins, A.J. Investigation of the potential of whips to injure horses. In Proceedings of the 11th International Conference of Racing Analysts and Veterinarians, Queensland, Australia, 18–24 May 1996.
28. National Health and Medical Research Committee. *Australian Code for the Care and Use of Animals for Scientific Purposes*, 8th ed.the Code; National Health and Medical Research Committee: Canberra, Australia, 2013. Available online: https://www.nhmrc.gov.au/guidelines-publications/ea28 (accessed on 25 November 2016).
29. British Horseracing Authority. BHA Briefing: New Figures Show Whip Offences Continued to Fall in 2015. Available online: http://www.britishhorseracing.com/wp-content/uploads/2016/01/BHA-BRIEFING-2015-Whip-Data-14-01-16.pdf (accessed on 15 July 2016).
30. Murrihy, R. Chief Steward NSW Racing, Sydney, NSW, Australia. 21 April 2016.
31. Van Gestel, M. Deputy Chairman of Stewards—Operations, NSW Racing, Sydney, NSW, Australia, 29 June 2016.
32. Racing NSW. Annual Report 2014. Available online: http://www.racingnsw.com.au/site/_content/document/00001257-source.pdf (accessed on 24 June 2016).
33. Van Gestel, M. Deputy Chairman of Stewards—Operations, NSW Racing, Sydney, NSW, Australia, 13 May 2016.
34. Australasian Legal Information Institute. Prevention of Cruelty to Animals Act 1979. Available online: http://www.austlii.edu.au/au/legis/nsw/consol_act/poctaa1979360/s4.html (accessed on 11 July 2016).

35. British Horseracing Authority. Guide to Procedures and Penalties 2016. Available online: http://www.britishhorseracing.com/wp-content/uploads/2014/03/guide-procpen-2016.pdf (accessed on 15 July 2016).
36. International Federation of Horseracing Authorities. International Agreement on Breeding, Racing and Wagering 25 January 2016. Available online: http://www.horseracingintfed.com/resources/2015Agreement.pdf (accessed on 8 July 2016).
37. International Group of Specialist Racing Veterinarians (IGSRV). WhipChip the Electronic Whip. Available online: http://igsrv.org/whipchip:-the-electronic-whip (accessed on 15 July 2016).
38. Law Council of Australia. Policy Statement–Rule of Law Principles. Available online: http://www.lawcouncil.asn.au/lawcouncil/images/LCA-PDF/a-z-docs/PolicyStatementRuleofLaw.pdf (accessed on 4 July 2016).
39. McGreevy, P.D.; Ralston, L. The distribution of whipping of Australian Thoroughbred racehorses in the penultimate 200 m of races is influenced by jockeys' experience. *J. Vet. Behav.* **2012**, *7*, 186–189. [CrossRef]
40. McGreevy, P.D.; Caspar, G.L.; Evans, D.L. A pilot investigation into the opinions and beliefs of Australian, British, and Irish jockeys. *J. Vet. Behav.* **2013**, *8*, 100–105. [CrossRef]
41. Van Gestel, M. Deputy Chairman of Stewards—Operations, NSW Racing, Sydney, NSW, Australia, 17 May 2016.
42. Racing Australia Media Release. Letter from Racing Australia Chairman—Sent To All Jockeys Today 19 November 2015. Available online: http://www.racingnsw.com.au/default.aspx?s=article-display&id=19005 (accessed on 23 June 2016).
43. Racing Australia Media Release. Whips. 23 October 2015. Available online: http://www.racingaustralia.horse/uploadimg/media-releases/Racing%20Australia%20Media%2023%20October%202015.pdf (accessed on 23 June 2016).
44. Racenet.com.au. Berry's Penalty the Severest of Whip Infringers. 2 April 2016. Available online: https://www.racenet.com.au/news/121649/Berry%E2%80%99s-penalty-the-severest-of-whip-infringers (accessed on 24 June 2016).
45. G1X. Stewards to Get Cracking on the Whip Rule. Available online: https://www.g1x.com.au/news/racing/stewards-to-get-cracking-on-the-whip-rule (accessed on 23 June 2016).
46. Racing Australia Media Release 24 June 2016. Whip Rules. Available online: http://www.racingaustralia.horse/uploadimg/media-releases/Racing%20Australia%20Media%2024%20June%202016.pdf (accessed on 30 July 2016).
47. Knight, P.K.; Hamilton, N.A. Handedness of whip use by Australian jockeys. *Aust. Vet. J.* **2014**, *92*, 231–234. [CrossRef] [PubMed]
48. Grandin, T. *Improving Animal Welfare: A Practical Approach*; CAB International: Oxfordshire, UK, 2010; p. 2.

2

Prepartum Lying Behavior of Holstein Dairy Cows Housed on Pasture through Parturition

Christa A. Rice, Nicole L. Eberhart and Peter D. Krawczel *

Department of Animal Science, University of Tennessee, Knoxville, 2506 River Dr. 258 Brehm Animal Science Knoxville, Knoxville, TN 37996, USA; ckurman@utk.edu (C.A.R.); neberhar@vols.utk.edu (N.L.E.)
* Correspondence: krawczel@utk.edu

Academic Editor: Clive J. C. Phillips

Simple Summary: Dairy cows experience meaningful biological changes during gestation that impact cow comfort and alter behavior, particularly during late gestation and leading up to calving. The housing environment can also have a positive or negative effect on cow comfort. Pasture access allows cows the freedom of movement and an ability to express natural grazing and resting behaviors. After observing cows housed on pasture during the late gestation and calving periods, this study found that lying behaviors only differed on the day of calving and the day prior to calving. Additionally, the proportion of time spent lying per hour decreased in the hour prior to calving compared to 6 h prior to parturition. The altered lying behaviors and activity observed in the hours before calving may indicate a decrease in cow comfort experienced by the cow during parturition. However, discomfort is typical of parturition. These data suggest that cows were able to express natural behaviors associated with calving and pasture when provided an adequate environment for cows during the prepartum period.

Abstract: Utilizing pasture-based systems may increase cow comfort during late gestation and calving as it lacks the constraints of confinement housing. The objective of this study was to quantify lying behavior and activity of Holstein dairy cows housed on pasture during the 6 d before calving. Sixteen Holstein dairy cows were moved to pasture 3 weeks before their projected calving date. Data loggers were attached 14 d prior to projected calving date. Behavior was evaluated 6 d before calving for all cows (n = 16) and 6 h prior to calving for a subset of cows (n = 6) with known calving times. Data loggers recorded at 1-min intervals to determine lying time (h/d and %/h), lying bouts (n/d and n/h), lying bout duration (min/bout), and steps (n/d and n/h). A repeated measures analysis of variance with contrasts was performed to determine if lying behaviors and activity differed between baseline and day or hour of interest. Lying time was greater 6 d prior to calving compared to the day of and before calving. Cows had longer lying bouts 6 d prior to calving compared to day of calving. Cows spent less time lying in the hour before calving compared to 6 h prior to parturition. The lack of change in behavior and activity during the 7 d prior to calving may indicate that pasture provided an adequate environment for cows during the prepartum period but did not impact cow welfare in the hours leading up to calving.

Keywords: transition cow; lying behavior; pasture

---

1. Introduction

A dairy cow experiences inherent stressors during the late stages of gestation, however, providing an appropriate housing environment may help alleviate the behavioral impact of these stressors. Housing options for late gestation dairy cows in the United States vary widely by producer [1]. This variation in housing indicates that there exists no consensus on the best approach for managing dairy cows during this stage of lactation.

In 2014, dairy cow management data was collected via interviews and questionnaires from dairy producers [1]. Pasture access for dry cows was common with 72.3% of operations allowing dry cows access to pasture (11.3% of those operations used pastures as a primary housing system for dry cows) and 34.0% of dry cows having some access to pasture [1]. For operations relying on indoor housing for dry cows, 18.2% of operations used tie stalls or stanchions with no outside access, 9.5% utilized freestalls without outside access, and 20.3% used freestalls with outside access [1].

The current understanding of dairy cow behavior surrounding calving is based on data from confinement housing [2–4]. Cows housed in freestalls spent approximately 10–11 h per d lying which was consistent over 10 d before and after calving [2]. Cows housed in individual calving pens 4 d prior to calving exhibited a decrease of 104 min/d in lying time and an increase of 0.8 bouts/d in lying bouts when compared to the 24 h before calving [3]. A competitive feeding situation tended to decrease lying time during the week before calving with competitively fed cows lying for 494 min/d vs. non-competitively fed cows which spent 641 min/d lying [4]. The number of standing bouts increased to 17.3 bouts/d during the 24-h period before and after calving relative to the mean number of bouts during the 10 d before and after calving (11.7–13.1 bouts/d) [2].

The greatest changes in behavior around calving occurred in the 24 h before and after parturition. Lying time decreased by approximately 2 h and standing bouts increased during this period [2] compared to pre- and post-calving periods. Cows housed in individual calving pens increased their lying bouts and overall activity during the 6 h before calving [3]. Calving difficulty also influenced behavior during this time; cows who calved unassisted increased their lying bout frequency over the 6 h before parturition, but cows who needed assistance only increased their lying bout frequency 2 h before parturition [5].

Many dairy producers have begun to recognize the benefits of pasture at a variety of stages of the lactation cycle. These benefits include decreased lameness and reduced restriction to natural behaviors, such as lying down, by freestall hardware [6,7]. Lying or resting behavior is an essential component of cow comfort and is an indicator of animal welfare. Cows housed indoors spend approximately 12 h per day lying or resting [8] and cows housed on pasture spend approximately 9 h per day engaging in lying behaviors [9]. Overall, cows prefer soft bedding surfaces and stall dimensions that minimize contact with the hardware of freestalls [10–12]. Pasture offers a soft surface and is free of restrictive freestall hardware, making it easier for cows to lie down comfortably. This suggests pasture may alleviate some of the inherent discomfort during the late gestation period, as indicated by the laterality that occurs throughout this period [13]. More importantly for the context of this study, the pasture-based system may reflect the preference of prepartum dairy cows. When given a choice between a freestall barn or pasture mid-lactation, cows spent less than 30% of their time on pasture between the morning and evening milkings, but spent 90% of their time on pasture at night [14]. Since the cows in this study were mid-lactation animals, it is likely that their feeding and milking routines affected where they chose to spend time, especially during the day [14]. Despite the confounding effects of feeding and milking routines, these data support the previous hypothesis that pasture may be more comfortable for cows than confinement systems, making it an ideal option for housing cows in the prepartum period because cows often experience stress and discomfort during this time [15].

Despite the common use of pasture in the dry period, to our knowledge, all research to date on lying behavior in transition cows has been performed in a confinement or freestall setting while preference for pasture was only assessed in mid-lactation animals. The objective of our study was to quantify the lying behavior and activity of Holstein dairy cows housed on pasture during the 6 d before calving and for the 6 h before calving for a subset of cows. It was hypothesized that lying behavior and activity would be unaffected before calving due to the improved ability of free movement provided by the pasture.

2. Materials and Methods

2.1. *Animals, Housing, and Measurement*

All cows were housed at the University of Tennessee's East Tennessee Research and Education Center (Walland, TN, USA). Sixteen Holstein dairy cows going into their second or greater lactation with a mean 305 ME of 12,380 kg $\pm$ 702.1 kg and mean 245 $\pm$ 5.5 DIM during previous lactations were moved from freestall housing to pasture approximately 3 weeks before their expected calving date. Behavior and activity data were collected from the 16 cows for 6 d before calving. A subset of 6 cows from the original 16 cows with known calving times were used to determine behavior 6 h before calving. The study took place from March to May with cattle on pasture being maintained as a dynamic group. The cows were housed in the same pasture, but not simultaneously. As groups of cows approached day 42 (42 days into the dry period), they were moved from the freestalls to the pasture. Pasture population ranged from a minimum of 2 cows to a maximum of 18. Cows were moved to pasture in groups of 10 (including cows not enrolled in the study) on a weekly basis as they were close to calving and removed from the pasture group immediately after calving. Cows in the pasture were checked approximately eight times per day by farm staff.

The pasture was 7.9 acres of orchard grass and fescue. Cows were moved a distance of 167 m by way of a fenced drover's lane from the barn to the pasture. Supplemental feed offered within the pasture consisted of round baled wheat, rye grass, or Sudan grass, and refusal TMR from the lactating herd. Round bales were fed as needed on a concrete feed pad which was 6 $\times$ 15 m with six concrete feed bunks and a hay rack measuring 2 $\times$ 3 m. Water was available ad libitum from an automated waterer surrounded by gravel located near the center of the pasture. Cows were able to seek natural shade in two areas (0.07 acres and 0.27 acres, respectively). Cows were checked multiple times throughout the day and during feeding. The University of Tennessee IACUC (protocol #1982-0111) approved all animal procedures.

2.2. *Data Collection*

IceTag data loggers (IceRobotics Ltd., Edinburgh, Scotland, UK) were attached to the cow's hind legs approximately 14 d before the expected calving date to collect data on lying time (min/d), number of lying bouts (bouts/d), duration of lying bouts (minutes/bout), and steps/d. The use of IceTag loggers for this type of research was previously validated [16]. A lying bout was defined as $\geq$2 min of uninterrupted lying [17]. For the subset of six cows, data loggers were used in the same manner and proportion of the hour spent lying (%/h), number of lying bouts (bouts/h), and steps/h were evaluated for 6 h prior to calving. Because some lying bouts lasted longer than an hour, they were excluded from the hourly analysis. For the daily analysis, data was summarized with day beginning at midnight.

2.3. *Data Analysis*

The individual cow was the experimental unit for all analyses. A repeated measures analysis of variance with contrasts using SAS (v9.3, SAS, Inc., Cary, NC, USA) was performed to determine if lying behaviors and activity differed between baseline and day or hour of interest. For evaluation of the daily means, the behavioral responses on day-6 was used as a baseline and the mean response of each subsequent day was compared against it. For the evaluation of hourly means, the behavioral responses of lying time (min/h) and steps/h were compared to the response 6 h prior to calving. Significance was declared at $p \leq 0.05$.

3. Results

3.1. Lying Time

Mean daily lying time of cows was (mean ± SE) 10.3 ± 0.3 h/d in the 6 days prior to calving. Lying time decreased to 9.3 ± 0.6 the day before calving and to 8.0 ± 0.8 the day of calving ($p = 0.006$ and $p = 0.009$, respectively; Figure 1a). Cows in the subset spent a mean proportion of 24.0% ± 4.1% of each hour lying during the 6 h before calving. Proportion of lying decreased in the hour prior to calving (11.8% ± 5.6%, $p = 0.04$; Figure 1b).

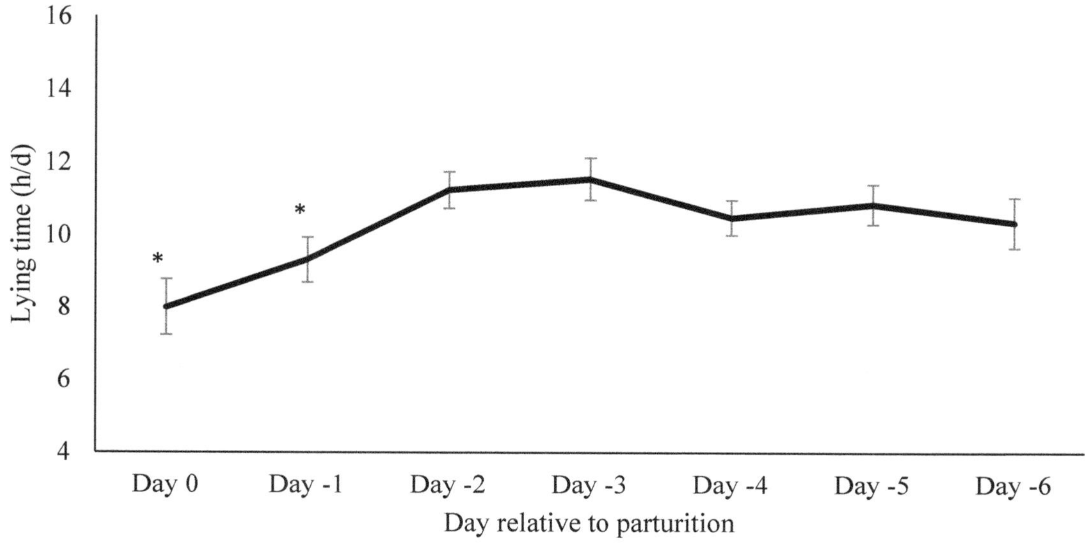

(a)

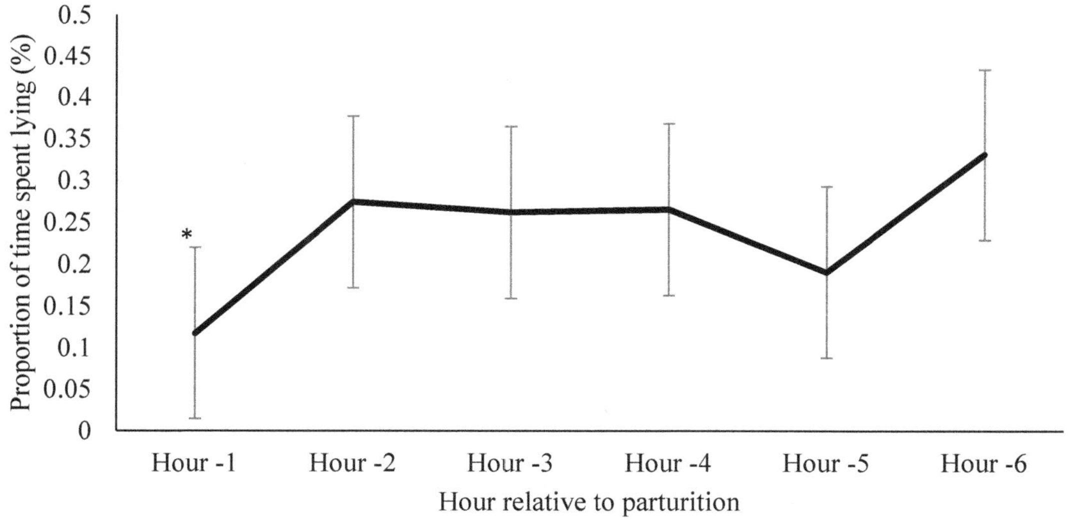

(b)

Figure 1. (**a**) Lying time per day relative to parturition. Changes in average lying time in minutes one week before calving in 16 cows. All days were compared to day-6 as a baseline (mean ± SE). * indicates $p < 0.01$. On the days of and before calving cows spent less time lying down compared to d-6. * indicates $p < 0.05$; (**b**) Proportion of time spent lying per hour relative to parturition. Mean hourly lying time 6 h before calving in six cows (mean ± SE). All hour intervals were compared to 6 h prior to calving as a baseline. In the hour before calving (hour-1), cows spent less time lying than 6 h prior to calving ($p = 0.0360$). * indicates $p < 0.05$.

3.2. Number of Lying Bouts

Cows in the current study had a mean daily number of lying bouts of 10.0 $\pm$ 0.4 n/d throughout the study. Lying bouts did not differ in the days leading up to parturition ($p \geq 0.07$; Figure 2a). For the subset of cows, mean number of lying bouts during the 6 h before calving was 11.8 $\pm$ 2.2 bouts/h. The number of lying bouts did not differ in the hours leading up to parturition ($p \geq 0.2987$; Figure 2b).

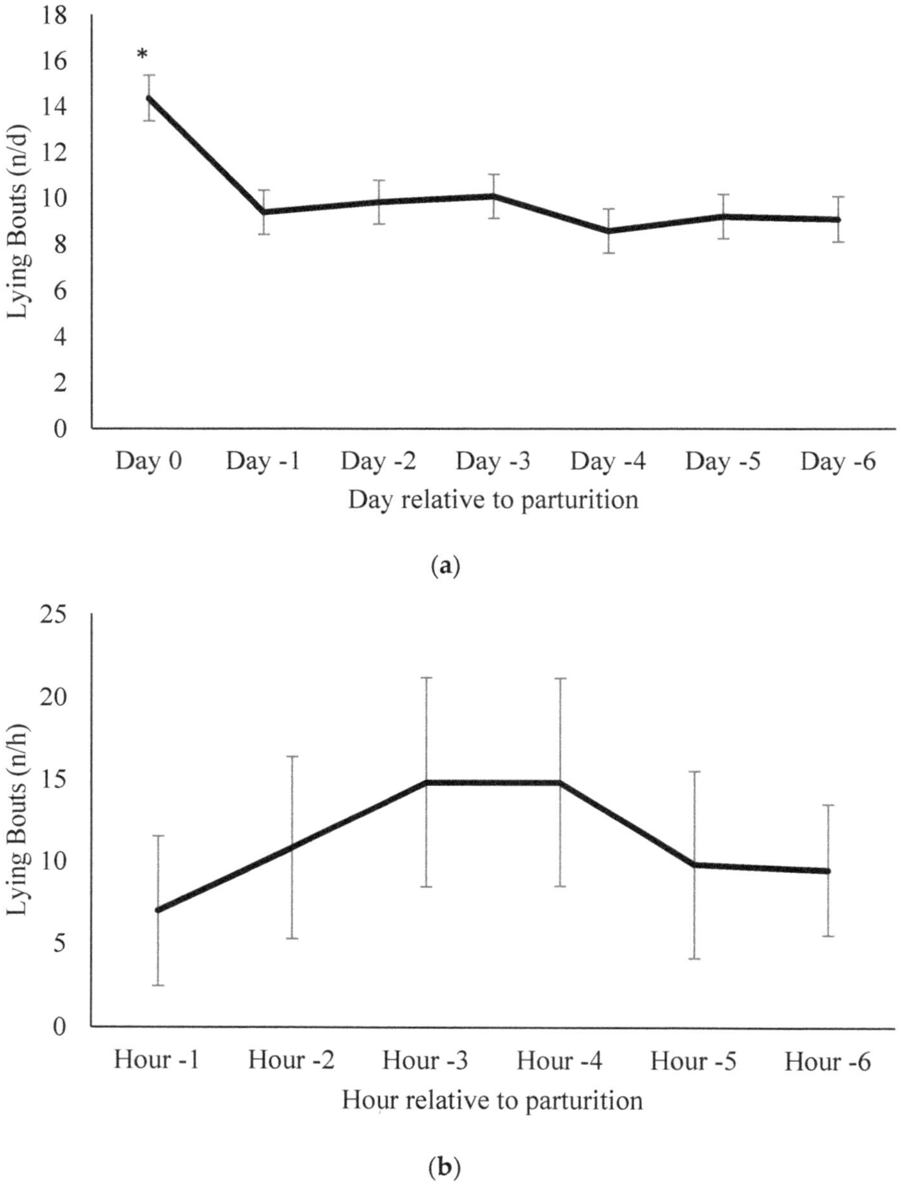

Figure 2. (**a**) Lying bouts per day relative to parturition. Mean number of lying bouts one week before calving in 16 cows (mean $\pm$ SE). All days were compared to day-6 as a baseline. On the day before calving cows had a greater number of lying bouts compared to day-6. * indicates $p < 0.05$; (**b**) Lying bouts per hour relative to parturition. Mean hourly lying bouts 6 h before calving in six cows (mean $\pm$ SE). There were no differences in lying bouts between the baseline (−6 h) and the hours leading up to parturition.

3.3. Lying Bout Duration

Mean lying bout duration in the days leading up to calving was 96.9 ± 14.9 min/d. Lying bout duration was shorter the day of calving compared to day-6 ($p < 0.0001$; Figure 3).

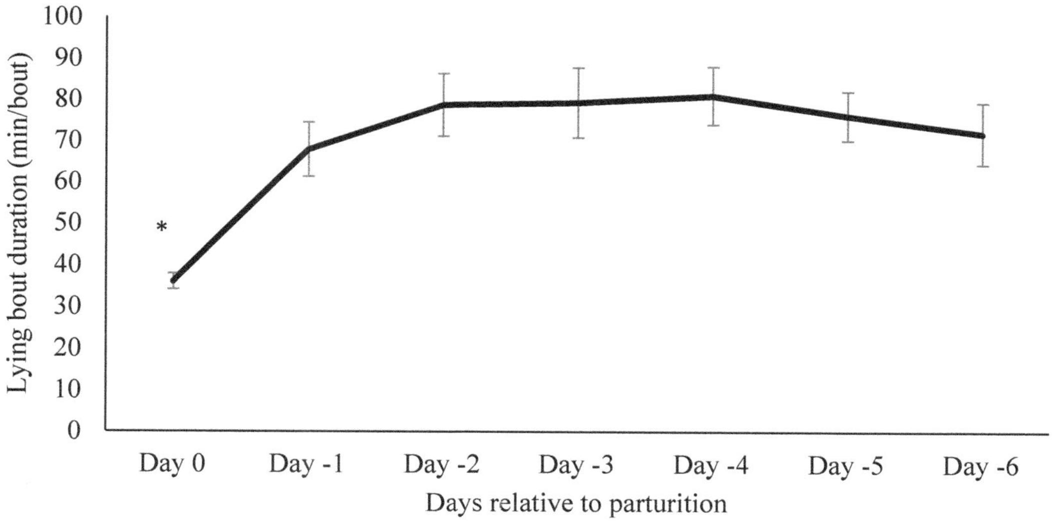

Figure 3. Lying bout duration in the days leading up to parturition. Mean lying bout duration in minutes one week before calving in 25 cows (mean ± SE). All days were compared to day-6 as a baseline. Lying bouts were shorter on the day of calving compared to day-6. * indicates $p < 0.05$.

3.4. Steps

Cows engaged in a mean of 3369.1 ± 76.0 steps/d during the days before calving. The number of steps did not differ between the days before calving ($p \geq 0.48$; Figure 4a). During the 6 h before calving for the subset of cows there was an average of 142.8 ± 7.7 steps/h. Steps/h did not differ during the 6 h period before calving ($p \geq 0.18$; Figure 4b).

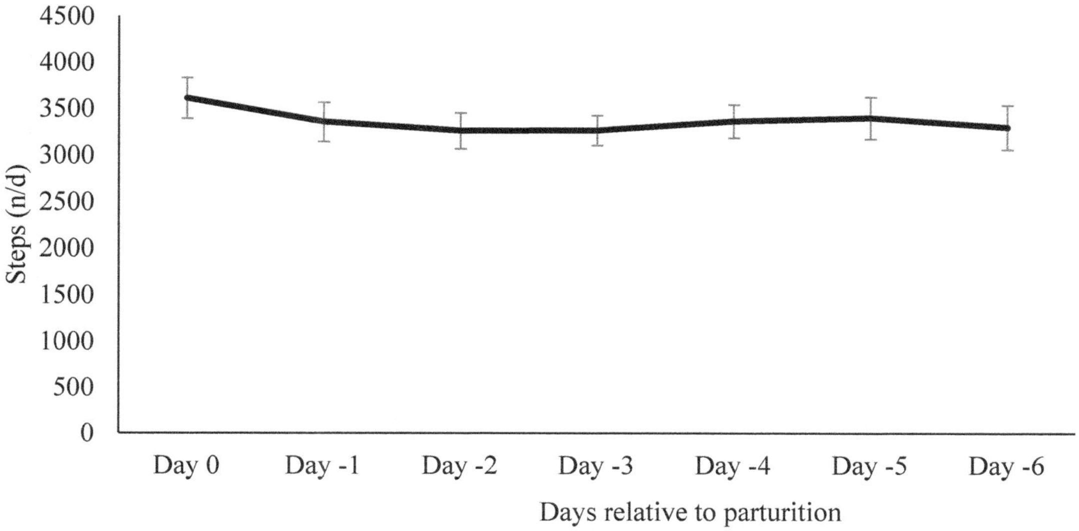

(**a**)

Figure 4. *Cont.*

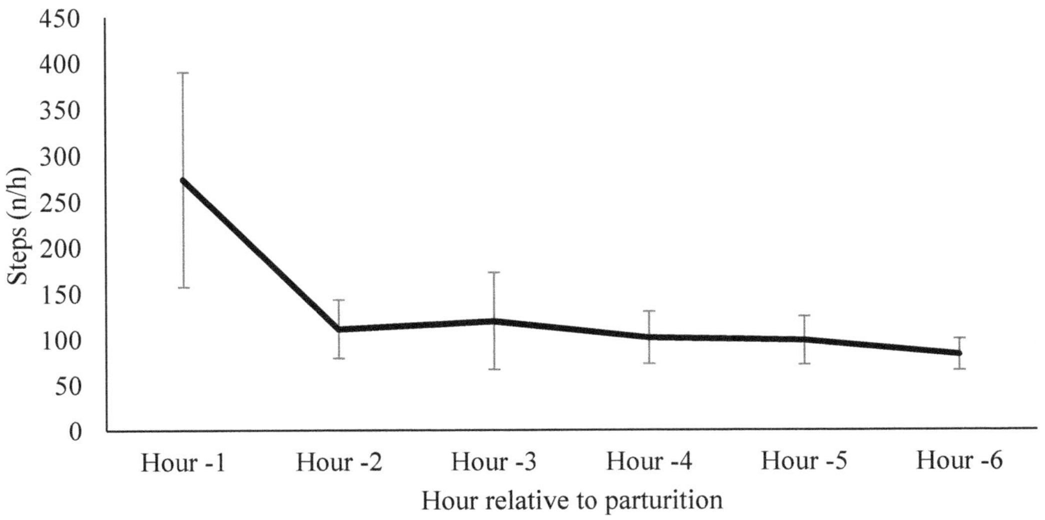

(**b**)

Figure 4. (**a**) Steps per day relative to parturition. Mean number of steps one week before calving in 16 cows (mean ± SE). All days were compared to day-6 as a baseline. There were no differences in the number of steps in the 7 d before calving compared to day-6; (**b**) Steps per hour relative to parturition. Mean number of steps 6 h before calving in six cows (mean ± SE). All hour intervals were compared to the baseline (6 h before calving). There were no differences in activity in the hours leading up to parturition.

4. Discussion

4.1. Lying Time

The observed lying time response was consistent with lying time reported for primiparous and multiparous cows fed in non-competitive conditions, where lying time ranged from approximately 600 to 650 min/d [4]. In contrast, values from the current study were considerably lower than the mean responses of 738 min/d and 985 min/d reported for cows housed in confinement over the last 10 or 4 d before calving [2,3]. When comparing freestall and pasture, cows on pasture had the lowest daily lying times (approximately 660 min/d) while cows confined to freestalls had the highest daily lying times (around 780 min/d) [14]. These data indicate that the lying times observed in the current study are within the normal range for cows on pasture.

Jensen observed that cows lay down less during the 6 h period before calving when compared to 2–4 days before calving [3]. Although cows on the previous study had greater mean daily lying time than cows in the current study, these data may suggest that pasture had no negative effect on prepartum cow comfort in conjunction with reduced lying times.

4.2. Number of Lying Bouts

Mid to late lactation cows on pasture had more lying bouts/d (15.3 bouts/d) than prepartum cows in the current study (10.0 ± 0.4 bouts/day) [6]. However, the previous study focused on cows with lameness, indicating that the greater number of lying bouts per day could have been due to decreased comfort associated with the lameness. Similarly, during the day before calving, confinement housed dairy cows increased their standing bouts which was attributed to discomfort and restlessness during parturition [2]. The lack of increase in lying bouts in the current study during the day before calving indicate that cows housed on pasture were less restless and uncomfortable during this time than cows housed in confinement during the same period. This suggests future work on the merits of prepartum housing may need to focus on lying bouts rather than lying time to assess cow comfort.

4.3. Lying Bout Duration

Lying bout duration was generally close to 60 min/bout across several studies of behavior of cows approaching parturition within confinement housing (63 min/bout, [2]; 57 min/bout, [3]; 50 min/bout, [5]). While the bout duration in the current study was considerably longer, it was within the range of 43 dairy farms in British Columbia, Canada (50 to 118 min/bout) where lying behavior was assessed in mid-lactation cows [8]. This may indicate that even though cows were approaching calving they did not experience enough discomfort to affect lying bout duration. It has been hypothesized that a reduction in lying time due to discomfort from events, such as an uncomfortable lying surface, was driven by a reduction in lying bouts rather than changes in bout duration [11].

4.4. Steps

Cows did not alter activity in the week prior to calving, which was in contrast to data obtained from animals housed in a confinement setting where cows increased activity during the day before calving [3]. Similarly, no change in the number of steps for cows housed on pasture in the 6 h prior to calving may indicate that pasture housed cows experience less restlessness before calving. Pain may cause an increase in activity since acute pain such as that associated with the induction of mastitis caused an increased number of steps [18]. However, cows on pasture spend more time grazing than cows in confinement spend feeding, suggesting that activity differences between confinement and pasture may not be similar.

5. Conclusions

Cows housed on pasture may experience reduced restlessness and discomfort associated with calving. When on pasture, cows did not show signs of discomfort, such as increased activity or increased number of lying bouts. Cows on pasture may also be more comfortable standing, as indicated by lower lying times when compared to literature values for cows housed in confinement. Further research to compare groups of cattle housed on pasture and confinement is needed in order to determine if pasture provides improved comfort for prepartum dairy cattle.

Acknowledgments: This project partially supported by USDA Hatch Funds.

Author Contributions: Christa Rice collected and analyzed data and wrote the manuscript. Nicole Eberhart edited the manuscript, assisted in writing, analyzed data, and prepared the manuscript for publication. Peter Krawczel was the primary investigator, designed the study, collected data, and assisted in manuscript editing.

Conflicts of Interest: The authors declare no conflict of interest.

References

1. USDA. *Dairy 2014, "Dairy Cattle Management Practices in the United States, 2014"*; #692.0216; USDA–APHIS–VS–CEAH–NAHMS: Riverdale, MD, USA, 2016.
2. Huzzey, J.M.; von Keyserlingk, M.A.G.; Weary, D.M. Changes in feeding, drinking, and standing behavior of dairy cows during the transition period. *J. Dairy Sci.* **2005**, *88*, 2454–2461. [CrossRef]
3. Jensen, M.B. Behaviour around the time of calving in dairy cows. *Appl. Anim. Behav. Sci.* **2012**, *139*, 195–202. [CrossRef]
4. Proudfoot, K.L.; Veira, D.M.; Weary, D.M.; von Keyserlingk, M.A.G. Competition at the feed bunk changes the feeding, standing, and social behavior of transition dairy cows. *J. Dairy Sci.* **2009**, *92*, 3116–3123. [CrossRef] [PubMed]
5. Miedema, H.M.; Cockram, M.S.; Dwyer, C.M.; Macrae, A.I. Behavioural predictors of the start of normal and dystocic calving in dairy cows and heifers. *Appl. Anim. Behav. Sci.* **2011**, *132*, 14–19. [CrossRef]
6. Hernandez-Mendo, O.; von Keyserlingk, M.A.G.; Veira, D.M.; Weary, D.M. Effects of pasture on lameness in dairy cows. *J. Dairy Sci.* **2007**, *90*, 1209–1214. [CrossRef]
7. Krohn, C.C.; Munksgaard, L. Behavior of dairy-cows kept in extensive (loose housing pasture) or intensive (tie stall) environments 2. Lying and lying-down behavior. *Appl. Anim. Behav. Sci.* **1993**, *37*, 1–16. [CrossRef]

8. Ito, K.; Weary, D.M.; von Keyserlingk, M.A.G. Lying behavior: Assessing within- and between-herd variation in free-stall-housed dairy cows. *J. Dairy Sci.* **2009**, *92*, 4412–4420. [CrossRef] [PubMed]
9. O'Driscoll, K.; Lewis, E.; Kennedy, E. Effect of feed allowance at pasture on lying behaviour and locomotory ability of dairy cows. *Appl. Anim. Behav. Sci.* **2015**, *166*, 25–34. [CrossRef]
10. Tucker, C.B.; Weary, D.M.; Fraser, D. Effects of three types of free-stall surfaces on preferences and stall usage by dairy cows. *J. Dairy Sci.* **2003**, *86*, 521–529. [CrossRef]
11. Tucker, C.B.; Weary, D.M. Bedding on geotextile mattresses: How much is needed to improve cow comfort? *J. Dairy Sci.* **2004**, *87*, 2889–2895. [CrossRef]
12. Tucker, C.B.; Weary, D.M.; Fraser, D. Influence of neck-rail placement on free-stall preference, use, and cleanliness. *J. Dairy Sci.* **2005**, *88*, 2730–2737. [CrossRef]
13. Tucker, C.B.; Cox, N.R.; Weary, D.M.; Spinka, M. Laterality of lying behaviour in dairy cattle. *Appl. Anim. Behav. Sci.* **2009**, *120*, 125–131. [CrossRef]
14. Legrand, A.L.; von Keyserlingk, M.A.G.; Weary, D.M. Preference and usage of pasture versus free-stall housing by lactating dairy cattle. *J. Dairy Sci.* **2009**, *92*, 3651–3658. [CrossRef] [PubMed]
15. Grummer, R.R. Impact of changes in organic nutrient metabolism on feeding the transition dairy-cow. *J. Anim. Sci.* **1995**, *73*, 2820–2833. [CrossRef] [PubMed]
16. Munksgaard, L.; Reenen, C.G.; Boyce, R. Automatic monitoring of lying, standing and walking behavior in dairy cattle. *J. Anim. Sci.* **2006**, *84*, 304.
17. Bewley, J.M.; Boyce, R.E.; Hockin, J.; Munksgaard, L.; Eicher, S.D.; Einstein, M.E.; Schutz, M.M. Influence of milk yield, stage of lactation, and body condition on dairy cattle lying behaviour measured using an automated activity monitoring sensor. *J. Dairy Res.* **2010**, *77*, 1–6. [CrossRef] [PubMed]
18. Siivonen, J.; Taponen, S.; Hovinen, M.; Pastell, M.; Lensink, B.J.; Pyorala, S.; Hanninen, L. Impact of acute clinical mastitis on cow behaviour. *Appl. Anim. Behav. Sci.* **2011**, *132*, 101–106. [CrossRef]

3

A Decade of Progress toward Ending the Intensive Confinement of Farm Animals in the United States

Sara Shields [1,*], Paul Shapiro [2] and Andrew Rowan [1]

1 Humane Society International, 1255 23rd Street, Northwest, Suite 450, Washington, DC 20037, USA; arowan@humanesociety.org
2 Humane Society of the United States, 700 Professional Drive Gaithersburg, MD 20879, USA; pshapiro@humanesociety.org
* Correspondence: sshields@hsi.org

Academic Editors: Marina von Keyserlingk and Clive J. C. Phillips

Simple Summary: Over the past ten years, unprecedented changes in the way farm animals are kept on intensive production facilities have begun to take hold in the U.S. veal, egg and pork industries. Propelled by growing public support for animal welfare, the Humane Society of the United States (HSUS) has successfully led the effort to transition farms from using restrictive cages and crates to more open aviary and group housing systems that offer the animals far more freedom to express natural behavior. This paper describes the background history of the movement, the strategy and approach of the campaign and the challenges that were overcome to enable this major shift in farming practices. The events chronicled are set within the context of the larger societal concern for animals and the important contributions of other animal protection organizations.

Abstract: In this paper, the Humane Society of the United States (HSUS) farm animal protection work over the preceding decade is described from the perspective of the organization. Prior to 2002, there were few legal protections for animals on the farm, and in 2005, a new campaign at the HSUS began to advance state ballot initiatives throughout the country, with a decisive advancement in California (Proposition 2) that paved the way for further progress. Combining legislative work with undercover farm and slaughterhouse investigations, litigation and corporate engagement, the HSUS and fellow animal protection organizations have made substantial progress in transitioning the veal, pork and egg industries away from intensive confinement systems that keep the animals in cages and crates. Investigations have become an important tool for demonstrating widespread inhumane practices, building public support and convincing the retail sector to publish meaningful animal welfare policies. While federal legislation protecting animals on the farm stalled, there has been steady state-by-state progress, and this is complemented by major brands such as McDonald's and Walmart pledging to purchase only from suppliers using cage-free and crate-free animal housing systems. The evolution of societal expectations regarding animals has helped propel the recent wave of progress and may also be driven, in part, by the work of animal protection organizations.

Keywords: ballot initiatives; animal welfare; gestation crate; battery cage; veal

---

1. Introduction: The Nature of the Animal Protection Movement from 1980 to 2000

1.1. Henry Spira's Influence

Strategy in the modern animal protection movement was greatly influenced by Henry Spira, a particularly effective activist, teacher and writer with a background in civil rights and other social justice causes. Influenced by Peter Singer (the author of the book Animal Liberation), after taking the philosopher's New York University class in 1974, Spira advanced the movement by carrying it beyond leafleting and protests. He insightfully narrowed the focus to very specific targets with obvious vulnerabilities, for example animal experiments that had questionable benefits to society, exposing them to the public in highly visible ways. An early target was the American Museum of Natural History, where cats were being intentionally blinded in experiments on their sexual behavior. Henry took note of key leverage points: taking the problem directly to the researcher's funding bodies, coaxing the support of local politicians and enticing the media to spotlight the "cat-torture" being funded with taxpayer dollars. After early successes, including halting the museum's cat experiments, he quickly moved on to larger targets, sending letters to the executives of multi-billion dollar corporations, including cosmetics giants Revlon and Procter & Gamble, and securing in-person meetings. He was keenly aware of the sensitivity of a brand's image to associations with the inhumane treatment of animals and had an acute sense of social attitudes and where they overlapped with the priorities of the animal rights movement, focusing on the suffering of species kept as pets, as the public already valued kindness to dogs and cats. When his concerns were dismissed, he forced attention to the issue by taking out full page ads in the New York Times, linking companies to perceived cruelty in experimentation and vivisection. He was a pioneer of shareholder advocacy; buying enough stock in companies to propose a shareholder resolution. He was the first to take on agribusiness in a big way, meeting with Perdue in 1987 and McDonald's in 1989 about the treatment of animals used in farming, an area that seemed insurmountable at the time. Spira influenced the style and approach of the whole next generation of animal advocates [1].

1.2. Farm Animals and Early Legislative Initiatives

The animal protection movement of the 1980s had focused on laboratory animals and animal testing, as well as dogs, other companion animals and wildlife. However, it was widely acknowledged that in terms of the numbers of animals affected, animals in agriculture dwarfed other human uses; while it was reported that there were about 11 million rodents, rabbits, dogs, cats and primates being used in laboratories in the United States [2] (p. 60) in 1982, 39 million cattle and calves, 83 million hogs [3] and over 4 billion broiler chickens were slaughtered for human food consumption [4] in the same year.

Because there were few legal protections for animals on the farm, it was natural to start working toward the enactment of new laws. In Europe, farm animal welfare legislation was gaining traction, with restrictions on permanent tethering of pigs in 1991 [5] and calves in 1997 [6]. In 1999 an EU-wide ban on conventional cages for egg-laying hens was set to be phased in by 2012 [7,8]. However, in the United States, early federal legislation specific to farm animals was largely unsuccessful.

On 6 June 1989, the U.S. House of Representatives, Subcommittee on Livestock, Dairy and Poultry held a joint hearing on H.R. 84, the Veal Calf Protection Act. Veal crates confine newly-born calves side-by-side in rows of stalls, typically measuring 66–76 cm (26–30 in) wide by 168 cm (66 in) long. The calves are tethered to the front with a chain or a rope. The calf's movement is limited to a few steps forward or backward, lying down and standing up, but the calf is unable to turn around for the 16–18 weeks that he is confined, until he is led out for slaughter. The stalls are convenient for the producer and maximize space utilization, but veal crates are particularly restrictive for a young bovine, who is normally a playful, energetic, social animal [9]. Early polling showed the public was more concerned about veal calves than other farm animals [10], and the issue became emblematic of intensive farming.

The principle sponsor of H.R. 84 was Congressman Charlie Bennett of the third District of Florida. Although the bill did not ban the use of veal crates, instead only mandating their minimum size, it met substantial opposition from the American Veal Association, certain animal scientists and cattle producers, who argued that the bill would put veal producers out of business, that it was based on emotion, that further research was necessary and that the bill would set a "dangerous precedent" for all of animal agriculture. Dr. Stanley Curtis from the University of Illinois called the law "naïve" and stated that "no behavioral need has been scientifically established for veal calves or any other animal" [11]. Despite this, there was also considerable support for the bill, including testimony from the HSUS's (Humane Society of the United States) vice president of Bioethics and Farm Animals, Dr. Michael Fox. However, the legislation was blocked from moving forward [11].

Given the strength and political influence of the industry lobby, it became clear that success in the United States would be more likely at the state level, where the public could weigh in using the ballot initiative process. This, too, was not easily achieved at first. In fact, there had been an attempt in Massachusetts in 1988 to use a citizen's initiative petition to establish some protections for veal calves. This initiative was unsuccessful (losing by a vote of one third to two thirds) [10] in part because the major newspapers in Massachusetts did not support the measure, and the campaign was poorly planned and implemented. However, animal protection organizations became more adept at carefully selecting and winning initiative and referendum campaigns on behalf of animals. A successful ballot initiative in 1990 in California to prevent the trophy hunting of mountain lions (Proposition 117) marked the beginning of an era of direct democracy in animal advocacy, bringing the issues to the people and circumventing elected officials who appeared to be beholden to hunting and farming interests. Within 10 years of Proposition 117 in California, 33 state questions, proposals, measures, propositions and amendments, mostly protecting wildlife, but also prohibiting cockfighting and horse slaughter, were proposed, debated and presented to the public for voting. Of the 33 proposed, 21 had passed in favor of animal welfare [12].

2. The 2002 Florida Gestation Crate Ballot Initiative

The First Confinement Ban

The first successful U.S. legislation to prohibit the use of an intensive confinement system on the farm was proposed for a public vote in Florida in November of 2002 [13]. Through a constitutional amendment, the citizen-led initiative aimed to prohibit the confinement of pigs in gestation crates during pregnancy. Also called sow stalls, these metal enclosures set on concrete floors measure approximately 2 ft (0.6 m) wide and 7 ft (2.1 m) long, or slightly larger than the sow's own body. On conventional, industrial farming operations, sows are kept in stalls for breeding and the following gestation period. They are lined in rows to maximize the number of breeding females that can be kept under one roof. While they are confined to the crate, each sow is able to take a step forward and backward and lie down and stand up, but she cannot turn around for the 114 day length of her gestation.

The Florida ballot initiative was organized by the HSUS working with Farm Sanctuary. Farm animal protection organizations throughout the nation rallied behind the effort, and volunteers from other states flew to Florida to help gather signatures for Amendment 10; those who could not be there were watching closely from across the nation. Under the proposed change, Article X, Section 21 of the Constitution of Florida was to state "Inhumane treatment of animals is a concern of Florida citizens ... It shall be unlawful for any person to confine a pig during pregnancy in an enclosure, or to tether a pig during pregnancy, on a farm in such a way that she is prevented from turning around freely".

With over two million votes cast, the initiative passed with 55% in favor. The vote was a historic win and established a model for progress at the state level. The initiative took effect on 5 November 2008, six years after it was passed, giving farmers time to transition to alternative systems.

3. Contemporary Farm Animal Protection Work at the HSUS

3.1. Launch of the Campaign

In June 2004, Wayne Pacelle was appointed by the HSUS Board of Directors as President and Chief Executive Officer (CEO) of the organization. Under Pacelle, farm animals would be a priority issue, and support for the work was greatly expanded. The investigations unit was bolstered; a litigation department was created; and fundraising efforts were accelerated. The HSUS grew in size and impact through the amalgamation of other leading animal protection organizations, including the Fund for Animals and the Doris Day Animal League.

The HSUS had long been an advocate and supporter of sustainable agriculture and had a section especially devoted to the topic. The section was headed by Dr. Michael Appleby (a world famous ethologist and poultry specialist) at the time Pacelle became CEO. Within months of Pacelle's appointment, he created a Factory Farming Campaign (FFC) to supplement the work of Dr. Appleby's Farm Animals and Sustainable Agriculture (FASA) section. The balance between the two sections would be the basis for future successes, combining rigorous desk research, information gathering and fact checking with advocacy and direct action.

The FFC campaign was ambitious, instructed by Pacelle to err on the side of making mistakes based on action taken, rather than opportunities missed by being too passive. It was also forward oriented, setting goals based on where the campaign envisioned animal agriculture 20 years into the future. The work was guided by Spira's approach, whose "Ten Ways to Make a Difference" [1] (pp. 184–192) was hung on the office wall in Gaithersburg, Maryland. The campaign work was set out in four pillars: public policy, corporate engagement, litigation, and investigations, all of which were to play major roles in the significant advancements for farm animals that unfolded over the next decade.

3.2. The Scientific Basis for Farm Animals' Campaign Work

From the beginning, the work of the HSUS on farm animal issues was firmly grounded in science. Farm animal welfare, as a scientific discipline, was formalized after publication of the 1964 book, Animal Machines, by Ruth Harrison [14], which was the first major critical work examining the treatment of animals in what she called "factory farming". Harrison was concerned by the advent of intensive agriculture, and after publication of her book, a government committee was convened to examine the matter, chaired by professor F. W. Rogers Brambell [15]. The committee's report launched the systematic investigation of farm animal welfare and formulated the initial version of the "five freedoms". The committee recommended that "An animal should at least have sufficient freedom of movement to be able without difficulty, to turn round, groom itself, get up, lie down and stretch its limbs" [16] (Paragraph 37).

Following the Brambell report, there was a call for scientific investigation into the welfare of farm animals in order to better inform public policy decisions in the United Kingdom [17] using rigorous experimentation and objective data, and the idea quickly took root throughout Europe. In 1976, the Council of Europe agreed the "Convention on the Protection of Animals Kept for Farming Purposes" [18], which influenced subsequent legislation and the research behind it [8]. Applied ethology (research on the welfare of animals used for farming and other purposes) has been ongoing ever since.

The idea that animals have behavioral needs (deeply engrained, ancestral behavioral patterns) in addition to their basic requirements for feed, water and shelter became a central tenant of the field as the research advanced [19,20]. Tangible physical and mental consequences manifest if animals are confined so tightly that normal movement is thwarted. For example, sows kept in gestation crates have lower bone strength and muscle weight when they are not permitted regular exercise [21,22] and more often show abnormal, repetitive behavior, such as stereotypic bar-biting [23,24], although other factors such as their concentrated feed also play an important role [25]. Even when not strictly required for survival, the motivation to perform some behavior remains strong, even in the commercial production environment. Sows are driven to wallow and root [26]; calves play and groom themselves [9]; and chickens dustbathe [27], perch [28] and search for secluded nesting sites when ready to lay an egg [29]. Deprived, restrictive environments cannot meet these behavioral needs. Evidence mounted showing the harm of a barren environment, devoid of interest or natural stimuli, on complex, intelligent, social species and the importance of providing an enriched environment for normal neural development [30] and the prevention of abnormal behavior [15,31].

At the HSUS, the Farm Animal Welfare (FAW, renamed from FASA) section developed a full library of white papers, covering every key animal welfare issue by species and topic, appropriately referenced and fact checked. These papers are still regularly accessed to inform the organization and are freely available to the public. While the science identifies numerous animal welfare issues in farming deserving of attention, the FFC narrowed the campaign focus in order to better make progress (a Spira tactic), choosing to work on a few intensive confinement issues, which were easy for the public to understand and support (another Spira tactic).

4. The 2006 Arizona Gestation Crate Ballot Initiative

Proposition 204

Following the victory in Florida, Arizona became the first state-level legislative endeavor for the FFC. Arizona was not a large veal producer in 2006, but it did have a dairy industry [32], and so, a ban on veal crates could potentially prevent the establishment of a more prominent veal confinement sector. Arizona had a good track record on ballot measures for animals, voting in 1994 to ban trapping and in 1998 to ban cockfighting. While Florida had two small gestation crate facilities, Arizona had a more significant pork industry [33] with Hormel production operations. The potential for improving the lives of a substantial number of animals played heavily into the decision. There were also good allies on the ground. Working in a coalition (Arizonans for Humane Farms), the HSUS combined forces with Farm Sanctuary, the Animal Defense League of Arizona and the Arizona Humane Society to mobilize the signature gathering effort. In Arizona, signature gatherers had to be voters [34], so volunteers could not be recruited from outside the state. The campaign focused on the fact that compassion is a universal value, appealing to all Arizonans, and enlisted a prominent conservative Republican sheriff, Joe Arpaio, from Maricopa County, to be a spokesperson for the campaign in television ads. Those opposing the measure seemed to have been caught off guard during the Florida signature-gathering effort, but were much more organized in Arizona, calling themselves the "Campaign for Arizona Farmers & Ranchers", raising over a million dollars, posting large yellow and black "HOGWASH" signs along Arizona highways and even trying to refer a counter measure to the ballot.

Proposition 204, the "Humane Treatment of Farm Animals Act", stipulated that "A person shall not tether or confine any pig during pregnancy or any calf raised for veal, on a farm, for all or the majority of any day, in a manner that prevents such animal from: 1. Lying down and fully extending his or her limbs; or 2. Turning around freely" [35]. It won in a landslide vote, 62% for and 38% against. The ban was set to take effect 31 December 2012.

In January of 2007, two of the nation's largest veal producers, Strauss Veal and Marcho Farms, announced they would abandon veal crates. Later that year, the American Veal Association's board of directors unanimously approved a new policy to move the entire U.S. veal industry to group housing within ten years [36]. The animal protection movement had changed an entire industry.

In February of the year the gestation crate ban was set to take effect, Hormel subsequently pledged to transition not only its Arizona facilities, but also its farms in Colorado and Wyoming, setting a goal of 2018 [37] and bringing the total to over 50,000 sows who would be freed from their gestation stalls by this one company alone [38]. The alternative system that producers began to explore, group housing, provided more space and permitted the sows to walk, socialize and lie down comfortably.

5. The Power of Undercover Investigations

5.1. Hallmark/Westland

In the fall of 2007, an investigator working for the HSUS documented inhumane treatment of downed dairy cows, those too weak or metabolically taxed to walk, at a slaughterhouse in Chino, California [39]. Plant workers at the Hallmark/Westland facility were filmed using a forklift to forcibly move cows who could not rise to their feet, dragging them with chains, kicking them, spraying high-pressure water hoses into their nostrils and shocking them with electric prods, all in an effort to get them to stand long enough for the U.S. Department of Agriculture (USDA) veterinary inspector to pass them for slaughter. The HSUS released the footage in January of 2008, and the widespread press coverage that followed sparked national public concern [40]. The reaction was unprecedented and surprised even animal advocates, who were accustomed to such footage being dismissed. Farm Sanctuary had been conducting investigations since 1986, and thirteen previous complaints of animal mishandling at the same plant over the preceding decade by the Inland Valley Humane Society and the Society for the Prevention of Cruelty to Animals seemed to have hardly been noticed; but after the HSUS investigation, two of the plant's employees were arrested and convicted on animal cruelty charges (this was unprecedented: the first time slaughterhouse employees had ever been convicted of animal cruelty at a slaughterhouse), and the USDA initiated the largest meat recall in U.S. history (143 million pounds of beef). The HSUS brought a federal False Claims Act lawsuit alleging the facility's owners had defrauded the U.S. Government by selling USDA beef from cruelly-treated cows in violation of its federal contracts. The U.S. Department of Justice joined in HSUS's claims and jointly prosecuted the case, which culminated in the largest judgment for animal abuse ever entered in a U.S. court: over $150,000,000. Some of the success of the investigation was due to the fact that the Hallmark/Westland plant was a major supplier of beef to the USDA's Commodity Procurement Branch, which provides beef to the National School Lunch Program. This intensified the public's reaction; parents across the nation lamented the feeding of meat from sick animals to school children. The Hallmark/Westland incident and subsequent investigations appears to have heightened the public's sensitivity to farm animals.

5.2. Widespread Objectionable Practices

Follow-up investigations at livestock auctions confirmed that the mishandling of downed animals at Hallmark/Westland was not a one-off occurrence [41]. Animal advocates argued it was part of a larger, systemic problem of relatively common inhumane practices in agriculture. However, the common defense from industry spokespersons has been that inhumane behavior was rare because it was in the producers' economic best interest to treat their animals well [42]. However, the Hallmark/Westland facility had passed two previous USDA audits (the more recent awarding a flawless report) [43] and had been designated "Supplier of the Year" by USDA in 2004–2005 [44].

Investigations had become a powerful method for exposing inhumane practices. Technology had advanced to the point that tiny cameras could be hidden, enabling hired investigators to record the daily operations of the facility, legally obtaining footage (previous activists had resorted to the risky tactic of trespassing with camcorders at night). Between 2001 and 2017, there were over 50 investigations of farms, auctions and slaughterhouses in the United States and Canada by a number of groups including The HSUS, Mercy for Animals (MFA) and Compassion Over Killing (COK). The footage obtained from these investigations showed everything from poor or unskilled animal handling (resulting in distress or injury to the animals), failed euthanasia attempts, neglect and willful cruelty, all set in standard animal housing and transport conditions. Where criminal animal cruelty was suspected, animal protection groups gave the footage to departments of agriculture and district attorneys, resulting in convictions, lost contracts, legal complaints and even the revision of laws. When released to the public, social media allowed rapid sharing of the footage, with some videos garnering over a million views. Investigations were picked up by major media channels including the Cable News Network (CNN) [45], the New York Times [46], and the Washington Post [47]. Each investigation further offered the opportunity for public discussion of the state of farm animal production in America. In response to the profusion of investigations, agricultural industries in several states began backing legislation to criminalize the taking of photos or video on a farm without the owner's consent [48] (so-called "Ag-Gag" bills that have subsequently been challenged under the First Amendment of the U.S. Constitution [49,50]).

On the one year anniversary of the Hallmark/Westland investigation, the popular industry trade magazine, Meatingplace, published an overview of how the recall permanently changed the meat business, writing "From regulations to school lunch suppliers to video recordings of operations, the process of taking meat products from pasture to plate will never be the same" [51].

6. California's Proposition 2

6.1. The Campaign

Following the passing of Proposition 204 in Arizona, Smithfield Foods, the nation's largest pork producer, announced in January of 2007 that it would phase out gestation crates at all of its company-owned sow farms within ten years. In Oregon [52] and Colorado [53], state legislatures passed bans on gestation crates in June 2007 and May of 2008, respectively. The ban in Colorado included veal crates. In early 2008, the PEW commission on Industrial Farm Animal Production (chaired by former Kansas Governor John Carlin) released an extensive $2\frac{1}{2}$-year study. One of its recommendations was to "Phase out the most intensive and inhumane production practices within a decade ... " [54]. Around this time, some food companies began pledging to move their supply chains away from intensive confinement practices (including Safeway, Burger King, Bon Appétit, Wolfgang Puck and Whole Foods). Against this backdrop, the HSUS began to qualify a farm animal welfare measure for the ballot in California.

While the pork and veal sectors in California were small, the egg industry was the fifth largest in the country, with 19 million hens [55]. In conventional, industrial egg production, laying hens are confined to small, wire cages of five to eight birds. The cages are lined in rows and stacked into tiers four, sometimes five, high. Using such battery cages, a single barn may hold hundreds of thousands of hens together under one roof. The U.S. trade industry association for the egg industry, the United Egg Producers (UEP), recommendation for space allowance for a white hen was then, and remains now, 67 in^2 (432 cm^2) [56]. A single sheet of notebook paper in the United States is 94 in^2 (603 cm^2), and campaigners used this comparison to demonstrate to consumers how little space birds were provided in the commercial egg industry.

Among animal advocates, there was a prevailing concern that it would be difficult to get people to care about chickens; in previous pre-ballot statewide polling, a measure to protect hens alone was less popular than polling language on pigs and calves [57]. However, the FFC suspected that most people just never really thought about chickens. Investigations in the 1990s revealed poor welfare of hens in battery cages, but this was before the rapid dissemination of investigative footage (the early videos were distributed on VHS tapes). FFC's leadership reasoned that it was not that people did not care, they simply did not know, and when given the chance to help all three species at once, voters would approve such a measure, as polling in California later showed [58]. They further wagered that if people could see for themselves, through video and photo evidence, the conditions in which hens in the egg industry were commonly kept, they would support reform.

In 2008, The HSUS, together with other organizations including Farm Sanctuary, the Animal Protection and Rescue League, Compassion Over Killing and the San Francisco Society for the Prevention of Cruelty to Animals, launched a ballot initiative in California that would provide gestating sows, calves raised for veal and egg-laying hens more living space. The language of the proposition was simple. It included a prohibition on tethering or confining "any covered animal, on a farm, for all or the majority of any day, in a manner that prevents such animal from:

(a) Lying down, standing up, and fully extending his or her limbs; and
(b) Turning around freely".

Fully extending limbs was further defined as "extending all limbs without touching the side of an enclosure, including, in the case of egg-laying hens, fully spreading both wings without touching the side of an enclosure or other egg-laying hens" [59]. For six months, thousands of volunteers gathered nearly 800,000 signatures from residents to place the measure on the ballot.

While most measures gain access to the ballot by hiring firms to pay signature gatherers, the animal welfare coalition in California relied primarily on volunteers. The volunteers stood outside animal shelters, pet supply stores and anywhere else they thought voters sympathetic to animals would likely be gathered. The top volunteer personally gathered more than 5000 signatures, largely by drawing in passersby to watch his three very well-trained border collies perform tricks. The effort quickly became the largest mobilization of public support in the history of the humane movement.

On 9 April 2008, Debra Bowen, the Secretary of State for California, certified that the initiative would appear on the November ballot, after concluding proponents of the measure had gathered more than enough voter signatures. Officially, it was titled the "Standards for Confining Farm Animals Act" [60], although the official legal name it would bear post-election was the "Prevention of Farm Animal Cruelty Act". It was listed as Proposition 2 on the ballot.

The initiative was endorsed by a long list of businesses, farmers, politicians (including Assemblyman Mark Leno and state Senator Carole Migden), veterinarians (including support from the California Veterinary Medical Association) and other organizations (including the Center for Food Safety, the Center for Science in the Public Interest, the Consumer Federation of America, Clean Water Action, the Sierra Club, the United Farm Workers and the Union of Concerned Scientists). Individual donors gave thousands of dollars to support the YES! on 2 campaign.

Those organizing in opposition to the measure included the UEP, the Pacific Egg & Poultry Association, the California Farm Bureau and multiple large egg producers, with production both inside and outside of the state, banding together as "Californians for SAFE Food". They advanced the arguments that the measure would jeopardize food safety and public health, heighten the risk of salmonella and avian flu outbreaks, increase the price of eggs and leave consumers with fewer choices [61]. They also predicted that it would dismantle the egg industry in California and that Proposition 2 was a disguised attempt to prevent people from eating meat. HSUS attorneys discovered that the American Egg Board, a federally-created organization overseen by the USDA, was planning to direct $3,000,000 of its funds to oppose Proposition 2 in California. Federal law prohibits such political activity by the Board, as HSUS made clear in a lawsuit filed to block that donation. A federal court in San Francisco agreed the transaction would be illegal and ordered the Board not to make it [62].

It was difficult to argue that the price of eggs would not increase when hens were given more room in cage-free systems, but previous analyses had suggested the increase would be modest [63], particularly when expressed on a per-egg basis or in the context of consumers' monthly food budget. However, looking more deeply into potential price increases, HSUS staff discovered an apparent egg producer attempt to drive up the price of eggs nationwide over many years. In part, this involved animal welfare standards developed to allow egg producers to reduce the total U.S. hen population and thus reduce egg output, thereby driving up the price of eggs. HSUS took these concerns to the Department of Justice and the Federal Trade Commission (FTC). A few months later, purchasers of eggs filed a federal antitrust class action lawsuit that is still pending [64].

The health and disease arguments against the measure however, were largely unsupported. Reviewing the literature on the incidence of salmonella (the number one food safety issue in egg production [65,66]) on cage and cage-free farms, FAW found that the largest study ever conducted, an EU baseline survey of more than 5000 operations in two dozen countries, concluded that it was the large, cage facilities, and not the cage-free farms, which had higher salmonella risk [67]. (In hindsight, it is noteworthy that the largest egg recall in the history of the United States, in which nearly 2000 people were sickened by *Salmonella* enteritidis, would be traced back to a large battery cage company in Iowa just two years after Proposition 2 [68].)

There was also debate regarding rural communities and the environment. HSUS identified a major egg producer dumping millions of gallons of liquefied manure from caged hens into a multi-acre lagoon right next to long-time residents. The HSUS filed an environmental lawsuit on behalf of the neighbors, which led to a 2011 jury verdict in favor of the neighbors' nuisance claims and an award of over half a million dollars in damages to them [69].

The YES! on 2 campaign utilized multiple media outlets including television advertising, radio, Facebook and Twitter. There were public discussions at universities and major fundraising events. Every part of the state was covered, with full-time organizers in the major media markets, including agricultural areas such as Fresno. In addition, the campaign produced an online animated video that went viral, with farm animals parodying Stevie Wonder's famous song, Superstition, reworked to be about the need to support Proposition 2. HSUS staff, volunteers and even Pacelle himself knocked on doors to ask people to vote yes on Proposition 2. The Oprah Winfrey Show devoted an entire episode to the effort in October.

Mercy for Animals released undercover videos from a major West Coast egg producer and distributor, just weeks before the vote. In addition to battery cages, the video also showed a worker stamping on a sick hen. The footage further bolstered the campaign, and there appeared to be a palpable change in social climate in favor of animals.

On 4 November 2008, Prop 2 campaigners gathered together to watch the voting results. In the end, it was a landslide victory; Proposition 2 passed with 63.5 percent of the votes in favor and 36.5 percent against. There were 11 other measures on the statewide ballot, but none received more yes votes than Proposition 2. It won majorities in 47 of 58 counties, across genders, and in all age, education and ethnic groups polled [70]. It was even favored among rural voters, including in some of the largest agricultural counties. The act was written into California's Health and Safety Code [59]. It remains one of the most important advancements for farm animals in the history of U.S. animal protection work, easing the way for subsequent public and corporate policy and standards setting work.

6.2. The Ensuing Legal Activity

Animal advocates consistently maintained that Proposition 2 established behavioral standards that can be easily understood. By the terms of the law, hens must be able to fully spread both wings without touching the side of their enclosure or another egg-laying hen. In the official ballot voting guide, the argument in opposition to the measure went further, stating that Proposition 2 "effectively bans 'cage-free' eggs, forcing hens outdoors for most of the day" [60]. After the initiative passed, however, some members of the egg industry wanted to continue using cage systems.

Three different lawsuits were filed on the grounds that the new law did not provide precise space requirements. In 2010, JS West, a large California egg producer that had installed "colony cages" supplying 116 in^2 (749 cm^2) of space per bird, sought a declaration that these new cages were acceptable under Proposition 2 in Fresno County court. The court dismissed the case in 2011, finding that JS West had not pled sufficient facts to establish an actual or present controversy [71]. In 2012, egg producer William Cramer and the Association of California Egg Farmers (ACEF) filed two separate lawsuits in state and federal court, respectively, arguing, among other things, that Proposition 2 was unconstitutionally vague. However, the courts did not agree, ultimately ruling against the plaintiffs. The Cramer court's 2012 decision noted that Proposition 2 was not vague because it "establishes a clear test that any law enforcement can apply, and that test does not require the investigative acumen of Columbo to determine if an egg farmer is in violation of the statute". The decision was upheld by the Court of Appeals in 2015 [72]. Additionally, the state court dismissed the ACEF case in 2013, with leave to amend [73]. ACEF voluntarily dismissed the case in 2014. The HSUS was a party in all three of these cases, with its Animal Protection Litigation team joining pro bono attorneys to defend Proposition 2.

Two years after Proposition 2 passed, state Assemblyman Jared Huffman sponsored AB 1437, a bill requiring that eggs sold in the state of California, regardless of where they were produced, come from conditions commensurate with the behavioral standards set forth in Proposition 2. The sales ban passed in 2010 and was signed into law [74] by then Governor Arnold Schwarzenegger.

This, too, was challenged in the courts. In 2014, the Attorney General of Missouri, Chris Koster, along with attorneys general from Nebraska, Alabama, Kentucky and Oklahoma and the Governor of Iowa, Terry E. Branstad, filed a lawsuit challenging AB 1437 under the Commerce Clause of the U.S. Constitution and on federal preemption grounds. Their district court lawsuit was dismissed in

October of the same year, when the court ruled that the state governments did not have standing in the matter to sue on behalf of all citizens of their states. Plaintiffs appealed the case to the 9th circuit in November of 2016. That court affirmed the lower court's decision, specifically noting that because California egg farmers were subject to the same rules as egg farmers from other states, the law was not discriminatory [75].

7. The 2010 Ohio Ballot Initiative

Countermeasures

The state-by-state legislative model continued, with wins in Maine [76] and Michigan [77] in 2009. The farm animal protection movement had become a formidable force, enough so that producer organizations and politicians were paying attention. The threat of a potential ballot initiative was enough leverage to initiate serious discussions with state legislators. However, the campaign took a different course when it reached Ohio. Ohio ranked second in egg production [78] and ninth in hogs [79] and was thought to have a significant veal production sector, as well.

Pacelle and HSUS staff began conversations with the Ohio Farm Bureau, the state's Veterinary Medical Association, the Cattlemen's Association and the Pork and Poultry Councils in February of 2009. The HSUS made it clear that a Proposition 2-style ballot initiative was being contemplated for 2010 in Ohio, but also that the preference was to negotiate a legislative compromise and avert a costly and divisive ballot initiative campaign, with progress in Colorado serving as a model. After initial conversations, the HSUS waited for a response.

On March 16, the HBO documentary Death on a Factory Farm was released. The film chronicled an undercover investigation of a Wayne County, Ohio, farm where, among typical animal housing and handling concerns, a producer was filmed using a forklift to raise heavy sows by a noosed logging chain, killing the pigs by slow hanging. After being hoisted off the ground, sows were shown struggling and kicking in mid-air for almost five minutes while employees stood by watching. During the ensuing trial for animal cruelty charges, the Ohio Pork Producers Council put up $10,000 in legal fees for the defense. Of the ten charges filed, only one resulted in a conviction, and it was unrelated to the hanging of sows. The farm manager who was filmed throwing baby piglets was fined $250 and sentenced to a training course on the proper handling of hogs.

In defensive mode, the politicians in the General Assembly of Ohio proposed a preemptive counter measure to the HSUS's proposed initiative. This was a constitutional amendment to create a Livestock Care Standards Board in the state tasked with writing guidelines for the care of livestock and poultry. The HSUS opposed the amendment, asserting a livestock board would simply codify industry norms, and continued with plans to go to the ballot. However, the amendment (Issue 2) passed in November 2009.

The coalition of animal protection groups, this time "Ohioans for Humane Farms", would have to gather 402,275 valid signatures of registered Ohio voters from 44 of 88 counties to place the measure on the ballot. The initiative was written to require the newly-formed Ohio Livestock Care Standards Board [80] to ban cage confinement practices. However, in order to address other major welfare issues, it also included a ban on the transport or sale of "downer cows" (as seen in the Hallmark/Westland investigation) and a ban on the use of strangulation as a "euthanasia" method (as spotlighted in the Death on a Factory Farm documentary).

The threat of the ballot initiative provided further impetus for continued negotiation, and to avoid bitter political conflict, discussions resumed. While volunteers gathered signatures to put the potential measure on the ballot, Gov. Ted Strickland tried to find a way to avert the measure while still passing farm animal welfare reforms. In a series of negotiations between Strickland, Pacelle and the Ohio Farm Bureau, a deal was brokered in July 2010. In exchange for the HSUS dropping ballot plans, the Ohio Farm Bureau and the farm trade associations agreed the Livestock Care Standards Board would promulgate rules to include the downer cattle and humane euthanasia provisions, as well as a moratorium on the construction of new battery cage facilities and the eventual phase out of gestation and veal crates. Pacelle also secured in the agreement provisions addressing other pressing animal protection issues, including recommendations to the state legislature to prohibit the sale and/or possession of dangerous exotic animals, including big cats, bears, primates and others.

It was a tragic turn of events that prompted the Ohio legislature to enact the ban on the acquisition of exotic pets in June of 2012. In Zanesville, the troubled owner of a menagerie of large carnivores and primates took his own life after opening the cages to set free approximately 50 animals. To protect the public, authorities shot 18 Bengal tigers, 17 lions, as well as wolves, grizzly bears and other animals.

8. The Federal Egg Bill

Partnering with Egg Producers

In 2011 when ballot initiative signature drives were underway in Washington and Oregon, the UEP approached the HSUS to initiate a conversation regarding the prospect of a federal bill to define minimum space requirements for egg-laying hens. This was a surprising turn of events because previous battery cage campaigns had produced a highly polarized, adversarial relationship between the two organizations.

After much private deliberation, a deal was announced on July 7 in a joint press conference. The HSUS and the UEP would work together in the U.S. Congress to pass "The Egg Products Inspection Act Amendments", which would set minimum space requirements and phase out the use of barren battery cages over the next 15 years, along with numerous other reforms.

It was a compromise agreement that banned battery cages, but permitted the use of enriched colony cages of the sort JS West had installed in California. However, the bill required nearly double the space per bird of a typical battery cage, 124 in^2 (800 cm^2). Importantly, the deal would also mandate labeling on all egg cartons nationwide, informing consumers of the housing system used to produce the eggs, with descriptions including "eggs from caged hens" on cartons of eggs originating from colony cage systems. In exchange, the HSUS put on hold efforts to qualify ballot measures in the Northwest. Pacelle explained to the volunteers collecting signatures in Oregon and Washington that a ballot initiative is only an option in around half of U.S. states. Many of the top egg producing states do not allow the process; at the time, federal legislation appeared to be the only option for setting minimum space standards for every hen in the nation.

The "Egg Products Inspection Act Amendments" were introduced in Congress twice, first in 2012 [81] and then again in 2013 [82]. The effort was praised as a rare coalescing of disparate interests in politics. However, the 2012 House bill was referred to an agricultural subcommittee, where it failed to advance. In the Senate, Sen. Dianne Feinstein (D-Calif.) offered it as an amendment to the 2012 Farm Bill, but it was not one of those debated on the floor. In 2013, the amendments again did not advance, blocked by the beef and pork lobbies from farm states, who maintained that it would set a precedent for the on-farm federal regulation of the rest of animal agriculture.

9. Corporate Policy

9.1. Engaging with Major Brands

During the legislative campaigning, the HSUS's renamed Farm Animal Protection (FAP) section engaged in a corporate outreach campaign, raising the battery cage and gestation crate issues with major pork and egg buyers, including restaurants, grocery stores, food service companies, fast-food chains, hotels, cruise lines and other segments of the food retail sector. Companies varied widely in their response. Some were interested in positioning themselves progressively on animal issues, while those at the other end of the spectrum were indifferent or, occasionally, extremely wary. The process started with a courteous letter to the CEO. Sometimes, the initial contact was ignored or politely dismissed, but usually, it was passed down the chain to (depending on the company) the egg or pork buyers, communications team or public relations department. Some companies would agree to a meeting right away, but in other instances, it took further action to initiate a meaningful conversation.

When friendly requests for a meeting were brushed aside, other avenues to advance the issue were chosen. Following again in the footsteps of Spira, the HSUS started buying sufficient shares in publically-traded companies to introduce shareholder resolutions. Federal rules require ownership of a minimum of $2000 of a company's stock for 12 consecutive months in order to introduce shareholder proposals. The FAP team began attending dozens of shareholder meetings, making the case that poor animal welfare policies put shareholders at a risk. These proposals rarely received majority support among shareholders (though even if they did, they were simply non-binding advisory proposals anyway), but simply having the issue on the proxy raised the stakes substantially, generating news attention and an audience with the key leaders in a company.

Given that brand image is vitally important to large companies, another way that activists engaged a company's attention was to link cruelty on farms (documented in undercover videos) to the retailers they supplied. When corporations were shown the conditions in which animals in their own supply chain were raised, it would often prompt more serious concern and discussion, particularly if a reporter called asking questions. If a petition were launched, it could quickly amass hundreds of thousands of online supporters. The HSUS also asked the Securities Exchange Commission and the Federal Trade Commission to investigate companies' potentially misleading claims about animal welfare. Frequently, when companies received inquiries from those agencies, they quickly reconsidered their strategy. Investigations, media interest and pubic support improved the willingness of companies to draft or strengthen their animal welfare policies.

The HSUS staff traveled across the country to meet with company officials to discuss farm animal welfare. In the early years of the campaign, the meetings were often largely focused on convincing the company that intensive confinement issues were legitimate concerns, deserving of their attention. These meetings also helped build trust and a good working relationship. As corporate awareness of changing state laws and consumer interest in farm animal welfare issues increased, the nature of the conversation changed. Companies began to focus on how to obtain animal products produced in alternative systems that could replace crates and cages.

One of the initial obstacles was lack of supply, but as conversations with producers continued, it was clear that, due to the complex nature of the supply chain, producers were not always aware of the growing demand. The HSUS worked with brands to clearly communicate that they wanted cage-free eggs and crate-free pork, to send a clear signal with a well-publicized animal welfare policy. Once the brand made the public commitment and spoke to their suppliers to arrive at a reasonable timeline in which to implement the switch, producers began to invest in the infrastructure. It was not easy, particularly with pork: companies were buying finished product from suppliers who bought from a processing facility who in turn were purchasing from farmers raising the pigs they bought as piglets from breeding facilities (where the mother sows were either in gestation crates or group housing systems). The key was getting communication that reached through the supply chain all the way down to the farm and breeding farm level.

Discussions sometimes took years. The first meeting with Walmart, the largest grocer in the nation, took place before Proposition 2 was even on the ballot in California and went on for another decade. Conversely, the fast-food chain Burger King became an animal welfare leader relatively quickly, declaring in 2012 that it would use only cage-free eggs in all of its restaurants within five years. Early commitments were often only small steps forward. For instance, Kraft Foods switched one million eggs to cage-free in 2011. While switching a million eggs made an impressive public statement, it was a very small portion of the company's total egg usage (USDA estimates are that Heinz, which bought Kraft in 2015, uses 313 million eggs annually [83]). Some companies switched their eggs in just one product line; Campbell Soup first phased in cage-free eggs for its Pepperidge Farm brand, and Unilever pledged to switch all of its eggs in Hellman's light mayonnaise to cage-free in 2010. However, the Unilever announcement was significant because of the prominent cage-free messaging on the product label and in print and television advertisements. Hyatt hotels begin listing "cage-free eggs" directly on their room service menus. All of these announcements were celebrated by the HSUS and other animal protection groups, such as MFA and Compassion in World Farming (CIWF) that were also meeting with major brands. These developments laid the groundwork for further, more sweeping commitments.

9.2. The Rise of Corporate Social Responsibility

The legislative changes, undercover videos, media and growing number of announcements by major brand names cultivated continued progress. The concept of cage-free eggs and crate-free pork production began to sink deeper into public awareness, and it was apparent that companies were following the issue more closely. Mitigating risk and protecting brand reputation was a factor, but more often than not, there seemed to be genuine animal welfare concern. Many companies were eager to use their buying power to transform the industry.

By about 2010, there started to be a noticeable difference in the composition of the company representatives who came to the table to discuss farm animal issues. Increasingly, it was the Corporate Social Responsibility (CSR) team, along with the purchasing department, and sometimes the company's executives, as well. CSR teams were tasked with understanding current issues of concern to consumers and building support for social, environmental and animal welfare commitments. They were increasingly serving as resources for the decision-makers in the company. Brands became much more likely to accept that they had ethical and social obligations, and their shareholders were increasingly including sustainability issues in their investment strategies [84]. This made movement on farm animal welfare much easier to facilitate.

As the work advanced, public commitments were often announced in joint press releases with the HSUS. Companies began to ask what else they could do about animal welfare. Rather than the HSUS contacting companies to solicit meetings, major brands started reaching out to the HSUS and other animal protection organizations.

10. The 2012 Gestation Crate Announcements

10.1. McDonald's

McDonald's was a pivotal company, in terms of both its buying power and its high profile brand name. This made it a target for animal activists, including Spira, for decades preceding FAP's work. In a 1980s leafleting campaign, British activists had challenged the corporation on a range of issues including cruelty to animals. McDonald's sued, and a court trial ensued, which became known as the "McLibel" case. It was the longest court case in British history. Although the final ruling was generally favorable to McDonald's (concluding that the defendants had not proven many of their claims), the extraordinary publicity surrounding the case was not. McDonald's was portrayed as a bully trying to stifle freedom of speech [85], while Chief Justice Roger Bell, who presided over the case, ruled that McDonald's was, in fact, "culpably responsible for cruel practices" involved in raising pigs, chickens and laying hens, specifically acknowledging the issue of severe restriction of movement [86]. People for the Ethical Treatment of Animals (PETA) continued the campaign in the United States with billboards in Chicago (McDonald's headquarters) and media advertisements.

In 1999, McDonald's hired Dr. Temple Grandin, a world-famous designer of livestock handling systems, to implement industry-leading slaughterhouse audits in the United States that improved the welfare of millions of cattle and pigs [87]. In 2000, they also worked with U.S. animal welfare scientists to increase cage-space allowances for hens in the company's supply chain [88], which was industry-leading at the time (although they continued to permit battery cages).

The HSUS first met with McDonald's in 2005 when FAP was formed, but the conversations did not lead to tangible progress until 2012, when HSUS was assisted by shareholder activist Carl Icahn. Animal protection groups had been meeting with Bob Langert, then Vice President of Sustainability at

McDonald's, but Icahn made a call directly to Don Thompson, the President and CEO. Icahn's call helped elevate the issue to top priority, and McDonald's agreed soon after to rid its supply chain of gestation crates [89] (pp. 33–39).

The McDonald's pledge came in two phases: a February 2012 joint press release with the HSUS, in which the company required all of its U.S. pork suppliers to outline their plans to phase out gestation crates, and a ten-year timeline, announced at the end of May. While the HSUS argued for a more rapid transition, the organization eventually agreed that such an enormous industry-wide shift required a longer timeframe.

10.2. The Companies That Followed

The year 2012 was pivotal for the corporate outreach campaign on gestation crates. Using established relationships with most of the major U.S. food brands, FAP circled back to each firm, holding up the McDonald's commitment as an example. If McDonald's, a price-sensitive company that sold a considerable amount of pork (one percent of the U.S. supply [90]), could do it, so could others. Further gestation crate commitments began to follow rapidly. In March, Wendy's (the second largest U.S. hamburger chain) and Compass Group (the largest food service company in the world) publicized their commitments. Burger King made an announcement in April, and Denny's committed in May. Many more followed, including Sonic, Kroger, Sodexo, TrustHouse, Sears Holdings, CKE Restaurants, Heinz, Kraft, Aramark, Wienerschnitzel, ConAgra, Target, IHOP and Appleby's, among many others. Within three years of McDonalds' move, nearly 60 major U.S. companies followed. The timelines varied, but were usually set to be complete within either five or ten years, by 2017 or 2022.

Smithfield Foods was one of McDonald's top suppliers, and by 2012, they were 38 percent into their transition to group housing [91]. However, with the continual retailer announcements and with the ongoing pressure of undercover investigations of hog farms being released by the HSUS, MFA and COK, other major pork producers also started transitioning to group housing. Hormel (makers of the iconic SPAM product) made its pledge in early 2012, and Cargill, Hatfield and Tyson all made public commitments within two years. However, these producers had a mix of company-owned pigs and contract growers, and the Hormel pledge did not extend to contract farms. In contrast, Smithfield asked its contract producers to switch (under threat of loss of contract extensions). Contract producers accounted for 40 percent of the company's sow herd (2100 farms across 12 states) [92].

A further challenge to the work on gestation crate policies continued to be that producers who use group housing systems were sometimes still confining sows in crates during breeding and, in some cases, for several more weeks, until the confirmation of pregnancy. Under this type of management system, sows would still be kept in stalls for up to six of their 16 weeks of gestation. Despite these caveats, the growing shift in the industry meant that well over a million sows would be provided with more living space (Table 1).

Table 1. Producer group housing pledges [93–100], including the total number of company-owned, company-managed and contract produced sows [38,101].

Company	Pledge Date	Number of Sows	Transition Time-Line		Time in Individual Crates during and Following Breeding
			Company Owned Farms	Contracted Producers	
Smithfield	2007	880,000	2017	2022	Individual stalls until confirmed pregnant
Hormel	2012	52,000	2018	Not included	Not available
Cargill	2014	175,000 *	2015	2017	28–42 days
Clemens	2014	55,100	2017	2022	7–10 days on company owned farms; up to 42 days in contract production
Tyson	2014	62,500	Does not own	32% group housing as of March 2016	Not available

* In 2015, JBS USA Pork acquired Cargill's U.S. pork business. The number of sows reported for Cargill is based on 2015 figures, rather than 2016.

11. The Evolving Social Consciousness

11.1. Financial Support for U.S. Animal Protection in the 21st Century

Financial support for animal protection in the United States has grown substantially in inflation-adjusted dollars since the beginning of the 20th century. In 1910, McCrea published an analysis of the animal movement in 1909 reporting that there were around 500 societies who raised approximately $0.5 per capita in the United States (in 2008 dollars) to support their activities [102]. Today, a reasonably accurate picture of the finances of the animal protection movement can be obtained by analyzing the 990 files maintained by Guidestar in its database. The animal protection organizations (those classified as D20 in the IRS database and now numbering around 21,000 are raising around $3 billion (in 2008 dollars) from donations and program service revenue. This constitutes around $9.5 per capita in the United States. In other words, animal protection groups raise almost 20-times as much money today as they did just over 100 years ago. The indications are that public support is still growing faster than inflation.

11.2. Further Barometers of Public Sentiment

A real sense that societal views regarding animals were changing began to take hold. In 2014 the 50th U.S. state, South Dakota, passed felony-level penalties for animal cruelty. In March of 2015, Ringling Brothers announced it would retire its performing elephants from traveling shows, surprising animal advocates who had protested the circus for years. A year later, still using wild animals including tigers and lions, the circus announced it was simply going out of business altogether. In November of 2015, the National Institutes of Health announced an official plan to move the remaining government-owned chimpanzees from research laboratories into sanctuaries [103], marking the end of an era of chimpanzee experimentation in government-funded disease research.

Perhaps one of the greatest reflections of the state of public sentiment toward animal exploitation and injustice came mid-summer of 2015, when an American dentist from Minnesota, Walter Palmer, hunted and killed Cecil, a well-known lion from Hwange National Park in Zimbabwe. Cecil was famous around the world because he had been studied by Oxford University's Wildlife Conservation Research Unit (WildCRU) and had been satellite-tracked since 2009. The hunt was allegedly illegal. Cecil was shot with a bow and arrow outside of park boundaries, wounded and tracked for 11 h before he was finally killed, then skinned and beheaded [104,105].

The hunt concluded on July 2, but Palmer was not named as the hunter until July 27. His naming was followed by a massive, global reaction. Protests were held in front of Palmer's dental clinic, and U.S. talk show host Jimmy Kimmel did a monologue on ABC (The American Broadcasting Company) the next day. The story was a top trending topic on social media, and the WildCRU website received 4.4 million visits, causing it to crash, along with Oxford University's site [105]. Various petitions were started, appealing to the White House, the U.S. Fish and Wildlife Service and the president of Zimbabwe, totaling more than two million signatures between them all [106]. Animal protection groups, including the HSUS, the Animal Legal Defense Fund, the International Fund for Animal Welfare and the Born Free Foundation, worked to convince major airlines around the world such as Delta, United and American Airlines to ban the shipment of big game trophies, thus closing options for hunters to bring back their spoils. The killing of Cecil the lion highlighted the growing social sensitivity to the issue of big game hunting. The team at Oxford argued that it was the largest public reaction in the history of wildlife conservation [105].

A further indicator that society's sensitivity to animal maltreatment was changing came on 17 March 2016. SeaWorld, the ocean-themed amusement park, declared that it would no longer breed captive orcas, making its current population the last generation in the park, and that it would end the public shows involving orcas. Perceptions of captive whales had changed following the release of the 2013 documentary Blackfish, which chronicled the death of a trainer, Dawn Brancheau, at SeaWorld Orlando. She was killed by Tilikum, a wild-caught killer whale who had had a troubled history in captivity. The documentary was serially rebroadcast by the major news outlet CNN, in itself a sign of changing public sentiment. SeaWorld's stock fell; attendance slumped; and the CEO resigned. SeaWorld's new CEO, Joel Manby, and the HSUS's Pacelle shared a friend in common, former congressman John Campbell, who brokered a meeting between the two parties. After negotiations with the HSUS, SeaWorld made its pivotal commitment, which Pacelle wrote about in a postscript of his newly-released book The Humane Economy [89]. Tilikum died at Sea World in January of 2017.

The changing perception of animals and of society's duties toward them likely aided the work to protect farm animals. A survey of 798 U.S. households, published in 2014, found that almost half of the respondents (46%) were somewhat or extremely concerned about the welfare of U.S. livestock animals [107]. A Gallup poll released in May 2015 found the number of people who believed that "animals should have the same protection from harm and exploitation as people" had increased from 25% in 2003 and 2008 polls to 32% in the 2015 poll [108]. Further survey work reported that people concerned about farm animal welfare were frequently younger and more often female [107].

The societal change was reflected in major news announcements, the public's reaction to them, media interest and personal conversations. Ideas about animal protection appeared to have shifted from the margins to mainstream.

12. The Demise of the Battery Cage in America

12.1. Freeing the Hens

Until 2015, the battery cage had been the standard form of egg production for more than 60 years. However, on 1 January 2015, California's Proposition 2 and AB 1437 took effect, and by law, all shell eggs produced or sold in the state had to be compliant with the new specifications, effectively preventing the sale of eggs produced in conventional battery cages. Further, since approximately 20 million eggs were being imported into California every day [109], there were reverberations around the country, particularly in the Midwest's top egg producing states.

In February, Sodexo, a major food service manager at thousands of college cafeterias, universities, hospitals, and corporate dining centers across the country, extended its previous commitment to switch all of its shell eggs to cage-free by announcing it would transition its liquid egg supply, as well, bringing its total to the equivalent of 239 million shell eggs used a year. Compass Group and Aramark had already made similar announcements in 2015.

In March, Steve Easterbrook took over as CEO of McDonald's. The company was in a slump, and Easterbrook was brought in to turn it around. One decisive move that he made was to launch the all-day breakfast menu, which set the chain to increase egg sales, then already at two billion annually [110].

Easterbrook, who was from the United Kingdom where McDonald's eggs were sourced from free-range facilities, made another pivotal change. He decided to address animal welfare concerns and draw customers back into its restaurants by pledging, on 9 September 2015, to move to 100 percent cage-free eggs in both United States and Canadian supply chains within 10 years. Again, McDonald's would change an entire industry.

Like the 2012 gestation crate announcement, McDonald's decision to go cage-free sparked an unprecedented chain reaction among other brands. In the following months, nearly a hundred other major companies enacted similar purchasing policies (Appendix A).

Egg producers began to transition to cage-free housing. Aviary systems, now firmly established in European countries, were being sold in the United States by equipment manufacturers boasting 25 years of management experience in alternative housing. Cage-free egg production was already increasing rapidly (Figures 1 and 2), but in October of 2015, Cal-Maine Foods and Rose Acre Farms announced a joint venture to establish a cage-free operation in Texas with a capacity for 2.9 million hens. Hickman's Eggs announced it would expand cage-free production by two million hens, and Herbruck's Ranch, a major supplier to McDonald's with 8.5 million hens, has not built a new cage facility since

2005 [111]. Many other producers were installing more cage-free production capacity. The industry trade publication Feedstuffs reported that the supply of cage-free eggs was becoming more reliable, and the American Egg Board divulged that the percentage of cage-free eggs available on the market had jumped more than 60% from 2014 to 2015 [112].

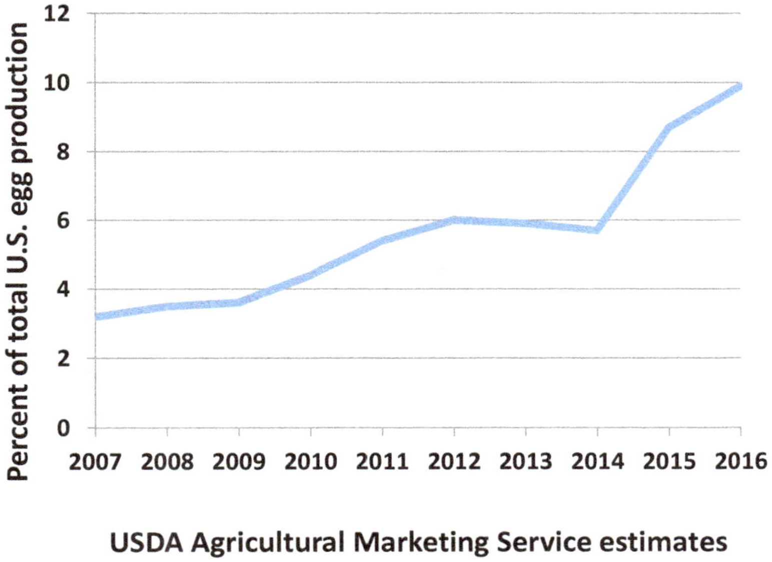

Figure 1. Percent cage-free egg production in the United States.

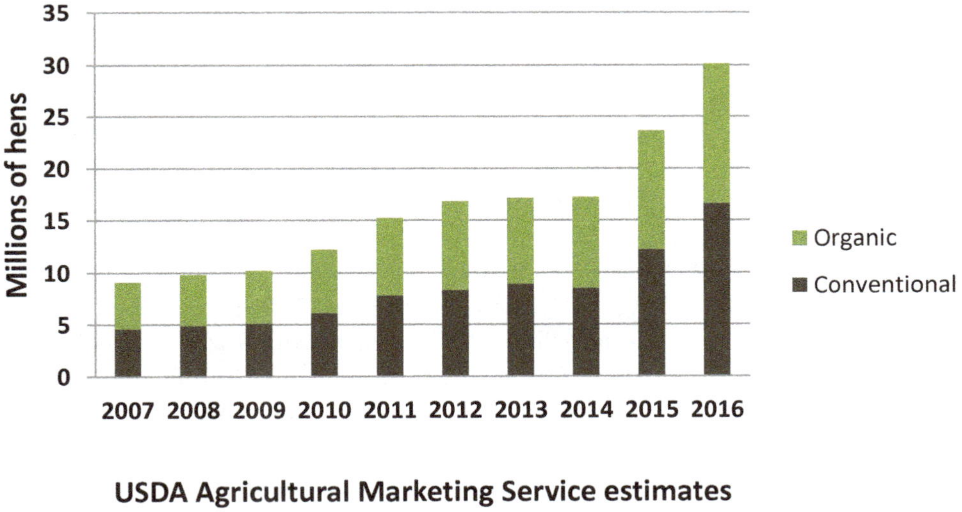

Figure 2. Number of cage-free laying hens in the United States.

12.2. Pressuring the Holdouts

Restaurants, food service providers and hotels were switching (Appendix A), but the grocery stores took longer. They were reluctant to stop selling battery cage eggs, reasoning that consumers should be offered a choice. From the HSUS's perspective, consumers were not making an informed choice, since the cartons do not indicate that the eggs come from caged hens. Some cartons even depict bucolic imagery despite the eggs being from caged hens. HSUS argued that cage-free should be the new baseline, with choices extending to free-range and pasture-based systems.

Early in 2015, the HSUS began focusing on Costco, a membership-only warehouse club that sold 2.9 billion eggs per year. In 2007, the club had committed to going 100 percent cage-free, but had failed to commit to a time frame. It was also unwilling to engage in dialog on this issue, so the HSUS began to apply pressure. Over the summer, the HSUS released undercover video footage of a Pennsylvania supplier to Costco, showing hens crowded into battery cages. Eggs from such hens were subsequently packaged and sold in cartons depicting pictures of hens roaming in outdoor pastures. Again, The HSUS engaged with the FTC and asked the agency to take action to stop the egg company from using images that could be misleading, but by the time the FTC investigated, the supplier had already abandoned the use of that particular packaging. The HSUS also filed a shareholder proposal shortly after breaking the investigation, asking Costco to disclose the risks to investors in the stores' supply chain. The HSUS launched a website, "CagedForCostco.com" and enlisted the help of celebrities. Ryan Gosling, Brad Pitt and Bill Maher spoke out, asking Costco to commit to a timeline for sourcing from cage-free suppliers, generating substantial media attention. Costco finally recommitted to going cage-free in December of 2015, only a few months following the McDonald's announcement.

While campaigning against Costco and meeting with other major egg retailers, the HSUS continued to focus on the one company with the buying power to put an end to the debate: Walmart. The company was the largest retailer in the world, with nearly 260 million customers per week in 11,535 stores world-wide [113]. Walmart sells 25% of all groceries in the United States. Meetings with Walmart had indicated there was some willingness to join the growing movement to go cage-free, but the company's reputation for being the lowest on costs was an obstacle. In dialog with the company, the HSUS emphasized that Walmart's scale made it the most important player in agriculture, that the public wanted supply chains to reflect their values and that Walmart could set the corporate standard. Discussions were affable, but progress was limited. The company had, in May 2015, released a sweeping, but general animal welfare policy. It was a good first step, but without a timeline for instituting a no-cage policy, it would not lead to real change.

After working internally with suppliers, Walmart took the next step on 5 April 2016, announcing that both Walmart and Sam's Club (the company's warehouse cost club retailer) in the United States would set a goal of transitioning to a 100 percent cage-free egg supply chain by 2025. This move settled the debate; clearly there was no future for cages in the U.S. egg industry.

By 25 April 2016, 14 out of the 15 top grocery stores had announced timelines to go completely cage-free, as well. The last grocery store, Publix, finally announced in July, after the HSUS created a YouTube video parody of battery cage confinement, "Why 20 people are stuck in an elevator" and aired it as a television commercial.

In August of 2016, the USDA estimated that it would require over 50 billion eggs to meet the requirements in all of the public pledges [83]. According to the USDA's calculation, over 200 million hens would have to be cage-free to meet this demand, or about 70% of the nation's total flock. USDA's data suggest that so far, the transition has kept pace with demand (Figure 3).

It appeared that few people in America had not noticed the change. Media stories about big name brands pledging to go cage-free were published regularly and covered in major news outlets [114–116]. National Public Radio covered the trend [117–119], and Fortune Magazine featured a front page story on Easterbrook titled "Inside McDonald's Bold Decision to Go Cage-free" [120]. However, it was not all good press, with some articles pointing out the management challenges in cage-free production, consumer confusion about what cage-free really meant and the logistics of switching such an enormous supply in a limited time-frame [121,122]. Naturally, the transition will be closely watched by many different stakeholders.

In March of 2016, McDonald's received the Henry Spira Corporate Progress Award from the HSUS. It is hoped that Spira would be pleased with the progress since his first meetings with McDonald's in 1989, were he still alive today. Spira's groundwork made way for the sea change that is so celebrated now, and his approach is firmly embedded in the history of the humane movement.

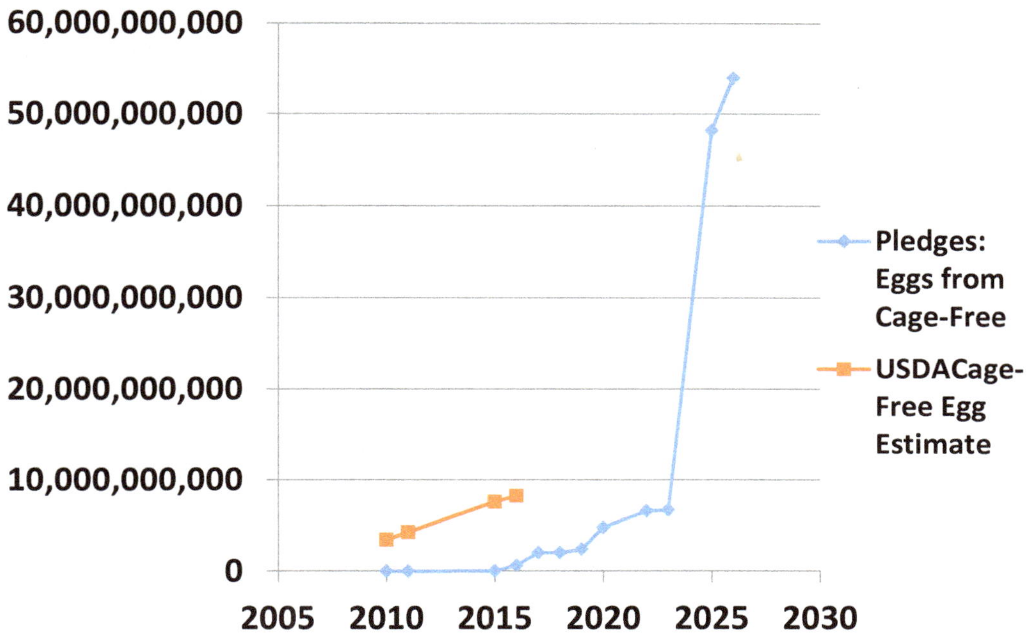

Figure 3. U.S. cage-free egg production: actual and pledged (billions; total U.S. egg production ca. 80 billion [123]). Calculated by year from USDA estimates [83] and company announcement dates provided in Appendix A.

13. Conclusions: Looking Ahead

Over the past decade, the plight of farm animals has worked its way into common discourse in legislatures, court rooms, universities, investment firms, business conferences, family dinners and the farms themselves. The entrenched cultural disregard for farm animal well-being has been replaced with a sincere reexamination of the way we raise animals for food, as part of a larger societal discussion entwined with other agricultural sustainability topics, such as the environment, global warming, antibiotic overuse, rural communities, human health and nutrition.

With the debate over cage confinement essentially settled in the United States, there was little organized opposition to the ballot initiative in Massachusetts, which passed in November 2016 [124] and will prohibit cages and crates in the state, as well as the sale of eggs, veal and pork from caged animals. The HSUS has begun work on new issues, turning back to the scientific literature for guidance, and initiating a major campaign on the welfare of broiler chickens. The confinement work is expanding into new regions of the world. Humane Society International (HSI), with 12 international hubs on six continents already in place, has plans to extend its presence further over the coming five years. HSI Farm Animals has already had tremendous early success. Since July of 2016, numerous companies, including Compass Group, Burger King and Sodexo, have extended their animal welfare policies globally, and after talks with HSI, Mexico-based Grupo Bimbo (the largest bakery company in the world) announced it would switch completely to cage-free eggs. In India, HSI working with the Federation of Animal Protection Organizations (FIAPO) is challenging battery cages in the courts, with consolidated state cases now moved to the High Court of Delhi. After working with HSI in Brazil, the world's largest pork producer, JBS, announced that it had phased out the use of gestation crates at all company-owned facilities in 2016. These are just a few examples, with further announcements coming regularly.

Reflecting back, future generations will likely be surprised that there was ever a time when it was considered acceptable to confine a calf or a pig so tightly that he or she could not even turn around for weeks or cage a bird so tightly that she could not spread her wings. The events that have led to the dismantling of cages and crates might then be viewed within the larger societal context of progress on a range of social and moral issues [124]. The farm animal welfare reforms over the last decade in the United States run parallel to other shifting societal norms, such as the Supreme Court extending the right of gay couples to marry and the first time in U.S. history that a woman was chosen to be the presidential candidate of a major political party. It may one day appear that these changes would have been inevitable with time, but it is the dedicated work of social reform advocates over many years driving the change, as the campaign to end the intensive confinement of farm animals clearly shows.

Author Contributions: Andrew Rowan conceived of the idea for the paper and contributed to its drafting. Sara Shields led the writing, and Paul Shapiro contributed and added key details.

Conflicts of Interest: The authors declare no conflict of interest.

Appendix A

Table A1. A partial list of company cage-free pledges based on USDA estimates [83] and organized by the date of each public pledge.

Company	Pledge Date	Annual Egg Usage	Transition Time-Line
Red Robin	25 May 2009	2,000,000	2010
Hellman's Mayonnaise	24 February 2010	350,000,000	2020
Hyatt Hotels	24 May 2011	3,000,000	2011
Burger King	25 April 2012	1,200,000,000	2017
Au Bon Pain	21 January 2013	41,000,000	2017
Marriott	27 January 2013	25,600,000	2015
Delaware North	22 October 2014	370,000,000	2016
Sodexo	19 February 2015	239,000,000	2020
Aramark	12 March 2015	215,000,000	2020
Centerplate	23 March 2015	600,000	2020
King's Food Markets	5 April 2015	35,000,000	2025
Otis Spunkmeyer	5 April 2015	100,000,000	2023
Hilton Hotels	6 April 2015	4,100,000	2017
WinCo Foods	25 April 2015	85,100,000	2025
Revolution Foods	1 July 2015	3,000,000	2018
Cheesecake Factory	28 July 2015	6,000,000	2020
McDonald's	9 September 2015	2,000,000,000	2025
Starbuck's	1 October 2015	195,000,000	2020
TGI Friday's	27 October 2015	7,000,000	2025
Kellogg's	29 October 2015	400,000,000	2025
Panera Bread	5 November 2015	120,000,000	2020
Taco Bell	16 November 2015	130,000,000	2016
Jack in the Box/Qdoba	20 November 2015	108,000,000	2025
General Mills	25 November 2015	331,000,000	2025
Costco	2 December 2015	2,900,000,000	2025
Royal Caribbean	3 December 2015	6,900,000	2022
Dunkin' Donuts	7 December 2015	390,000,000	2025
Subway	8 December 2015	110,000,000	2025
Arby's	9 December 2015	12,000,000	2020
Einstein Bros	14 December 2015	9,000,000	2020
Flowers Foods	14 December 2015	100,000,000	2025
Peet's Coffee	14 December 2015	3,000,000	2020
Caribou Coffee	15 December 2015	4,000,000	2020
Shake Shack	15 December 2015	2,000,000	2016
Carnival Corp.	21 December 2015	7,000,000	2025
Nestle	22 December 2015	185,000,000	2020
Wendy's	4 January 2016	890,000,000	2020
Quizno's	7 January 2016	45,000,000	2025
Denny's	14 January 2016	400,000,000	2026
Mondelez	16 January 2016	68,000,000	2020

Table A1. *Cont.*

Company	Pledge Date	Annual Egg Usage	Transition Time-Line
California Pizza Kitchen	20 January 2016	3,000,000	2022
Taco John's	20 January 2016	8,000,000	2025
Target	20 January 2016	1,590,000,000	2025
P.F. Chang's/Pei Wei	23 January 2016	14,000,000	2025
Schwan Food Co	29 January 2016	37,000,000	2020
White Castle	29 January 2016	12,000,000	2025
Sonic	1 February 2016	155,000,000	2025
Starwood Hotels	1 February 2016	5,800,000	2020
BJ's	9 February 2016	294,000,000	2022
Trader Joe's	12 February 2016	638,400,000	2025
Cracker Barrel	15 February 2016	220,000,000	2026
Applebee's	18 February 2016	4,000,000	2025
CVS Health	18 February 2016	71,100,000	2025
Golden Corral	18 February 2016	20,000,000	2026
IHOP	18 February 2016	214,000,000	2025
Black Bear Diner	19 February 2016	20,000,000	2025
Bloomin' Brands	22 February 2016	1,300,000	2025
Ahold	23 February 2016	1,085,000,000	2022
Krystal Burger	23 February 2016	11,000,000	2026
Albertsons	1 March 2016	3,073,000,000	2025
Kraft Heinz	1 March 2016	313,000,000	2025
WAWA	1 March 2016	69,100,000	2020
Delhaize America	2 March 2016	1,757,000,000	2025
Kroger	4 March 2016	3,942,000,000	2025
Bob Evans	4 March 2016	100,000,000	2025
The Fresh Market	8 March 2016	236,600,000	2025
Chick-fil-A	9 March 2016	57,000,000	2026
Aldi	10 March 2016	1,260,000,000	2025
Schnuck's	16 March 2016	144,200,000	2025
Sprouts Market	22 March 2016	324,800,000	2022
Basha's	23 March 2016	151,200,000	2017
Shoney's	23 March 2016	36,000,000	2025
Raley's	24 March 2016	170,800,000	2020
PepsiCo (Quaker)	28 March 2016	10,000,000	2020
Super Value	29 March 2016	670,000,000	2025
Weis Markets	29 March 2016	232,400,000	2026
Darden Restaurants, Inc.	30 March 2016	10,000,000	2018
Stater Bros.	30 March 2016	236,600,000	2025
Walgreens	30 March 2016	76,500,000	2025
Meijer	1 April 2016	313,600,000	2025
Lowes Food Stores	2 April 2016	131,600,000	2025
Smart & Final, Inc.	2 April 2016	365,400,000	2025
Ingles Markets	4 April 2016	282,800,000	2025
Krispy Kreme	4 April 2016	620,000,000	2026
Save Mart Supermarkets	4 April 2016	303,800,000	2025
Snyders-Lance	4 April 2016	100,000,000	2025
Walmart	5 April 2016	11,500,000,000	2025
Fairway Foods	5 April 2016	21,000,000	2025
King Kullen	7 April 2016	51,800,000	2025
Gelson's Markets	8 April 2016	36,400,000	2020
Giant Eagle	8 April 2016	294,000,000	2025
Boyer's Food Market	9 April 2016	25,200,000	2026
Brookshire Grocery Co.	11 April 2016	210,000,000	2025
SpartanNash	11 April 2016	224,000,000	2025
LeBrea Bakery	12 April 2016	2,500,000	2016
Vallarta Supermarkets	13 April 2016	65,800,000	2025
Southeastern (Winn-Dixie)	15 April 2016	1,068,200,000	2025
Tops Markets	16 April 2016	229,600,000	2025

Table A1. *Cont.*

Company	Pledge Date	Annual Egg Usage	Transition Time-Line
Walt Disney	16 April 2016	57,600,000	2016
H-E-B	18 April 2016	429,800,000	2025
Wegman's	19 April 2016	126,000,000	2025
Wakefern (ShopRite)	22 April 2016	490,700,000	2025
C&S Wholesale Grocers	25 April 2016	427,000,000	2025
Woodman's Market	26 April 2016	22,400,000	2025
Dollar Tree/Family Dollar	28 April 2016	250,000,000	2025
7-Eleven	3 May 2016	118,500,000	2025
Allegiance Retail (Foodtown)	3 May 2016	119,000,000	2022
Dollar General	5 May 2016	320,000,000	2025
Rite Aid	6 May 2016	42,800,000	2025
Heinen's	9 May 2016	30,800,000	2020
Price Chopper/Market 32	9 May 2016	189,000,000	2025
Carl's Jr/Hardee's	10 May 2016	59,000,000	2025
Dairy Queen/Orange Julius	16 May 2016	40,000,000	2025
Superior Grocers	16 May 2016	63,000,000	2025
Bojangles	26 May 2016	92,000,000	2025
Northgate Gonzalez	27 May 2016	56,000,000	2025
Grocery Outlet	1 June 2016	210,000,000	2025
US Foods	1 June 2016	1,115,000,000	2026
Dierberg's	2 June 2016	35,000,000	2025
Associated Food Stores	10 June 2016	594,600,000	2025
CraftWorks Restaurants	15 June 2016	2,500,000	2022
Jerry's Enterprises	20 June 2016	36,400,000	2025
Sysco	20 June 2016	1,274,000,000	2026
IGA, Inc.	21 June 2016	870,000,000	2025
Rouses Markets	26 June 2016	61,600,000	2025
Foodland Super Market	29 June 2016	53,200,000	2025
Reasor's	29 June 2016	26,600,000	2025
BiRite Foodservice	1 July 2016	13,000,000	2026
Niemann Foods	1 July 2016	61,600,000	2025
Martin's Super Markets	8 July 2016	30,800,000	2025
Harmon's	11 July 2016	22,400,000	2020
Festival Foods (WI)	15 July 2016	33,600,000	2025
Publix Super Markets	15 July 2016	1,579,200,000	2026
Homeland Stores	20 July 2016	84,000,000	2025
Papa John's	20 July 2016	39,000,000	2016
Mi Pueblo Food Center	22 July 2016	26,600,000	2025
Sedano's	22 July 2016	47,600,000	2026
Casey's General Stores	26 July 2016	80,300,000	2025
Barilla America, Inc.	31 July 2016	38,000,000	2020
Buehler Food Markets	2 August 2016	21,000,000	2025
Focus Brands	5 August 2016	134,000,000	2026
Key Food Stores	8 August 2016	280,000,000	2025
Compass Group	15 September 2016	377,600,000	2019
PAQ, Inc. (Food 4 Less)	12 October 2016	58,800,000	2025

References

1. Singer, P. *Ethics into Action: Henry Spira and the Animal Rights Movement*; Rowman & Littlefield Publishers, Inc.: Lanham, MD, USA, 1998.
2. U.S. Congress. Office of Technology Assessment. *Alternatives to Animal Use in Research, Testing, and Education*; U.S. Congress. Office of Technology Assessment: Washington, DC, USA, 1986.
3. United States Department of Agriculture. *Livestock Slaughter Annual Summary: 1982*; United States Department of Agriculture (USDA), Statistical Reporting Service: Washington, DC, USA, 1983.

4. United States Department of Agriculture. *Quick Stats*; United States Department of Agriculture (USDA), National Agricultural Statistics Service: Washington, DC, USA, 1982.
5. European Union. *Council Directive 91/630/EEC of 19 November 1991 Laying Down Minimum Standards for the Protection of Pigs*; European Union: Brussels, Belgium, 1991.
6. European Union. *Commission Decision of 24 February 1997 Amending the Annex to Directive 91/629/EEC Laying Down Minimum Standards for the Protection of Calves*; European Union: Brussels, Belgium, 1997.
7. European Union. *Council Directive 1999/74/EC of 19 July 1999 Laying Down Minimum Standards for the Protection of Laying Hens*; European Union: Brussels, Belgium, 1999.
8. Appleby, M.C. The European Union ban on conventional cages for laying hens: History and prospects. *J. Appl. Anim. Welf. Sci.* **2003**, *6*, 103–121. [CrossRef] [PubMed]
9. Phillips, C. *Cattle Behaviour and Welfare*, 2nd ed.; Blackwell Publishing: Oxford, UK, 2002.
10. Rowan, A.N. Animal well-being: Key philosophical, ethical, political, and public issues affecting food animal agriculture. In *Food Animal Well-Being*; United States Department of Agriculture and Purdue University: Indianapolis, IN, USA, 1993; pp. 23–35.
11. United States Congress. *Veal Calf Protection Act: Joint Hearing Before the Subcommittee on Livestock, Dairy, and Poultry and the Subcommittee on Department Operations, Research, and Foreign Agriculture of the Committee on Agriculture, House of Representatives*; U.S. Government Printing Office: Washington, DC, USA, 1989.
12. Rowan, A.N.; Rosen, B. Progress in animal legislation: Measurement and assessment. In *The State of the Animals III*; Salem, D.J., Rowan, A.N., Eds.; Humane Society Press: Washington, DC, USA, 2005; pp. 79–94.
13. Centner, T.J. Limitations on the confinement of food animals in the United States. *J. Agric. Environ. Ethics* **2010**, *23*, 469–486. [CrossRef]
14. Harrison, R. *Animal Machines: The New Factory Farming Industry*; Vincent Stuart Ltd.: London, UK, 1964.
15. Broom, D. A History of Animal Welfare Science. *Acta Biotheor.* **2011**, *59*, 121–137. [CrossRef] [PubMed]
16. Brambell, F.W.R. *Report of the Technical Committee to Enquire into the Welfare of Animals Kept Under Intensive Livestock Husbandry Systems*; Presented to Parliament by the Secretary of State for Scotland and the Minister of Agriculture, Fisheries and Food by Command of Her Majesty; Her Majesty's Stationary Office: London, UK, 1965.
17. Keeling, L.J. Healthy and happy: Animal welfare as an integral part of sustainable agriculture. *Ambio* **2005**, *34*, 316–319. [CrossRef] [PubMed]
18. Council of Europe. *Convention on the Protection of Animals Kept for Farming Purposes*; Council of Europe: Strasbourg, Belgium, 1976.
19. Duncan, I.J. Behavior and behavioral needs. *Poult. Sci.* **1998**, *77*, 1766–1772. [CrossRef] [PubMed]
20. Jensen, P.; Toates, F.M. Who needs 'behavioural needs'? Motivational aspects of the needs of animals. *Appl. Anim. Behav. Sci.* **1993**, *37*, 161–181. [CrossRef]
21. Marchant, J.N.; Broom, D.M. Effects of dry sow housing conditions on muscle weight and bone strength. *Anim. Sci.* **1996**, *62*, 105–113. [CrossRef]
22. Schenck, E.L.; McMunn, K.A.; Rosenstein, D.S.; Stroshine, R.L.; Nielsen, B.D.; Richert, B.T.; Marchant-Forde, J.N.; Lay, D.C., Jr. Exercising stall-housed gestating gilts: Effects on lameness, the musculo-skeletal system, production, and behavior. *J. Anim. Sci.* **2008**, *86*, 3166–3180. [CrossRef] [PubMed]
23. Arellano, P.E.; Pijoan, C.; Jacobson, L.D.; Algers, B. Stereotyped behaviour, social interactions and suckling pattern of pigs housed in groups or in single crates. *Appl. Anim. Behav. Sci.* **1992**, *35*, 157–166. [CrossRef]
24. Vieuille-Thomas, C.; Pape, G.L.; Signoret, J.P. Stereotypies in pregnant sows: Indications of influence of the housing system on the patterns expressed by the animals. *Appl. Anim. Behav. Sci.* **1995**, *44*, 19–27. [CrossRef]
25. Terlouw, E.M.C.; Lawrence, A.B.; Illius, A.W. Influences of feeding level and physical restriction on development of stereotypies in sows. *Anim. Behav.* **1991**, *42*, 981–991. [CrossRef]
26. Stolba, A.; Wood-Gush, D.G.M. The behaviour of pigs in a semi-natural environment. *Anim. Prod.* **1989**, *48*, 419–425. [CrossRef]
27. Olsson, I.A.S.; Keeling, L.J. Why in earth? Dustbathing behaviour in jungle and domestic fowl reviewed from a Tinbergian and animal welfare perspective. *Appl. Anim. Behav. Sci.* **2005**, *93*, 259–282. [CrossRef]
28. Olsson, I.A.S.; Keeling, L.J. Night-time roosting in laying hens and the effect of thwarting access to perches. *Appl. Anim. Behav. Sci.* **2000**, *68*, 243–256. [CrossRef]

29. Rafael, F.; Appleby, M.C.; Hughes, B.O. Effects of nest quality and other cues for exploration on pre-laying behaviour. *Appl. Anim. Behav. Sci.* **1996**, *48*, 37–46.
30. Van Praag, H.; Kempermann, G.; Gage, F.H. Neural consequences of environmental enrichment. *Nat. Rev.* **2000**, *1*, 191–198. [CrossRef] [PubMed]
31. Lewis, M.H.; Presti, M.F.; Lewis, J.B.; Turner, C.A. The neurobiology of stereotypy I: Environmental complexity. In *Stereotypic Animal Behaviour: Fundamentals and Applications to Welfare*, 2nd ed.; Mason, G., Rushen, J., Eds.; CABI: Wallingford, UK, 2006; pp. 190–226.
32. United States Department of Agriculture. *Milk Cows and Production Final Estimates 2003–2007*; United States Department of Agriculture (USDA), National Agricultural Statistics Service: Washington, DC, USA, 2009.
33. United States Department of Agriculture. *Hogs and Pigs: Number of Operations and Inventory*; United States Department of Agriculture (USDA), National Agricultural Statistics Service, Arizona Agricultural Statistics: Phoenix, AZ, USA, 2005.
34. Arizona Revised Statutes. *Initiative, Referendum and Recall*; Arizona Revised Statutes: Phoenix, AZ, USA, 2005; Title 19, Section 19–114(A); Available online: http://law.justia.com/codes/arizona/2014/title-19/ (accessed on 4 October 2016).
35. Arizona Revised Statutes. *Cruel and Inhumane Confinement of a Pig during Pregnancy or of a Calf Raised for Veal*; Arizona Revised Statutes: Phoenix, AZ, USA, 1998; Title 13, Criminal Code, Chapter 29, Section 2910.07; Available online: https://www.animallaw.info/statute/az-initiatives-proposition-204-inhumane-confinement (accessed on 4 October 2016).
36. Smith, R. Veal group housing approved. *Feedstuffs*, 6 August 2007; p. 3.
37. Hormel Plans Phase-Out of Gestation Crates by 2017. Available online: http://www.nationalhogfarmer.com/animal-well-being/hormel-plans-phase-out-gestation-crates-2017 (accessed on 4 October 2016).
38. Freese, B. *Top 35 U.S. Pork Powerhouses® 2016*; Meredith Agrimedia: Des Moines, IA, USA, 2016.
39. United States Department of Agriculture. *Evaluation of FSIS Management Controls over Pre-Slaughter Activities*; Audit Report No. 24601-0007-KC; United States Department of Agriculture (USDA), Office of Inspector General Great Plains Region: Kansas City, MO, USA, 2008.
40. Continuing Problems in USDA's Enforcement of the Humane Methods of Slaughter Act. Available online: https://www.gpo.gov/fdsys/pkg/CHRG-111hhrg65127/html/CHRG-111hhrg65127.htm (accessed on 4 October 2016).
41. Letter to The Honorable Ed Schafer, Secretary, United States Department of Agriculture. RE: Humane Handling of Downed Animals at Slaughterhouses, Auctions, and Markets. 2 May 2008. Available online: www.humanesociety.org/assets/pdfs/farm/pacelle-to-usda-downers-05-02-08.pdf (accessed on 9 April 2017).
42. Review of the Welfare of Animals in Agriculture. Available online: https://www.gpo.gov/fdsys/pkg/CHRG-110hhrg39809/html/CHRG-110hhrg39809.htm (accessed on 4 October 2016).
43. Kucinich, D. *Opening Statement. Hearing on Adequacy of the USDA Oversight of Federal Slaughter Plants*; Domestic Policy Subcommittee, House Office Building: Washington, DC, USA, 17 April 2008.
44. Weiss, R. Video reveals violations of laws, abuse of cows at slaughterhouse. *The Washington Post*, 30 January 2008.
45. Fiegel, E. *Humane Society: Undercover Video Shows Alleged Abuse at Egg Farm*; CNN: Washington, DC, USA, 2010.
46. Kitroeff, N.; Mackey, R. Activists accuse Walmart of condoning torture of pigs by pork suppliers. *The New York Times*, 1 November 2013.
47. O'Keefe, E. Official blows whistle on food-safety agency. *The Washington Post*, 5 March 2010.
48. Shea, M. Punishing animal rights activists for animal abuse: Rapid reporting and the new wave of ag-gag laws. *Columbia J. Law Soc. Probl.* **2014–2015**, *48*, 337–371.
49. Landfried, J. Bound & gagged: Potential first amendment challenges to ag-gag laws. *Duke Environ. Law Policy Forum* **2012**, *23*, 377–403.
50. Liebmann, L. Fraud and first amendment protections of false speech: How United States v. Alvarez impacts constitutional challenges to ag-gag laws. *Pace Environ. Law Rev.* **2014**, *31*, 565–593.
51. Johnston, T. Hallmark/Westland: One year later. *Meatingplace*, February 2009; p. 10.
52. Prohibition against Restrictive Confinement. Available online: https://www.oregonlaws.org/ors/600.150 (accessed on 4 October 2016).

53. Confinement of Calves Raised for Veal and Pregnant Sows. Available online: https://www.animallaw.info/statute/co-farming-article-505-confinement-calves-raised-veal-and-pregnant-sows (accessed on 4 October 2016).
54. Putting Meat on the Table: Industrial Farm Animal Production in America. Available online: http://www.pewtrusts.org/en/research-and-analysis/reports/2008/04/29/putting-meat-on-the-table-industrial-farm-animal-production-in-america (accessed on 4 October 2016).
55. United States Department of Agriculture. *Chickens and Eggs 2007 Summary*; United States Department of Agriculture (USDA), National Agricultural Statistics Service: Washington, DC, USA, 2008.
56. United Egg Producers. *United Egg Producers Animal Husbandry Guidelines for U.S. Egg Laying Flocks*; United Egg Producers: Alpharetta, GA, USA, 2008.
57. Meadow, B.; Kannel, S.; Undem, T.; Crain, D. *Arizona Farm Animal Confinement Issues, Survey Research Analysis*; Prepared for the Humane Society of the United States; Lake Snell Perry Mermin Decision Research: Washington, DC, USA, 2005.
58. Lake Research Partners. *California: 800 Likely Voters in the 2008 Election*; Prepared for the Humane Society of the United States; Lake Research Partners: Washington, DC, USA, 2007.
59. Farm Animal Cruelty. Available online: https://leginfo.legislature.ca.gov/faces/codes_displayText.xhtml?lawCode=HSC&division=20.&title=&part=&chapter=13.8.&article= (accessed on 4 October 2016).
60. Prop 2 Standards for Confining Farm Animals Initiative Statute. Available online: https://ballotpedia.org/California_Proposition_2,_Standards_for_Confining_Farm_Animals_(2008) (accessed on 4 October 2016).
61. Nieves, E. Farm animal rights law would require more room to roam. *The Mercury News*, 28 July 2008.
62. Californians for Humane Farms v. Schafer, No. C08-03843 MHP, 2008 *Westlaw* 4449583. Available online: https://www.animallaw.info/case/californians-humane-farms-v-schafer (accessed on 4 October 2016).
63. Bell, D. A review of recent publications on animal welfare issues for table egg laying hens. Available online: https://www.google.ch/url?sa=t&rct=j&q=&esrc=s&source=web&cd=1&ved=0ahUKEwiQrard__DTAhXkKcAKHR0UAuoQFgglMAA&url=http%3A%2F%2Fanimalsciencey.ucdavis.edu%2Favian%2FWelfareIssueslayingHens.pdf&usg=AFQjCNH5Ro9QZSAlN8AuHSbqtbrez57_Ig&cad=rja (accessed on 4 October 2016).
64. Processed Egg Products Antitrust Litigation, No. 08-md-2002, 2016 Westlaw 5539592. Available online: http://www.leagle.com/decision/In%20FDCO%2020160907D87/IN%20RE%20PROCESSED%20EGG%20PRODUCTS%20ANTITRUST%20LITIGATION (accessed on 4 October 2016).
65. Patrick, M.E.; Adcock, P.M.; Gomez, T.M.; Altekruse, S.F.; Holland, B.H.; Tauxe, R.V.; Swerdlow, D.L. *Salmonella* Enteritidis infections, United States, 1985–1999. *Emerg. Infect. Dis.* **2004**, *10*, 1–7. [CrossRef] [PubMed]
66. Callaway, T.R.; Edrington, T.S.; Anderson, R.C.; Byrd, J.A.; Nisbet, D.J. Gastrointestinal microbial ecology and the safety of our food supply as related to Salmonella. *J. Anim. Sci.* **2008**, *86*, E163–E172. [CrossRef] [PubMed]
67. European Food Safety Authority. Report of the Task Force on Zoonoses Data Collection on the Analysis of the baseline study on the prevalence of Salmonella in holdings of laying hen flocks of Gallus gallus. *EFSA J.* **2007**, *97*, 1–84.
68. Centers for Disease Control. Multistate Outbreak of Human Salmonella Enteritidis Infections Associated with Shell Eggs (Final Update). 2 December 2010. Available online: www.cdc.gov/salmonella/2010/shell-eggs-12-2-10.html (accessed on 4 October 2016).
69. United States District Court. *Avila, et al. v. Olivera Egg Ranch, LLC, 2011 Westlaw 12882109*; United States District Court, Eastern District of California: Sacramento, CA, USA, 2011.
70. Meadow, B.; Ulibarri, J. *Post-Election Survey of 800 Californians Who Voted in the 2008 Election*; Lake Research Partners: Washington, DC, USA, 2008.
71. California Superior Court. *J.S. West Milling Co., Inc. v. State of California, et al., No. 10-CECG-04225*; California Superior Court: Fresno, CA, USA, 2011.
72. United States Court of Appeals. *Cramer v. Harris, 591. Federal Appendix 634*; United States Court of Appeals, Ninth Circuit: Los Angeles, CA, USA, 2015.
73. California Superior Court. *Association of California Egg Farms v. State of California, et al., No. 12-CECG-03695*; California Superior Court: Fresno, CA, USA, 2013.

74. Shelled Eggs; California Health and Safety Code, Division 20. Miscellaneous Health and Safety Provisions. Chapter 14, Section 25996. Available online: http://codes.findlaw.com/ca/health-and-safety-code/hsc-sect-25996.html (accessed on 4 October 2016).
75. United States Court of Appeals. *Missouri ex rel. Koster v. Harris, Petition for Cert. filed No. 16-1015, 847 Federal Reporter 3d 646, 650*; United States Court of Appeals, Ninth Circuit: San Francisco, CA, USA, 2017.
76. Cruelty to Animals; Maine Revised Statutes: Title 7, Agriculture and Animals, Part 9, Animal Welfare, Chapter 739: Section 4011. Available online: http://legislature.maine.gov/statutes/7/title7sec4011.html (accessed on 4 October 2016).
77. Animal Industry Act; Michigan Compiled Laws. Chapter 287: Section 287.746. Available online: http://www.legislature.mi.gov/(S(abtezdr0xncbzlfxmztnr0e4))/mileg.aspx?page=GetObject&objectname=mcl-act-466-of-1988 (accessed on 4 October 2016).
78. United States Department of Agriculture. *Chickens and Eggs 2009 Summary*; United States Department of Agriculture (USDA), National Agricultural Statistics Service: Washington, DC, USA, 2010.
79. Ohio Department of Agriculture. *2009 Annual Report*; Ohio Department of Agriculture: Reynoldsburg, OH, USA, 2009; Available online: http://www.agri.ohio.gov/divs/Admin/Docs/AnnReports/ODA_Comm_AnnRpt_2009.pdf (accessed on 4 October 2016).
80. Ohio Livestock Care Standards Board; Ohio revised codes. Title 9. Agriculture Chapter 904: Sections 904.01–904.09. Available online: http://codes.ohio.gov/orc/904 (accessed on 4 October 2016).
81. 112th Congress, Second Session. H.R. 3798. To Provide for a Uniform National Standard for the Housing and Treatment of Egg-Laying Hens, and for Other Purposes. 23 January 2012. Available online: https://www.congress.gov/bill/112th-congress/house-bill/3798/text (accessed on 4 October 2016).
82. 113th Congress, 1st Session. S. 820. To Provide for a Uniform National Standard for the Housing and Treatment of Egg-Laying Hens, and for Other Purposes. 25 April 2013. Available online: https://www.congress.gov/bill/113th-congress/senate-bill/820/text?format=txt (accessed on 4 October 2016).
83. United States Department of Agriculture. *Cage-Free Statistics from USDA*; United States Department of Agriculture (USDA), National Agricultural Statistics Service: Washington, DC, USA, 2016.
84. Sustainability Goes Mainstream: Insights into Investor Views. Available online: https://www.pwc.com/us/en/pwc-investor-resource-institute/publications/assets/pwc-sustainability-goes-mainstream-investor-views.pdf (accessed on 4 October 2016).
85. Summary of the Judgement (Read in Open Court) before The Hon. Mr Justice Bell. High Court of Justice, Queens Bench Division: London, Thursday, 19 June 1997. Available online: https://www.big-lies.org/general/ek2-mclibel-judgment-summary.html (accessed on 4 October 2016).
86. Armstrong, S. Bunfight. *The Guardian*. 13 September 1999. Available online: https://www.theguardian.com/media/1999/sep/13/marketingandpr.mondaymediasection (accessed on 4 October 2016).
87. Grandin, T. Effect of animal welfare audits of slaughter plants by a major fast food company on cattle handling and stunning practices. *J. Am. Vet. Med. Assoc.* **2000**, *216*, 848–851. [CrossRef] [PubMed]
88. Grandin, T. The McDonald's Effect. *Beef Magazine*. 1 February 2001. Available online: http://www.beefmagazine.com/mag/beef_mcdonalds_effect (accessed on 4 October 2016).
89. Pacelle, W. *The Humane Economy*; HarperCollins: New York, NY, USA, 2016.
90. Strom, S. McDonald's set to phase out suppliers' use of sow stalls. *The New York Times*, 13 February 2012. Available online: http://www.nytimes.com/2012/02/14/business/mcdonalds-vows-to-help-end-use-of-sow-crates.html (accessed on 4 October 2016).
91. Murphy, R. Smithfield Foods moves to end use of breeding crates on company farms. *Daily Press*, 7 January 2015. Available online: http://www.dailypress.com/news/isle-of-wight-county/dp-nws-smithfield-gestation-crates-progress-20150107-story.html (accessed on 4 October 2016).
92. Barclay, E. Smithfield prods its pork suppliers to dump pig crates. *National Public Radio*, 7 January 2014. Available online: http://www.npr.org/sections/thesalt/2014/01/07/260439063/smithfield-prods-its-pork-suppliers-to-dump-pig-crates (accessed on 4 October 2016).
93. Press Releases. Smithfield Foods Nears 2017 Goal for Conversion to Group Housing Systems for Pregnant Sows. Smithfield, VA. 4 January 2017. Available online: www.smithfieldfoods.com/newsroom/press-releases-and-news/smithfield-foods-nears-2017-goal-for-conversion-to-group-housing-systems-for-pregnant-sows (accessed on 9 April 2017).

94. Smithfield 2015 Sustainability & Financial Report. Housing of Pregnant Sows. Available online: www.smithfieldfoods.com/integrated-report/2015/animal-care/housing-of-pregnant-sows (accessed on 9 April 2017).
95. Hormel Foods. Animal Welfare. Available online: www.hormelfoods.com/About/CorporateResponsibility/Animal-Welfare (accessed on 9 April 2017).
96. Cargill. *Corporate Responsibility Report*; Cargill: Minneapolis, MN, USA, 2014.
97. Clemens Food Group Commitment to Animal Care. Available online: www.cfgsustainability.com/animal-welfare/at-the-farm.aspx (accessed on 9 April 2017).
98. Robert Ruth; Clemens Food Group. Personal communication, 2016.
99. Tyson Fresh Meats, Inc. Letter to Suppliers, dated 8 January 2014. Available online: https://www.google.ch/url?sa=t&rct=j&q=&esrc=s&source=web&cd=1&ved=0ahUKEwjg3ae4gPHTAhWHDcAKHQ6zC48QFggiMAA&url=http%3A%2F%2Fwww.tysonfoods.com%2F~%2Fmedia%2FCorporate%2FMedia%2FPosition%2520Statements%2FPork%2520Supplier%2520Letter%25201-8-14.ashx&usg=AFQjCNGA92Nq8T_TqDx-ypEKTv6LPvXH2g&cad=rja (accessed on 4 October 2016).
100. Tyson Foods Inc. Position Statement on Sow Housing Dated 23 March. Available online: www.tysonfoods.com/media/position-statements/sow-housing (accessed on 6 October 2016).
101. Top 25 U.S. Pork Powerhouses® 2015. Available online: http://www.porelia.com/wp-content/uploads/2016/09/2015PorkPowerhousesChartREV.pdf (accessed on 4 October 2016).
102. McCrea, R.C. *The Humane Movement: A Descriptive Survey*; Henry Bergh Foundation, Columbia University Press: New York, NY, USA, 1910.
103. National Institutes of Health (NIH), Office of Research Infrastructure Programs. NIH Plan to Retire All NIH-Owned and -Supported Chimpanzees. Available online: https://dpcpsi.nih.gov/orip/cm/chimpanzeeretirement (accessed on 9 April 2017).
104. Cecil the Lion Killing Sparks Outrage around the World. *CBS News*, 29 July 2015. Available online: http://www.cbsnews.com/news/cecil-the-lion-killing-sparks-outrage-around-the-world/ (accessed on 4 October 2016).
105. Macdonald, D.W.; Jacobsen, K.S.; Burnham, D.; Johnson, P.J.; Loveridge, A.J. Cecil: A moment or a movement? Analysis of media coverage of the death of a lion, *Panthera leo*. *Animals* **2016**, *6*, 26. [CrossRef] [PubMed]
106. Tan, A. American dentist who admitted killing Cecil the lion now hounded on social media. *ABC News*, 30 July 2015. Available online: http://abcnews.go.com/International/american-dentist-admitted-killing-cecil-lion-now-hounded/story?id=32757906 (accessed on 4 October 2016).
107. McKendree, M.G.S.; Croney, C.C.; Widmar, N.J.O. Effects of demographic factors and information sources on United States consumer perceptions of animal welfare. *J. Anim. Sci.* **2014**, *92*, 3161–3173. [CrossRef] [PubMed]
108. Riffkin, R. In U.S., more say animals should have same rights as people. *Gallup Social Issues*, 18 May 2015. Available online: http://www.gallup.com/poll/183275/say-animals-rights-people.aspx (accessed on 4 October 2016).
109. Wells, J. How California egg rules could affect everyone's breakfast. *NBC News*, 2 January 2015. Available online: http://www.nbcnews.com/business/consumer/how-california-egg-rules-could-affect-everyones-breakfast-n278531 (accessed on 4 October 2016).
110. Jargon, J.; Beilfuss, L. McDonald's continues with image shift with move to cage-free eggs in North America. *The Wall Street Journal*, 9 September 2015. Available online: https://www.wsj.com/articles/mcdonalds-to-source-cage-free-eggs-in-u-s-canada-1441798121 (accessed on 4 October 2016).
111. Brulliard, K. How eggs became a victory for the animal welfare movement. *The Washington Post*, 6 August 2016. Available online: https://www.washingtonpost.com/news/animalia/wp/2016/08/06/how-eggs-became-a-victory-for-the-animal-welfare-movement-if-not-necessarily-for-hens/?utm_term=.0ea04229cb03 (accessed on 4 October 2016).
112. Jolley, C. Tipping point reached on cage-free. *Feedstuffs*, 28 February 2016. Available online: http://feedstuffsfoodlink.com/blogs-tipping-point-reached-cage-free-10695 (accessed on 4 October 2016).
113. Walmart Press Release. *Commitment Represents Company's Continued Focus on Advancing Animal Welfare*; Walmart Press: Bentonville, AR, USA, 2016.

114. Gelles, D. Eggs that clear the cages, but maybe not the conscience. *The New York Times*, 16 July 2016. Available online: https://www.nytimes.com/2016/07/17/business/eggs-that-clear-the-cages-but-maybe-not-the-conscience.html (accessed on 4 October 2016).
115. Malcolm, H. Walmart's cage-free egg vow could cut prices, aid hens. *USA Today*, 7 April 2016. Available online: https://www.usatoday.com/story/money/2016/04/06/cage-free-eggs-expected-to-get-cheaper/82702828/ (accessed on 4 October 2016).
116. Colman, Z. The fight for cage-free eggs. *The Atlantic*, 16 April 2016. Available online: https://www.theatlantic.com/politics/archive/2016/04/a-referendum-on-animal-rights/478482/ (accessed on 4 October 2016).
117. Barclay, E. The year in eggs: Everyone's going cage-free, except supermarkets. *National Public Radio*, 30 December 2015. Available online: http://www.npr.org/sections/thesalt/2015/12/30/461483821/the-year-in-eggs-everyones-going-cage-free-except-supermarkets (accessed on 4 October 2016).
118. Charles, D. Most U.S. egg producers are now choosing cage-free houses. *National Public Radio*, 15 January 2016. Available online: http://www.npr.org/sections/thesalt/2016/01/15/463190984/most-new-hen-houses-are-now-cage-free (accessed on 4 October 2016).
119. Charles, D. What the rise of cage-free eggs means for chickens. *National Public Radio*, 27 June 2013. Available online: http://www.npr.org/sections/thesalt/2013/06/27/195639341/what-the-rise-of-cage-free-eggs-means-for-chickens (accessed on 4 October 2016).
120. Kowitt, B. Inside McDonald's bold decision to go cage-free. *Fortune*, 18 August 2016. Available online: http://fortune.com/mcdonalds-cage-free/ (accessed on 4 October 2016).
121. Lempert, P. Shift to cage-free eggs is likely to disappoint. *Forbes*, 8 May 2016. Available online: https://www.forbes.com/sites/phillempert/2016/05/08/shift-to-cage-free-eggs-is-likely-to-disappoint/#3e6846b85fd3 (accessed on 4 October 2016).
122. Chaussee, J. The insanely complicated logistics of cage-free egss for all. *Wired*, 25 January 2016. Available online: https://www.wired.com/2016/01/the-insanely-complicated-logistics-of-cage-free-eggs-for-all/ (accessed on 4 October 2016).
123. United States Department of Agriculture. *Chickens and Eggs 2016 Summary*; United States Department of Agriculture (USDA), National Agricultural Statistics Service: Washington, DC, USA, 2017. Available online: http://usda.mannlib.cornell.edu/usda/current/ChickEgg/ChickEgg-02-27-2017.pdf (accessed on 4 October 2016).
124. An Act to Prevent Cruelty to Farm Animals. Available online: https://malegislature.gov/Laws/SessionLaws/Acts/2016/Chapter333 (accessed on 4 October 2016).

4

Exploring the Role of Farm Animals in Providing Care at Care Farms

Jan Hassink [1,*], Simone R. De Bruin [2], Bente Berget [3] and Marjolein Elings [1]

1. Wageningen University and Research, P.O. Box 16, 6700 AA Wageningen, The Netherlands; marjolein.elings@wur.nl
2. National Institute for Public Health and the Environment, Centre for Nutrition, Prevention and Health Services, P.O. Box 1, 3720 BA Bilthoven, The Netherlands; simone.de.bruin@rivm.nl
3. Agderforskning, Gimlemoen, P.O.Box 422, 4604 Kristiansand, Norway; bente.berget@agderforskning.no

* Correspondence: jan.hassink@wur.nl

Academic Editor: Karen Thodberg

Simple Summary: This paper provides insight into the role of farm animals in farm-based programs and their importance to different types of participants. Farm animals provide real work, close relationships, challenging tasks and opportunities for reflection. They also contribute to a welcoming atmosphere for various types of participants.

Abstract: We explore the role of farm animals in providing care to different types of participants at care farms (e.g., youngsters with behavioural problems, people with severe mental problems and people with dementia). Care farms provide alternative and promising settings where people can interact with animals compared to a therapeutic healthcare setting. We performed a literature review, conducted focus group meetings and carried out secondary data-analysis of qualitative studies involving care farmers and different types of participants. We found that farm animals are important to many participants and have a large number of potential benefits. They can (i) provide meaningful day occupation; (ii) generate valued relationships; (iii) help people master tasks; (iv) provide opportunities for reciprocity; (v) can distract people from them problems; (vi) provide relaxation; (vii) facilitate customized care; (viii) facilitate relationships with other people; (ix) stimulate healthy behavior; (x) contribute to a welcoming environment; (xi) make it possible to experience basic elements of life; and (xii) provide opportunities for reflection and feedback. This shows the multi-facetted importance of interacting with animals on care farms. In this study the types of activities with animals and their value to different types of participants varied. Farm animals are an important element of the care farm environment that can address the care needs of different types of participants.

Keywords: farm animals; care farm; mental illness; youth; dementia

---

1. Introduction

The intensification and industrialization of livestock farming is under discussion in Western Europe. Environmental problems, homogenization of the landscape, outbreaks of animal diseases and poor animal welfare have created a negative image of the agricultural sector [1]. Increasing pressure on the agricultural sector and changing demands from society have changed the focus of an increasing number of farmers in Western Europe, generating increasing interest in multifunctional agriculture [2,3] involving the integration of new activities around the core of agricultural production [1,4]. Combining agricultural production with health and social care services is an innovative example of more sustainable rural development and multifunctional agriculture [5]. The combination of agricultural production and care and support services is known under different names, including care farming, green care and social farming [6].

Care farming, green care or social farming is a fast-growing sector across Europe [6,7]. It is an innovation at the crossroads of agriculture and healthcare, where the agricultural sector is actively involved in providing care to different client groups. Clients, or participants in the vocabulary of care farmers, are involved in agricultural production and/or farm-related activities. Care farms offer adult day care services, supported workplaces and/or residential places to clients with a variety of disabilities [8], including people with mental disorders, people with dementia, troubled youths and people with intellectual disability. The size of care farms in terms of their number of clients, number of animals, arable land and other measures differs widely across care farms [6,7]. Care farming has reached different stages of development in different countries. The Netherlands and Norway are leading in the number of care farms with approximately 1100 green care farms in each country [9,10]. Care farming is in line with the objectives of rural development policies as well as of health and social care policies. It promotes the wider idea of multifunctional agriculture, represents opportunities to reduce lack of services in rural areas and establishes new bridges between urban and rural areas [7]. It is also in line with the changing paradigms in the health sector; from an emphasis on disease, disease prevention and limitations caused by diseases, to a more and more a positive approach to health with an emphasis on health promotion, possibilities, and participation [11].

The combination of the personal and dedicated attitude of the farming family and other staff, carrying out meaningful farm-related activities, and an informal and open community in a green environment are seen as attractive aspects of care farms for various client groups [11], the perceived benefits being improved physical, mental and social well-being [12]. Activities involving farm livestock are important aspects of farm-based programs and they are expected to play an important role in stimulating the well-being of care farm participants, many of who prefer working with livestock to activities involving crops [13] and it is expected that interacting with farm animals is important to them [11,14].

2. Animals in Healthcare

During the 20th century, the presence of animals in institutional care settings increased, while different concepts of animal interventions, so-called Animal Assisted Interventions (AAI), were developed [15]. AAI refers to any intervention that intentionally includes or incorporates animals as part of a therapeutic or ameliorative process or environment [15,16]. AAI encompasses both Animal-Assisted Therapy (AAT) and Animal-Assisted Activities (AAA). Animal-assisted therapy (AAT) refers to goal-directed interventions with animals as an integral part of the treatment process for a particular client [17], while the term Animal-Assisted Activities (AAA) refers to a general category of interventions without a protocol [15]. Most studies involving AAI deal with interactions between humans and companion animals [15].

Numerous studies have shown that human-animal contact can enhance the physical, social and mental health of people with different types of disabilities. It has been suggested that contact with animals can reduce stress and anxiety, distract people from negative emotions, facilitate interpersonal relations and provide social support (e.g., [18–23]). Caring for animals can meet the basic social needs for caring for another living being and experiencing reciprocity [20,22]. Taking care of animals can boost people's confidence and self-esteem, and help create a more positive self-image. In interactions with animals people can experience basic elements of life [22,23]. Interacting with animals can provide feelings of safety and comfort, and allow people to display affection [20,24,25]. Developing a bond with animals is considered to be helpful in developing relationships with other people [24]. Several studies have shown benefits of animals for human health and wellbeing. However in the context of care farms and farm animals there is still much unknown regarding the benefits of animals for human health and wellbeing.

3. Animals at Care Farms

The role and effect of farm animals at care farms for different client groups is a relatively new area of research that requires further study. Care farms provide a setting that is somewhat different from the therapeutic healthcare settings that are usually studied, mainly based on small and stable social communities with flexibility in activities in nature-based settings. The animals on care farms are, generally speaking, used for production rather than for therapeutic purposes. As such, it is an environment that offers different types of activities and interaction between people and animals compared to other settings. In most cases, the focus is on productive farm work, like feeding the animals, cleaning stables and milking the cows. However, interacting with animals during these activities at care farms may also facilitate social and communicative contact with the animals [14,16].

In some cases, the care farmers have additional healthcare education [26], while others have no training or background in healthcare. In contrast with therapeutic settings involving only one animal species, there are usually different species of farm animals on care farms. Several studies have pointed at the important role of farm animals in making farms attractive to different types of client groups (e.g., [14,27–30]). Previous studies have shown that participants at care farms explicitly mentioned the importance of farm animals in the process of achieving personal and collective well-being [13,31]. Care farms often have a wide variety of animals, including chickens, pigs, sheep, horses and cows [13]. There are different types of care farms, including care farms focusing on agricultural production and efficient animal production, and care farms that focus on providing care [32]. The types of animal and the number of animals will vary between different types of care farms. In many cases, care farmers also keep small pets specifically for the participants [13]. Until recently, few studies have focused particularly on the role and effects of farm animals [14,21,27], and none of them have focused on the differences between species of farm animals with respect to their role and value for different types of care farm participants, which is why this study takes a closer look at the significance of animals in the care farm context [33]. Although this paper focusses on the perceived benefits of farm animals for different types of participants on care farms, there is also need to critically think about animals' potentially contested positions on farms and within therapeutic spaces [31]. Being involved in a care farming system may not be necessarily be beneficial for animals [31].

4. Aims

The aim of this paper is to provide insight into the activities of participants involving farm animals and the perceived benefits of interacting with different species of farm animals to different types of participants of care farms. We provide an overview of results of previous studies, focusing on the experiences of participants on care farms, including the interaction with farm animals. We then enrich these findings with the results of interviews and focus group meetings with care farmers, and interviews with different types of participants focusing on the role of (different species of) farm animals.

5. Methods

5.1. Study Design

We combined different data sources to generate insight into the activities, interactions and perceived benefits of farm animals to different types of participants. Whenever possible, we tried to specify these aspects for different species of animals. We started by conducting a literature review. Data from the studies involved were aggregated and provided us with a general understanding of the perceived benefits of care farm animals to different client groups. Secondly, we performed secondary analysis of data collected during six studies carried out between 2001 and 2015, all of which involved interviewing or observing care farmers and/or different types of participants. The authors of this paper were all involved in one or more of these studies. Since the aim of the study described in this paper was different from the aims of the studies from which the data were initially collected, we considered our approach to be a secondary data-analysis. We had direct access to the original data from interviews and observations of (family caregivers of) participants of care farms and care farmers. This second step allowed us to check our findings against those from the literature study. Also, we could add to and enrich existing findings by adding, for instance, findings involving client groups not yet described in existing literature or by adding quotations from respondents underlining the findings from the literature study. Thirdly, we organized focus group meetings with care farmers, allowing the care farmers to reflect on the findings from the first two steps and us to connect our findings to the practical experiences of care farmers in The Netherlands.

5.2. Literature Review

5.2.1. Search Strategy

Our literature review focused on English language papers involving on care farms. Because we expected the number of studies to be limited, we set no limit on the date range. The search was conducted in the Scopus database, using the following keywords: *care farm*, *green care*, *farming for health* and *social farm*. In addition, the authors manually added other studies of which they were aware.

5.2.2. Study Selection

Papers were eligible if they met the following predefined inclusion criteria: 1. The paper should focus on green care farms; 2. The paper described original research, including data about the role of farm animals for participants. From the 16 papers we found, six papers met the inclusion criteria. One reviewer (Jan Hassink) reviewed all the titles and abstracts of the papers that were extracted on the basis of their expected relevance. When considered relevant, the complete paper was retrieved and reviewed for its relevance again. Since we expected to be able to include only a limited number of studies, all eligible studies were included in our review, regardless of their quality.

5.3. Secondary Data-Analysis of Six Qualitative and Observational Studies

In all studies, participants were eligible for receiving care services of a care farm in line with the assessments of Dutch or Norwegian agencies tasked with evaluating care needs and eligibility of receiving care. For participants with mental illness, the focus of care was on rehabilitation and stimulation of participation in useful activities. For youth care clients, the focus of care was on stimulation of participation in activities in a productive agricultural setting. For people with dementia, the focus of care was on providing a structured and meaningful day (i.e., adult day services). Farmers and social workers on the care farm were reimbursed for delivering care by either the local government, national government or a health insurance company.

5.3.1. Study 1: Interviews with Care Farmers in The Netherlands

Between 2001–2003, Dutch care farms were selected with long-term experience with interaction between farm animals and participants, providing day care services to people with mental illness, children and youths with behavioral problems. Based on the information from the National Support Centre of care farms, an advocacy organization for green care farmers in The Netherlands, 20 farmers were selected from different regions in The Netherlands. Semi-structured face-to-face interviews of approximately 2 h were conducted, focusing on general characteristics of farm animals that are important to participants and differences between farm animals with respect to activities, characteristics and therapeutic qualities [34].

5.3.2. Study 2: Interviews with Youth Care Clients in The Netherlands

In 2013, semi-structured interviews were conducted with 11 young people who had completed a living and working program on an agricultural production-oriented youth care farm in The Netherlands [30]. The "living and working program" was developed for youngsters between the ages of 16 and 23 with severe social and mental health problems, varying from externalizing (acting out, e.g., aggression) to more internalizing problems (inward, e.g., anxiety and mood disorders, social withdrawal). The program aimed at providing young people with a normal working environment. The youngsters were not really motivated to participate in the farm program. However, if they had rejected the farm based program, they would have been referred to a compulsory program in a residential youth care institution or a judicial trajectory. The youngsters lived and worked on productive diary and intensive pig farms for a period of six months on an individual basis. They helped the farmers in their daily activities. The focus was on carrying out productive work on animal husbandry farms. The youngsters had the same working hours as the farmer. Unlike on most other care farms, there were no other participants on the farm, except for the youngsters placed on the farms. A detailed description of the program is provided by Schreuder et al. [30]. In the interviews, we focused on the experiences of the youngsters with the different elements of the living and working program. An important element of the program was taking care of the animals.

5.3.3. Study 3: Interviews with People with Severe Mental Illness in The Netherlands

In 2010, we conduced semi-structured interviews with 13 participants with severe mental problems, some of whom also had addiction problems [35]. They all attended day activities at care farms in The Netherlands. The aim of the interviews was to gather information about the rehabilitation process of participants with severe mental problems at care farms and conventional day activity centers, and the role of different aspects of working on a farm in this process. We asked the respondents to reflect on their rehabilitation experiences, the impact of the care farm on their physical, social and mental health, and the importance of the qualities of the care farms, one of which involved interacting with animals. The interviews were recorded and transcribed, allowing us to use the participants' own words in the analysis.

5.3.4. Study 4: Video Observations of People with Severe Mental Illness in Norway

Between 2004–2006, a study involving 35 adult participants with severe mental health problems was performed, the aim being to examine how the participants behaved when interacting with animals [27]. An ethogram and video records were used to study the working abilities (measured as intensity and exactness of the work with the animals) and different types of behavior they displayed when interacting with the animals. Video recordings were made near the start and end of a three-month intervention. Furthermore, the aim was to see if there were differences among different types of patients and whether their improved working abilities correlated with higher self-esteem, coping ability and quality of life, or lower levels of depression or anxiety. Finally, the physical distance between the participants and the animals was examined as an indicator of the level of fear towards the animals and to see whether closer proximity to the animals was related to active work and caring for the animals by the end of the intervention period [27].

5.3.5. Study 5: Interviews with People with Severe Mental Illness in Norway

In 2015, a study was performed looking into the experiences of people with mental health problems who took part in green care activities (including working and interacting with the animals) on care farms in Norway [33]. A hermeneutic phenomenological research design was applied and ten semi-structured interviews were conducted. The aim of study design was to capture the clients' experiences of the work and social interactions on the care farm and their perceptions regarding their personal health and daily functioning and regarding how the care farm context can motivate, engage and improve the way people participating in care farm activities function.

5.3.6. Study 6: Interviews with People with Dementia and Their Family Caregivers in The Netherlands

A qualitative descriptive study was performed between November 2012 and November 2013, [36], with the aim of generating insight into the characteristics of people with dementia and their family caregivers visiting adult day services centers, understanding the factors associated with initiating adult day services, understanding the factors associated with selecting the day service setting, and identifying the value of adult day services in terms of specific social participation domains. Semi-structured interviews were conducted with people with dementia living at home and their family caregivers. Adult day services at green care farms are offered to people with dementia living in the community. The aim is to provide a structured and meaningful day program to people with dementia and, in doing so, provide support and relief to family caregivers [37]. The participants were recruited, using purposeful sampling, via care professionals at ten green care farms and five regular adult day services facilities in The Netherlands. These professionals provided contact details of eligible participants with the permission of the participants involved. In addition, an invitation to participate in the study was placed on the website of a Dutch patient organization for people with dementia.

5.4. Focus Group Meetings

In the Autumn of 2016, four exchange and discussion meetings with care farmers and social workers on care farms were organized in four different regions (south, north, west and east parts) of The Netherlands about the qualities and valued/appreciated aspects of care farms for different types of participants of care farms. We organized these meetings to reflect on the results of scientific studies on care farms and to facilitate exchange of experiences. After the researchers (the first two authors of this paper) presented the main findings from scientific research on the value of care farms for different client groups, the researchers, care farmers and social workers together discussed and exchanged views and experiences related to the qualities of care farms. One of the topics that was discussed in-depth was the role and use of farm animals in farm-based programs for different types of participants.

All care farms that were members of the National Federation of Care Farms (the national advocacy organization of care farmers) were invited to take part in the meetings. In all, 60 care farmers took part in these focus group discussions. During each meeting, we divided the larger group into smaller subgroups of about five persons, which enabled each person to share experiences. The exchange and discussions in each group were facilitated by an external facilitator. The researchers and facilitator made notes of the outcomes of the group discussions.

5.5. Data Analysis

Data from the literature study were extracted in a structured way, including 1. study design; 2. study population; 3. type of activities with farm animals and 4. experienced value/benefits of farm animals. The most important findings from studies in the literature study were summarized.

For the secondary data from studies 1–6, we conducted a thematic analysis adopting a deductive approach. The original transcripts from the interviews were read and reread. Our analysis of the data was directed by the following main themes, which we agreed on beforehand: (i) activities with animals; (ii) the value of animals in general to different types of participants; (iii) the value of specific animal species to different types of participants. Additionally, on the basis of the themes that had emerged from the literature review, the main themes were further divided into subthemes. We used a similar approach on the data from the focus group meetings, writing reports of the meetings for further study. Whenever the data allowed us to, we specified the findings for different types of participants. To illustrate our findings, we included quotations from the original transcripts. The source of each quotation is identified by providing a description of the type of respondent. Also the notes of the focus group meetings with care farmers and social workers were thematically analyzed.

6. Results

6.1. Literature Review

Our findings are presented in three sections, corresponding to the three aims of our study, i.e., activities with animals, the value of animals in general to different types of participants and the value of specific animal species to different types of participants.

6.1.1. Activities with Animals on Care Farms

We found six scientific papers in English containing specific data about the benefits of interacting with farm animals experienced by participants of care farms. The characteristics of these studies, the activities on the farm and the benefits experienced, are presented in Table 1.

Table 1. Characteristics of the studies selected with the literature review.

Authors	Study Design	Activities	Study Population	Experienced Benefits of Animals
Pedersen et al., 2012 [14]	Interviews with 8 participants of care farms addressing the participants' experiences	Ordinary tasks in cow shed; Grooming, mucking, feeding, taking care of the calves, milking	People with mental illness: depression	Ordinary work, Being appreciated, Closeness, warmth and calmness, Distraction from difficulties; Flexibility in tasks, Ability to accomplish work tasks, tranquility; Getting energy
Kogstad et al., 2014 [38]	Observations and 2–4 interviews with 9 participants (17–25 year in age)	Ordinary tasks with different species of animals	Youth; drop-outs from school	Safe relationship; mastering of tasks, flexibility in tasks, silence, distraction from worries; giving care to other living beings; building motivation and confidence
Granerud and Eriksson 2014 [28]	Interviews with 20 current and former participants of care farms 22–55 years in age	Ordinary tasks with different species of animals	Long-term unemployed people with mental illnesses; some with addiction problems	Feelings of familiarity; Relationship without stigmatization or complications; Meaningfulness; Structure; Physical activity; Variation; Mastery of tasks
Iancu et al., 2014 [39]	Interviews with 14 participants on 13 care farms addressing the experiences with care farm program	Ordinary tasks with different species of animals	People with mental illnesses	Motivation, Responsibility, Working at your own pace, Choice in activities
Ferwerda et al., 2012 [29]	Interviews with 7 care farmers offering services to children with autism spectrum disorders (ASD)	Feeding, clean stables, milking goats, brushing and riding horse, walking dog, cuddle, playing, tell animal stories	Children with autism spectrum disorder	Social support, trustful relationship, living creature to tell stories to, conquer of fear
Gorman 2016 [31]	31 interviews with farmers and external organisations. Ethnographic research on one farm	Ordinary tasks with different species of animals	Diverse	Something to engage with and respond to; Shared relations, knowledge and experiences; Stimulating conversation; Feeling comfortable; level of ownership; place attachment; Physical and healthy activity in implicit way; Purposeful tasks; Screening out negative perceptions; Becoming care giver

In three cases, the participants were people with mental illnesses, in one case, drop-outs from school, in another case, children with disorders on the autism spectrum and, in one case, participants with different backgrounds. In two of the papers we found, interviews were conducted with care farmers, in the remaining four, with participants. On most of the care farms involved in these studies, participants interacted with production and companion animals. The participants of care farms are involved in both productive farming activities like feeding, cleaning the shed and milking cows, as well as activities associated with creating an intimate bond, like hugging and stroking.

6.1.2. Benefits of Farm Animals for Different Types of Participants in General

The studies illustrate the broad potential impact of farm animals on participants. The findings of the studies were clustered into 10 different categories of benefits (see Table 2). The animals are recognized as being the fabric of the care farm and playing an important role in providing structure and useful, meaningful and diverse activities to the participants. By interacting with/taking care of animals, participants can create a bond and experience warmth, closeness and security, while mastering tasks can improve their confidence, responsibility and personal growth. Interacting with animals can help participants forget their problems and become more relaxed, and allows them to experience reciprocity. The role of caregiver contrasts with their customary role as care recipient. Participants can choose different types of activities. Their experiences with animals allows them to interaction with other human beings. Animals can contribute to more healthy behavior from participants by stimulating physical exercise. Finally, the animals can provide a more welcoming environment.

Table 2. Perceived benefits of farm animals for different types of participants.

Meaningful day occupation	ordinary work, purposeful tasks, meaningfulness
Valued relationship	appreciation, closeness, warmth, safe, trustful, relationship without stigmatization or complications, social support, something to engage with, living creature to tell stories to
Mastery of tasks	ability to accomplish work tasks, building motivation and confidence, responsibility, conquer of fear
Reciprocity	giving care to other living being, becoming caregiver
Distraction from problems or difficulties	screening out negative perceptions, getting energy
Relaxation	tranquility, feeling comfortable
Tailored care/support	working at your own pace, choice in activities
Relationship with other human beings	shared relations, knowledge and experiences, stimulating conversation
Stimulating health behavior	physical and healthy activity in implicit way
Welcoming environment	feeling at home, place attachment

6.1.3. Values of Specific Animal Species for Different Types of Participants

None of the studies provided more detailed information about the value of specific animal species to for the different types of participants and about any differences between the different species.

6.2. Secondary Data-Analysis of Qualitative and Observational Studies and Focus Group Meetings with Care Farmers

Like the findings of the literature review, the findings of the secondary data analysis and focus group meetings are also presented in three sections, corresponding to the three aims of our study.

6.2.1. Activities with Animals on Care Farms

The care farmers who took part in the focus group meetings indicated that regular agricultural work should be the basis for the interactions with animals on care farms. This involves feeding, cleaning the stables and moving animals to the field or stable. The exact type of work depends on the type of species, as well as on the participants involved. Young people with behavioral problems, for instance, did all the regular work on dairy and intensive pig farms, which included feeding, cleaning the stables and moving pigs from one stable to another. It was intensive work on farms with a focus on agricultural production. People with mental problems, however, were generally placed on care farms with less focus on agricultural production. In addition to productive work, there was substantial time for stroking and hugging the animals. Farmers adapted the amount of work to the abilities of the participant. Also, for people with dementia, the activities focused less on agricultural production. For several people with dementia and their family caregivers, the presence of animals at the farm was one of the reasons to attend adult day services at a green care farm. Several family caregivers and people with dementia stated that they were fond of animals and were interested in feeding, watching, cuddling and grooming the animals, because they had or used to have pets themselves, or grew up or had worked on a farm. As such, they were used to working with, cuddling and/or taking care of animals.

6.2.2. Values of Farm Animals for Different Types of Participants in General

Meaningful Day Occupation

The care farmers who took part in the focus group meetings emphasized that actual useful work should be the basis of the interaction of participants with animals. In their view, interactions should develop in a natural way. During the work, there is time for hugging or stroking the animals. They were against creating artificial contacts with animals. Participants can develop a bond by taking care of the same animals for a longer period of time. The care farmers we interviewed expressed that taking care of the animals is conceived as real and useful work that provides the participants with a structured environment. In many cases, participants start by feeding the animals. The afternoon is often used to clean the stables. According to the care farmers, many participants do not like all the aspects of the work. They realize, however, that the work has to be done and that it is important to take good care of the animals and keep them healthy and happy.

> "We always start the day with feeding the animals. Some of the participants like to do other things, but I think it is important for their development that they also do tasks that have to be done on the farm. In the afternoon, they can choose what they like to do" (Farmer 1)

Also, participants of farm-based programs indicated that they saw working with the animals as useful work. Respondents with mental illnesses, for instance, mentioned that they found the work they were doing important. As one respondent stated:

> "It is normal and really useful work. We do things that have to be done. These animals are depending on you. You feed them and that is why they can exist" (participant with mental illness)

Also, for participants with dementia, taking care of the animals gives them the feeling that they still perform useful work. As one family caregiver put it:

> "But the green care farm . . . .I know he's good with animals; animals are fantastic for him. But in regular adult day services centers . . . .you won't find a dog; you won't find a cat. There is nothing. And here (at the green care farm), there are horses, you can walk or ride to a goat. You can feed the animals. Actually, they are turning people into a kind of farmers. That makes people think: I have a paid job again" (family caregiver of person with dementia on a waiting list for a care farm)

Valued Relationship

The care farmers indicated that farm animals are appealing to many participants, in particular younger animals and participants can develop a close relationship with them:

> "Giving the bottle to the calf is such an intimate and quiet job. It can help make children feel at home on the farm" (Farmer 2)

Animals invite people to take care of them. They can offer security, closeness and warmth when you cuddle them. According to the farmers, animals have no hidden agenda, they do not gossip and, for many participants, especially those struggling with relationships with human beings, communicating with and talking to animals can be a first step towards developing trust and building relationships, as is illustrated by the following quote:

> "One of our clients with autism is very much concerned with animals. In the institute, he is stressed and aggressive. On the farm he is much more relaxed. The animals make him quiet and offer him safety and structure. The animals don't give him the wrong stimuli. He only talks to persons that also have an interest in animals" (Farmer 3)

Two of the youngsters indicated they were surprised when they started to recognize the individual characteristics of different cows:

> "The cow is very special. Cows are like human beings, each cow has its own character. You get to know them. I never expected that. It was always the same cow that approached me when I entered the stable, and always the same cow that did not want to be milked by the robot" (young participant)

Two youngsters indicated that being responsible for and taking good care of the cows was important to them and helped them sustain the farm program, especially when the relationship with the farmer was not very good.

> "The farmer and I had not a good match.....The cows were important for me. Taking good care of the cows motivated me to continue the work and do it properly.....I have never been so close to a cow. It was very special and pleasant company. They never complain"

The care farmers, as well as many of the participants with mental illnesses themselves, emphasized the importance of having an intimate bond with an animal, especially when the participants had a bad time. And, for some of them, contact with animals was important in experiencing a safe way to build contact with another living creature:

> "When I feel a bit depressed, I stand next to the horse and she lays her nose on my arm and I brush her very gently. That gives me peace. And then I talk to her and she becomes more cheerful as well" (participant with mental illness)

> "They don't become angry with you, they don't say nasty things. I prefer animal company compared to humans. People can say things that hurt me" (participant with mental illness)

Mastery of Tasks

Another value of farm animals is that they can offer challenging and physically demanding activities, like cleaning the stables or moving pigs from one stable to another. Several of the care farmers indicated that animals can also challenge people's courage. It takes courage to work with big animals like cows or horses. Some animals can also do unexpected things, and the participants have to deal with this unpredictable behavior. Overcoming challenges can help them build self-esteem. As one farmer put it:

> "A horse is a very big animal for most participants. Riding a horse can be very challenging. But when you manage to lead the horse, it gives them such a good feeling. Many of them (children from a residential youth care organization) want to give some steering to their lives" (Farmer 5)

> "Working with cows can be challenging, when you have to catch a cow in the herd grazing in the field, you have to show that you are really strong" (Farmer 6)

Care farmers also indicated that the physicality of working with animals is stimulating. There is no discussion and it is a good motivator to become active.

> "when you come to feed the pigs in the morning, they start making so much noise that when you are still sleepy, you will wake up" (Farmer 7)

One of the farmers explained how he used this to stimulate a participant with mental problems to get out of his bed and come to the farm.

> "For him, it is always difficult to start in the morning. We hang a picture of Sue our pony at the end of the bed with the text you are only a few steps away from feeding her. This helps him start the day" (Farmer 8)

Also, the answers provided by the participants indicated that animals helped them master certain tasks. For instance, some of the youngsters, although they were not experienced when it came to working with cows, indicated that cows were very pleasant and special company to them. Being responsible for and taking good care of the cows was important to them and helped them sustain the farm program: "It was important for me that the farmer gave me the responsibility for taking care of the cows" (young participant)

As it was for the youngsters, working with big animals also could be challenging for participants with mental illnesses. However, when people manage to overcome that challenge, it can make them feel proud and improve their confidence and self-esteem.

> "I have always been fond of horses. Helping with the horses gave me a lot of confidence. That is only when I work with the horses. I do not know why, but I really like working with horses" (participant with mental illness)

For some participants with mental problems, it came as a surprise that they liked working with animals, as they had no interest in working with animals before.

> "I was never fond of animals. But I even learned riding a horse here. When it goes well with the riding, riding a horse gives me a very good feeling. I feel proud, like, "so you can do this"" (participant with mental illness)

Reciprocity

The interviews with the care famers showed that being at the farm give participants a positive experience and a better balance, as they not only receive care, but provide care as well.

Also, participants with mental illness indicated that taking care of animals gave them something in return. They experienced the positive response of the animal to the care they received, which gave the participants a good feeling. They felt that caring for the animals is special, because they are living beings.

> "I cheer up by working with the animals. When you take care of them, it gives me a good feeling. And they also give something in return. Some of them show it and they come to get hugged" (participant with mental illness)

> "If I were dealing with dead things, it would not give me the same sense of responsibility (laughing). Perhaps it would be easier to give a damn on a bad day. But I do; as long as you are dealing with live animals, you do care" (participant with mental illness)

Distraction from Problems or Difficulties

Care farmers indicated that the participants are encouraged to take care of another living being. As a result, they focus less on their own problems, which is especially important for participants with mental problems. As one farmer put it:

> "When you have a psychiatric problem, you realize that animals can become sick and how important it is to take care of them properly. You can only take care of an animal properly when you also take good care of yourself" (Farmer 9)

In addition, many participants with mental problems indicated that the physical work with the animals also distracted them from their own problems.

> "It is good to be active physically instead of thinking about my problems all the time" (participant with mental illness)

Relaxation/Rest

Farmers indicated that working with animals can make participants more relaxed. Some farmers specifically mentioned that cows have a very relaxing impact on participants. One of the farmers intentionally used cows specifically for participants suffering from depression.

> "Cows have a calmness around them, I let them lean against their warm body. It relaxes participants who are very busy" (Farmer 10)

> "Let a depressed client take care of the cows in the stable for a few days, especially brushing, and he or she will come to herself and feelings of sadness will be released" (farmer 11)

Some of the youngsters and participants with mental illnesses also indicated that working with cows made them feel relaxed and that interacting with animals had a calming effect. This gave them a good feeling:

> "It's so liberating...You are completely alone, and it's totally calm, and you can hear the birds, and it's just something about life. All the painful and negative thoughts disappear a little. They are put to the side. You get some kind of inner peace" (participant with mental illness)

> "I can tell the animals what bothers me, whole stories. And then I can cope with it and become myself a bit. So it's a kind of becoming more relax" (participant with mental illness)

Customized Care/Support

When there is sufficient diversity in activities and a participant takes part in a variety of activities, the care farmer and participant discover what kind of activities are appealing. For the care farmers, the challenge is to create the conditions in which a successful bond can be created. As one of the farmers indicated:

> "When you offer a rich context, and you are open to what is happening, you can find the right match" (Farmer 12)

Many participants with mental illnesses indicate that it is important for them to work at their own pace. When they have a bad day, the activities are adjusted accordingly.

For several people with dementia and their family caregivers, a green care farm was a more appealing environment than a regular long-term care institution, one of the reasons being that the farm-related activities are usually not available in regular adult day services centers, where activities are mostly organized indoors and are different in nature (e.g., craft work, playing games, gymnastics, memory training, etc.). The importance of taking part in animal-related activities and the types of preferred activities with animals varied. For some people, it was sufficient to watch the animals or to walk to the cages, stables and meadows. For others it, being actually involved in the work was more important. This reflects that at the farm participants had freedom of choice in the activities they wanted to do and what fitted their interests and preferences most:

> "We do have a pony, a donkey, sheep, hamster, rabbits, dog, cats and a potbellied pig. That's fun, but I don't do much with them. Every now and then, you know, in the winter time, when there is not so much to do in the green house or outside, I help them. Otherwise, other people take care of the animals. That is all well-organized" (person with dementia attending adult day services at a green care farm)

> "Yes, that's the kind of thing we are doing [at the green care farm], taking care of animals. Sometimes, it is playing with a rabbit or cuddling a guinea pig. But, in other instances, it is more about brushing the donkeys or goats, you name it, those kind of activities. I like that" (person with dementia attending adult day services at a green care farm)

Relationships with Other People

The interviews revealed that the shared interest in animals can stimulate contacts and conversation between people. Care farmers indicated that participants with a similar interest in the cows or horses developed new friendships and contacts based on their common interest. This could also happen with people outside the care farm. This is illustrated by one of the youngsters, who became really interested in horses and made new contacts with his uncle and aunt, as they also raised horses, and started helping them with the work.

A participant with mental illness indicated that it was difficult for her to go outside, because she felt that everybody was watching her. When she was walking with the dog of the farmer, it was easier, because she had something to do. It also helped her make contact with other people because other people walking their dogs started a conversation, sharing experiences of having a dog.

Stimulating Health Behavior

Taking care of farm animals involves physically demanding tasks. Many participants with mental illness mentioned that they become tired in a way that makes them feel good. As one of the participants put it:

> "It is a different type of getting tired than I am used to. It is more physical instead of mental" (participant with mental illness)

For several people with dementia, being around animals is a stimulus to become (physically) active and spend time outdoors. This was one of the reasons for initiating adult day services at care farms rather than in regular adult day care, as explained by one of the people with dementia:

> "I like it that the staff at the green care farm at certain moment . . . if they wouldn't do that, I would be chatting with others all day. Luckily they ask me every now and then, "Well, Mr. P. could you please help me with cleaning the cages of the animals or taking care of them?" Well, I am always prepared to do that. But I need it, that they ask me to do something" (person with dementia attending adult day services at a green care farm)

Welcoming Environment

Care farmers indicated that, in addition to farm animals, pets, like cats and dogs are also important to participants. They help create a familiar environment where the participants feel welcome. Because they walk around freely, it is easy for participants to make contact with them. One of the farmers gave an example where the cat was always sitting on the table in the morning, welcoming each participant and waiting for a stroke.

> "They really are part of the identity of the care farm and contribute to the informal atmosphere"

This is confirmed by some of the participants with mental illness. They also indicate this specific role of the animals on the farm, especially the dogs.

> "And the dogs make me feel welcome when I arrive here. They approach you. I was always afraid of dogs, but not for these two" (participant with mental illness)

Experiencing Basic Elements of Life

From the secondary data analysis, two new subthemes emerged, in addition to those that emerged from the literature review. From the interviews, it became clear that, thanks to the presence of animals, participants are also able to experience basic elements of life. Animals are part of everyday life. They stimulate all senses, they move around and each animal feels differently. Farm animals make life processes visible in a natural way. Participants see new animals being born, becoming ill and dying. Animals can also help start a discussion about sexuality.

> "Lambs in spring are the ultimate spring feeling; a bunch of jumping, running and playing lambs make everyone joyful and happy" (Farmer 13)

> "A rooster jumping on a chicken or a bull on a cow can lead to discussions about sexuality" (Farmer 11)

Mirroring

The second additional subtheme that emerged from the secondary data analysis was "mirroring". Care farmers said that they used the interactions between animals and the interactions between participants and animals to give participants feedback on their behavior. Some of the farmers used the interaction with animals to provide participants new insights into painful events in their lives. Many farm animals, like cows, live in herds and their interaction in the group can be a trigger for participants. As two farmers put it:

> "Cows live in a herd with strict rules and social order based on power, size and experiences. Participants often feel a bond with the cow that is lowest in order. This reflects their own experiences in life. This can be painful but also clarifying" (Farmer 13)

> "Separation of a calf from the cow is an emotional process for many participants with a psychiatric background. It is the recognition. It offers a good introduction to talk about their lives" (Farmer 14)

Other farmers used the interaction with farm animals to give participant direct feedback on their behavior and/or stimulate behavioral changes: "The horse is a mirror for our youngsters. When someone gets on the horse in a restless way and is shouting, the horse becomes nasty or unmanageable. At that moment, I have an introduction to discuss their behavior" (Farmer 15)

> "One of the boys once pinched the udder of the cow. The cow kicked him and the boy will never do it again. In that way, he also learns how he has to behave with others" (Farmer 16)

The participants also acknowledged that interacting with animals provided opportunities for reflection. Some youngsters with behavioral problems, for instance, did not really have a bond with the cows, but understood that the cows were really important to the farmer:

> "The cows were very big animals. I did not really have a bond with them but I felt that the cows were very important for the farmer and that he had a very close bond with them" (young participant)

Also, for participants with mental problems, animals can be a mirror reflecting difficult or painful experiences in their own lives. Different participants with mental illnesses gave examples of this mirror effect and indicated that the interaction with animals can also provide feedback on their own behavior.

> "Such a small innocent calf. You have to be quiet and I am quite a busy person. But now it is my calf. The calf was very shy but now it eats out of my hand. I just get so excited about it. That it trusts a stranger like me and that I can hug it and sucks my fingers. That it doesn't have to be afraid. Just like I was afraid of others. She teaches me how I can deal with myself" (participant with mental illness)

> "I like to tell the people about the butterfly garden. I tell them how the butterfly comes out of the cocoon; some of them never come out, others come out very slowly. I can build entire philosophies about this process" (participant with mental illness)

> "Sometimes I became reckless and then it went wrong. When the horse reacts in such a way it is your mistake, you have done something wrong. You can learn from this" (participant with mental illness)

6.2.3. Values of Specific Animal Species for Different Types of Participants

According to the care farmers, participants can establish a bond with all types of farm animals, including horses, cows, pigs, sheep, goats and chickens. In the focus group meetings, farmers expressed that in many cases they cannot predict which species particular participants will have a match with. Generally, larger animals are challenging and give participants a sense of pride when they can master them. They help develop self-esteem. Smaller animals are often used for experiencing security, warmth and connection. Cats and dogs are considered a special category of animals. Because they can walk around freely, it is easy for participants to make contact with them. For many participants, it is nice to be able to hug them and tell them their experiences. They really are part of the identity of the care farm and contribute to the informal atmosphere.

The respondents experienced some major differences between different species of farm animals. Cows are considered large animals with a warm, kind-hearted, dreamy character. Some care farmers said that cows make people calm and bring them into contact with their emotions. Horses, on the other hand, are considered large and versatile animals with which participants can form a close bond. Some of the farmers indicated that, for many participants, it is also challenging to work with horses and that some participants are afraid of horses. Pigs are considered cheerful, roguish animals that focus on food. Especially the small piglets are appealing to participants. Goats are considered curious animals with a focus on the participants. They are easy to stroke, but also jumpy and unpredictable. For some of the participants, this unpredictable behavior can be a challenge. Sheep are considered vulnerable animals that are not easy to caress. For many participants, lambs in spring are attractive and the ultimate expression of joy. Chickens being held in a non-productive way are nice to watch, especially when they can walk outside. They provide a cozy homelike atmosphere.

The interviews with the participants did not reveal a great deal of insight into the values of specific animals. It was only in the interviews with youngsters that some differences were mentioned. Some of the youngsters went to dairy farms, others to intensive pig farms. The experiences of the youngsters with the pigs were different than those with the cows. The interaction with farm animals was more important on dairy farms than on pig farms. The youngsters on dairy farms established a connection with individual cows and recognized differences in their characters. On intensive pig farms, the participants did not mention contacts with individual animals or character differences.

7. Discussion

For this study, we combined different data sources to gather insights into activities, interactions and perceived benefits of farm animals for different types of participants. Although knowledge about the value and impact of animals on people is still limited, our study has increased our insight into how different types of participants on care farms can benefit from the presence of and interaction with farm animals. Activities with animals and the value of animals to participants that were observed in literature were also observed in the data we included in the secondary data-analysis. Two additional types of values were revealed by the secondary data analysis. In addition to existing literature, the secondary data analysis provided some data about people with dementia that were not found in the literature. Although our study also provided additional information about the perceived benefits of interacting with different animal species for different participant groups, knowledge about what kind of animals provide the best match with whom remains limited.

Care farmers emphasized that useful work should be the basis of interacting with farm animals. This is a different approach to the one found in Animal-Assisted Therapy, where animals are part of the therapeutic process [23]. Other major differences between care farms and therapeutic settings are the role of the farmer as owner of the animals and mediator between participants and farm animals, the diversity of animals and activities on the care farm and the duration of the farm program. Although actual work, and not a therapeutic intervention with animals, is the basis, our study shows that, in such a productive setting, interaction with farm animals can also have diverse benefits for different types of participants. Our study revealed that participants can perform a variety of activities with animals. In general, they are involved in productive activities as well as more recreational activities. For youngsters with behavioral problems, the focus is mostly on productive work. For people with mental illness participants responded well to farm animals as well as pets, like cats and dogs. Flexibility in activities is important. For young people with behavioral problems and participants with mental illness, it is considered real work. For participants with dementia, animals provide a stimulus to remain active and spend time outdoors. For some participants with dementia, it is considered useful work, while, for others, animals are important to create a pleasant and welcoming atmosphere. In addition, we found differences in activities between different species of animals. Moreover, the participants responded well to farm animals as well as pets, like cats and dogs.

Our study shows that participants can develop a genuine and intimate connection with farm animals. We have shown examples where people formed a connection with large animals, like cows or horses, but also with smaller animals, like rabbits. Respondents also mentioned the importance of cats and dogs when it comes to developing an intimate connection and creating a homelike atmosphere. Our results are in line with earlier studies, that indicate that contact with farm animal can reduce stress, offer warmth and closeness, make participants more cheerful, help them forget their difficulties, overcome challenges and increase their coping skills [14,22,23,40]. Participants can use contact with animals to experience an intimate relationship with other living creatures. Young animals are especially appealing in this respect, while working with large animals can help participants master challenging tasks and thus improve their sense of pride, self-esteem and confidence.

Most studies report only a limited number of benefits of interacting with animals, because they often focus on specific goals and specific interventions (e.g., [22,23]). From different sources of information, we extracted 12 different categories of benefits of farm animals for different types of participants. What the three groups of participants included in this study had in common was that spending their days in a meaningful fashion was mentioned as an important value. Most of the data we found involved people with mental illnesses. Data for other types of participants is still limited. For participants with mental illness, all the categories listed above appear to be important. For youngsters with behavioral problems, animals are important because they provide real, useful and demanding work that stimulates responsibility, intimate contact and opportunities for reflection. For participants with dementia, animals are important because they are nice to watch, stimulate the participants to stay active and give them the experience of real work. At the same time, it should also be acknowledged that, since we relied mostly on secondary data, animals may also have additional values for the different types of participants included in this study. However, this was not yet supported by the data. Furthermore, in this study, we focused mainly on people with mental illnesses, youths and people with dementia. However, there are several other types of participants who visit care farms (e.g., people with learning disabilities, long-term unemployed, children with autism spectrum disorders) on whom we had no data. Therefore, for these types of participants, no conclusions can as yet be drawn.

Performing useful work and taking care of another living being are important. Care farmers should find a balance between offering useful work, giving participants the experience of being actual co-workers, and giving them the freedom to interact with the farm animals, to develop a deeper relationship and experience emotional support [14]. This balance is different for different types of participants. Based on this study, farmers, participants and their representatives and care professionals may be able to make more informed decisions about using farm animals in a care farming context, and in realizing the goals of different client groups in order to contribute to a more inclusive society.

Previous studies [14,22,23,40] revealed a smaller number of specific benefits of interacting with animals. Only a limited number of studies indicate that animals can provide opportunities for feedback and reflection [40], as was found in our study with regard to participants with mental problems and youngsters with behavioral problems. For people with mental problems, vulnerable animals in particular serve as a mirror for their experiences. Animals assist them in their healing process and recognition of painful episodes in their life. For youngsters with behavioral problems, animals can be a mirror to correct their behavior or learn from past mistakes. From the interviews with the youngsters with behavioral problems, we learned that working with cows and horses helped some of them to sustain the program, even when the relationship with the farmer was strained. The good relationship with the cows and horses was a kind of substitute for the relationship with the farmer. This importance of animals for adolescents lacking a good connection to adults is in line with earlier research [41].

Care farmers indicated that they cannot predict in advance with which animal a particular participant will have a good match. Having different species of animals will increase the likelihood that a good match can be found. For care farmers, the challenge is not only to focus on the work that has to be done on the farm, but also create the right context for different types of participants and recognize experiences with animals that can be beneficial to participants. Care farmers should also realize that the way they treat their animals can have a considerable impact on their relationship with participants. For most participants, taking good care of the animals means treating them well and with respect.

One subject that we did not include in this study is that of negative experiences and potential dangers when interacting with farm animals [31]. Future studies should also examine these aspects. Secondly, we did not examine the effect the interactions had on the animals involved, which is something future research will need to address, in particular in light of the ongoing social and political debate regarding animal welfare issues [42].

Although our study has increased our knowledge, at the same time we have to acknowledge that a major limitation of our study is that we heavily drew on secondary data. The main disadvantage of using secondary data is that the data were not collected to answer our specific research questions [43]. As such, we may not have a complete and balanced picture of the activities and values of animals for different client groups. For instance, the amount of data per group of participants varied and was limited for most groups of participants. Most data refer to participants with mental illnesses and, to a lesser extent, youths. Less information is available regarding other client groups. As a result, we need to be cautious when it comes to drawing sweeping conclusions. There is still much to discover. Another limitation is that most studies and original data we used involved the experiences of participants on care farms in general. Most studies did not specifically focus on the benefits of farm animals. Because the time available to conduct the interviews was limited, it is possible that essential information about the interaction with animals was not discussed. As such, this study should be seen as an exploratory first step, and more research is needed before we will be able to draw more far-reaching conclusions. Finally, it should be noted that we took a pragmatic approach to the review of the literature. Although we did not perform a systematic literature review, we nevertheless consider this a valuable part of our study. Literature assessing the values of animal species on care farms has not been summarized before. Our review therefore helped us to get an understanding of the state of this research field and to compare findings described in the literature with those from our secondary data analysis.

Previous studies have shown that care farms offer an appealing context for different types of participants [32,43]. Based on this study, we conclude that farm animals are important building stones for realizing the appreciated qualities of green care farms. Key qualities of care farms are: being part of a community; a personal, equal relationship with the farmer; useful, diverse work; and the green environment offering space and tranquility [32]. We have shown that farm animals play a crucial role in establishing these key qualities: our study showed how different types of participants take part in and appreciate the useful activities involving animals and how animals contribute to community building, a homelike atmosphere and a sense of belonging. Participants not only develop personal relationships with the farmer and other people, but also with specific farm animals. Finally, some of the respondents indicated they experience relaxation and tranquility when interacting with farm animals. These are qualities associated with the green environment.

8. Conclusions

To conclude, our study has shown that farm animals are important to different types of participants on care farms. They provide opportunities to participate in activities in a non-institutional setting where participants feel part of a farming community. Activities can be either productive farming activities or activities associated with creating an intimate bond. Although our study suggests that interacting with animals in a care farming context has several potential benefits for different types of participants, more research will be necessary to obtain a better understanding of how these different values can indeed contribute to participants' health and wellbeing.

Acknowledgments: The study on people with dementia was supported by the Dutch Ministry of Economic Affairs (through the Netherlands Organization for Health Research and Development (Grant 72801.0001) and the Dutch Alzheimer Society (Grant WE.03-2011-05)). The focus group meetings with care farmers were supported by the Dutch Ministry of Economic Affairs (through the Netherlands Organization for Health Research and Development (Grant 72801.0005).

Author Contributions: Jan Hassink was the principal author and provided data for the paper. Simone R. de Bruin, Bente Berget and Marjolein Elings provided data and assisted with the writing and editing of the paper.

Conflicts of Interest: The authors declare no conflict of interest.

References

1. Meerburg, B.G.; Korevaar, H.; Haubenhofer, D.K.; Blom-Zandstra, M.; van Keulen, H. The changing role of agriculture in Dutch society. *J. Agric. Sci.* **2009**, *147*, 511–521. [CrossRef]
2. Vereijken, P.H. Transition to multifunctional land use and agriculture. *NJAS* **2002**, *50*, 171–179. [CrossRef]
3. Wiskerke, J.S.C.; Bock, B.B.; Stuiver, M.; Renting, H. Environmental co-operatives as a new mode of rural governance. *NJAS* **2003**, *51*, 9–25. [CrossRef]
4. Ilbery, B.W. Farm diversification and the restructuring of agriculture. *Outlook Agric.* **1988**, *17*, 35–39.
5. Mölders, T. Multifunctional agricultural policies: Pathways towards sustainable rural development? *Int. J. Sociol. Agric. Food* **2013**, *21*, 97–114.
6. Hassink, J.; Van Dijk, M. *Farming for Health: Green-Care Farming across Europe and the United States of America*; Springer: Dordrecht, The Netherlands, 2006.
7. Di Iacovo, F.; O'Connor, D. *Supporting Policies for Social Farming in Europe: Progressing Multifunctionality in Responsive Rural Areas*; SoFar Project: Supporting EU-Agricultural Policies; Arsia: Firence, Italy, 2009.
8. Elings, M.; Hassink, J. Green care farms, a safe community between illness or addiction and the wider society. *J. Ther. Commun.* **2008**, *29*, 310–323.
9. Ernst and Young Advisory. *The Use of Day Care on Farms. Business Case Day Care on Farms*; Ernst and Young Advisory: The Hague, The Netherlands, 2012.
10. Norwegian Ministry of Agriculture and Food. *Strategy for Research and Research-Based Innovation 2007–2012*; Norwegian Ministry of Agriculture and Food: Oslo, Norway, 2012.
11. Hassink, J.; Elings, M.; Zweekhorst, M.; van den Nieuwenhuizen, N.; Smit, A. Care farms in The Netherlands: Attractive empowerment-oriented and strengths-based practices in the community. *Health Place* **2010**, *24*, 423–430. [CrossRef] [PubMed]
12. Hine, R.; Peacock, J.; Pretty, J. Care farming in the UK: Contexts, benefits and links with therapeutic communities. *Ther. Commun.* **2008**, *29*, 245–260.
13. Leck, Ch.; Evans, N.; Upton, D. Agriculture who cares? An investigation of "care farming" in the UK. *J. Rural Stud.* **2014**, *34*, 313–325. [CrossRef]

14. Pedersen, I.; Ihleback, C.; Kirkevold, M. Important elements in farm-animal assisted interventions for persons with clinical depression: A qualitative interview study. *Disabil. Rehabil.* **2012**, *34*, 1526–1534. [CrossRef] [PubMed]
15. Kruger, K.A.; Serpell, A. Animal-Assisted Interventions in Mental Health. In *Handbook on Animal-Assisted Therapy. Theoretical Foundations and Guidelines for Practice*, 3rd ed.; Fine, A.H., Ed.; Academic Press: San Diego, CA, USA, 2010; pp. 33–48. ISBN 978-0-12-381453-1.
16. Berget, B.; Pedersen, I.; Enders-Slegers, M.J.; Beetz, A.; Scholl, S.; Kovacs, G. Benefits of Animal-Assisted Interventions for Different Target Groups in a Green Care Context. In *Green Care. For Human Therapy, Social Innovation, Rural Economy, and Education*; Gallis, C., Ed.; Nova Science Publishers Inc.: New York, NY, USA, 2013; pp. 65–91.
17. Pet partners. Animal Assisted Activities (AAA) Animals-Assisted Therapy (AAT). Available online: http://www.deltasociety.org/Page.aspx?pid=320 (accessed on 18 May 2011).
18. Allen, K.; Balscovich, J.; Tomaka, J.; Kelsey, R.M. The presence of human friends and pet dogs as moderators of autonomic responses to stress in women. *J. Personal. Soc. Psychol.* **1971**, *61*, 582–589. [CrossRef]
19. Beetz, A.; Kotrschal, K.; Hediger, K.; Turner, D.; Uvnas-Moberg, K. The effect of a real dog, toy dog and friendly person on insecurely attached chilldren during a stressful task: An exploratory study. *Anthrozoös* **2011**, *24*, 349–368. [CrossRef]
20. Enders-Slegers, J.M.P. Een Leven Lang Goed Gezelschap: Empirisch Onderzoek Naar de Betekenis van Gezelschapsdieren Voor de Kwaliteit van Leven van Ouderen. Ph.D. Thesis, University Utrecht, Utrecht, The Netherlands, 2010.
21. Berget, B.; Ekeberg, O.; Pedersen, I.; Braastad, B.O. Animal-assisted therapy with farm animals for persons with psychiatric disorders: Effects on anxiety and depression. A randomized controlled trial. *Occup. Ther. Ment. Health* **2011**, *27*, 50–64. [CrossRef]
22. Bachi, K.; Terkel, J.; Teichman, M. Equine-facilitated psychotherapy for at-risk adolescents: The influence on self-image, self-control and trust. *Clin. Child Psychol.* **2011**, *17*, 298–312. [CrossRef] [PubMed]
23. Hauge, H.; Kvalem, I.L.; Berget, B.; Enders-Slegers, M.J.; Braastad, B.O. Equine-assisted activities and the impact on perceived social support, self-esteem and self-efficacy among adolescents—An intervention study. *Int. J. Adolesc. Youth* **2014**, *19*, 1–21. [CrossRef] [PubMed]
24. Martin, F.; Farnum, J. Animal-assisted therapy for children with Pervasive Developmental Disorders. *West. J. Nurs. Res.* **2002**, *24*, 657–670. [CrossRef] [PubMed]
25. Serpel, J.A. Guest editor's introduction: Animals in children's lives. *Soc. Anim.* **1999**, *7*, 87–94. [CrossRef]
26. Ihleback, C.; Ellingsen-Dalskau, L.H.; Berget, B. Motivations, experiences and challenges of being a care farmer—Results of a survey of Norwegian care farmers. *Work* **2016**, *53*, 113–121. [CrossRef] [PubMed]
27. Berget, B.; Skarsaune, I.; Ekeberg, Ø.; Braastad, B.O. Humans with mental disorders working with farm animals. *Occup. Ther. Ment. Health* **2007**, *23*, 101–117. [CrossRef]
28. Granerud, A.; Eriksson, B. Mental health problems, recovery, and the impact of green care services: A qualitative, participant-focused approach. *Occup. Ther. Ment. Health* **2014**, *30*, 317–336. [CrossRef]
29. Ferwerda-Zonneveld, R.T.; Oosting, S.J.; Kijlstra, A. Care farms as a short-break for children with Autism-Spectrum Disorders. *NJAS* **2012**, *59*, 35–40.
30. Schreuder, A.; Rijnders, M.; Vaandrager, L.; Hassink, J.; Enders, M.J.; Kenndy, L. Exploring salutogenic mechanisms of an outdoor experiential learning program on youth care farms in The Netherlands: Untapped potential? *Int. J. Adolesc. Youth* **2014**, *19*, 139–152. [CrossRef] [PubMed]
31. Gorman, R. Therapeutic landscapes and non-human animals: The roles and contested positions of animals with care farming assemblages. *Soc. Cult. Geopgraphy* **2017**, *18*, 315–335. [CrossRef]
32. Hassink, J.; Hulsink, W.; Grin, J. Care farms in The Netherlands: An underexplored example of multifunctional agriculture towards an empirically grounded organization-theory based typology. *Rural. Sociol.* **2012**, *77*, 569–600. [CrossRef]
33. Ellingsen-Dalskau, L.H.; Berget, B.; Pedersen, I.; Tenllnes, G.; Ihlebaeck, C. Understanding how prevocational training on care farms can lead to functioning, motivation and well-being. *Disabil. Rehabil.* **2016**, *38*, 2504–2513. [CrossRef] [PubMed]
34. Hassink, J. *De Betekenis van Landbouwhuisdieren in de Hulpverlening. Resultaten van Interviews met Professionals op Zorg- en Kinderboerderijen*; Rapport 45; Plant Research International: Wageningen, The Netherlands, 2002.

35. Elings, M.; Haubenhofer, D.; Hassink, J.; Rietberg, P.; Michon, H. *Effecten van Zorgboerderijen en Andere Dagbestedingsprojecten Voor Mensen Met een Psychiatrische en Verslavingsachtergrond*; Rapport 376; Plant Research International: Wageningen, The Neherlands, 2011.
36. De Bruin, S.R.; Stoop, A.; Molema, C.C.M.; Vaandrager, L.; Hop, P.; Baan, C.A. Green care farms: An innovative type of adult day service to stimulate social participation of people with dementia. *Gerontol. Geriatr. Med.* **2015**. [CrossRef] [PubMed]
37. De Bruin, S.R.; Oosting, S.J.; Tobi, H.; Blauw, Y.H.; Schols, J.M.G.A.; De Groot, C.P.G.M. Day care at green care farms: A novel way to stimulate dietary intake of community-dwelling older people with dementia? *J. Nutr. Health Aging* **2010**, *14*, 352–357. [CrossRef] [PubMed]
38. Kogstad, R.E.; Agdal, R.; Hopfenbeck, M.S. Narrative and natural recovery: Youth experience of social inclusion through green care. *Int. J. Environ. Res. Public Health* **2014**, *11*, 6052–6068. [CrossRef] [PubMed]
39. Iancu, S.C.; Zweekhorst, M.B.; Veltman, D.J.; van Balkom, A.J.; Bunders, J.F. Mental health recovery on care farms and day centres. A qualitative comparative study of users' perspectives. *J. Disabil. Rehabil.* **2014**, *36*, 573–583. [CrossRef] [PubMed]
40. Weigel, R.R. 4-H animal care as therapy for at-risk youth. *J. Ext.* **2002**, *40*, 1–3.
41. Ewing, C.A.; MacDonald, P.M.; Taylor, M.; Bowers, M.J. Equine-facilitated learning for youth with severe emotional disorders: A quantitative and qualitative study. *Child Youth Care Forum* **2007**, *36*, 59–72. [CrossRef]
42. Cornish, A.; Raubenheimer, D.; McGreevy, P. What we know about the public's level of concern for animal welfare in food production in developed countries. *Animals* **2016**, *6*, 74. [CrossRef] [PubMed]
43. Boslaugh, S. *Secondary Data Sources for Public Health: A Practical Guide*; Cambridge University Press: Cambridge, UK, 2007; ISBN 978-0-521-87001-6.

5

Factors that Influence Intake to One Municipal Animal Control Facility in Florida: A Qualitative Study

Terry Spencer [1,*], Linda Behar-Horenstein [2], Joe Aufmuth [3], Nancy Hardt [1], Jennifer W. Applebaum [4], Amber Emanuel [5] and Natalie Isaza [6]

[1] College of Medicine, University of Florida, Gainesville, FL 32611, USA; nhardt@gmail.com
[2] Colleges of Dentistry, Education, Veterinary Medicine, & Pharmacy, University of Florida, Gainesville, FL 32611, USA; Lsbhoren@ufl.edu
[3] George A. Smathers Libraries, University of Florida, Gainesville, FL 32611, USA; mapper@uflib.ufl.edu
[4] College of Liberal Arts and Sciences, University of Florida, Gainesville, FL 32611, USA; jennyapplebaum@ufl.edu
[5] College of Health & Human Performance, University of Florida, Gainesville, FL 32611, USA; amberemanuel@ufl.edu
[6] College of Veterinary Medicine, University of Florida, Gainesville, FL 32611, USA; isazan@ufl.edu
* Correspondence: tspencer@ufl.edu

Simple Summary: Animal shelters try to save homeless dogs and cats by returning lost pets to missing owners, adopting animals to new homes, and by reducing intake. We mapped the annual intake of one county animal shelter to discover where the homeless animals came from and selected one area of high-intake for stray adult dogs to study. We performed field interviews and reviewed available census and child-maltreatment data to create a theory about why so many stray dogs came from this study area. The study-area residents experience multiple socioeconomic challenges secondary to poverty including: interpersonal violence; housing instability; and lack of access to reliable transportation and communication services. Such factors lead residents to view domestic dogs not only as pets, but also as commodities that can add income to households, and often as burdens that results in pet abandonment. The community-specific data collected in this study can drive creation of strategic solutions for preventing pet abandonment and serve to reduce intake of stray dogs to the local animal shelter.

Abstract: This qualitative study identified a study area by visualizing one year of animal intake from a municipal animal shelter on geographic information systems (GIS) maps to select an area of high stray-dog intake to investigate. Researchers conducted semi-structured interviews with residents of the selected study area to elucidate why there were high numbers of stray dogs coming from this location. Using grounded theory, three themes emerged from the interviews: concerns, attitudes, and disparities. The residents expressed concerns about animal welfare, personal safety, money, and health. They held various attitudes toward domestic animals in the community, including viewing them as pets, pests, or useful commodities (products). Residents expressed acceptance as well as some anger and fear about the situation in their community. Interviewees revealed they faced multiple socioeconomic disparities related to poverty. Pet abandonment can result when pet owners must prioritize human needs over animal needs, leading to increased shelter intake of stray dogs. Community-specific strategies for reducing local animal shelter intake should address the issue of pet abandonment by simultaneously targeting veterinary needs of animals, socioeconomic needs of residents, and respecting attitude differences between residents and shelter professionals.

Keywords: animal shelters; GIS mapping; socioeconomic disparities; grounded theory; pet abandonment

---

1. Introduction

Mapping data with the aid of geographic information systems (GIS) is commonly-used to visualize relationships and analyze trends in industries such as, government, healthcare administration, and emergency management [1]. Animal shelter research has previously used this technology to characterize pet-adoption patterns and intake of dogs and cats [2–5]. In addition, GIS-mapping technology was previously used in Alachua County, Florida to: (a) identify neighborhoods with the greatest health disparities; (b) advocate for community outreach services; and (c) develop a mobile clinic and the Southwest Advocacy Group (SWAG) family resource center to deliver primary healthcare as a result [6].

In a survey of pet owners desiring to surrender their dogs to an animal shelter from a zip code associated with lower income addresses, researchers found that cost of veterinary care was a primary factor in their decision. This study also revealed secondary factors that influenced pet surrender, including: income, landlord issues, behavior issues, moving, lack of time, and new children [7]. In another study researchers surveyed pet-owners surrendering their large-breed dogs to identify strategies for keeping these harder-to-adopt dogs out of shelters. The study yielded no universally applicable intake-diversion strategies. This led the authors to conclude that community-specific solutions were needed to deter intake of large-breed dogs surrendered to animal shelters [8].

However, in addition to being surrendered by their owners, dogs enter municipal shelters impounded by animal control as homeless strays or as legal impounds during humane investigations or rabies quarantines. In fact, intake of strays typically represents the greatest volume of pets that enter municipal animal shelters [9]. No published studies to date have investigated community-specific factors that influence intake of stray dogs to an animal shelter.

The purpose of this study was to use qualitative-research methods to explore a community from which high numbers of stray dogs entered a local municipal animal shelter. The community was selected by visualizing annual shelter intake data on GIS maps. Researchers would then interview a sample of residents and explore community-specific factors that might influence the shelter's intake of stray dogs from the selected study area. The animal shelter selected for this project, operated by the Alachua County government, functions as animal control and is the only open-admission shelter in the county that accepts owner surrenders, strays, and legal impounds. There is one other limited-admission humane society in the county that transfers pets from this municipal shelter and accepts owner surrenders on a space-limited basis.

2. Materials and Methods

Research activities took place between January 2015 and January 2017. Investigators first worked with staff of the local animal shelter to improve the accuracy of their intake data and entry systems. Intake data reports were collected from the shelter's Chameleon Software (HLP, Inc., Littleton, CO, USA) between August 2015 and July 2016. Researchers standardized, or cleaned, intake addresses in order to match addresses to latitude and longitude geographic coordinates, a process called geo-coding. A GIS was used to spatially visualize the data. The geo-coded intake data was then aggregated to a uniform grid and overlaid with analyzed historical child maltreatment data to visually aid in study area selection. Researchers did not perform further spatial analysis of the data.

One study area was selected to investigate for this project. Multiple sources of information were used to explore the study area, including: GIS maps, public records, census data, field observations, and semi-structured interviews with residents of neighborhoods of interest [10]. Researchers analyzed the compiled data by applying a constructivist, grounded-theory (GT) approach as described by Charmaz [11] to identify any unique themes that arose from the collected data.

2.1. *Animal Intake Data Collection, Cleaning, and Geo-Mapping*

A total of 3747 intake records were collected throughout the one year study period, of which 846 records were eliminated through the data cleaning and geocoding process, leaving 2901 clean records. The cleaning process removed incomplete addresses, addresses outside of the county, and addresses related to veterinary care facilities or government related buildings. Addresses not located during the geocoding process were also removed. The clean data percentage of the final retained geo-coded data available to map was 77.42%. Lastly, the 2901 records were aggregated to 1608 unique addresses and summarized by species categories. Environmental Science and Research Institute's (ESRI) Geocoder accessed through ArcGIS Online was used to geocode the addresses [12]. The geo-coded coordinates were then spatially visualized using ESRI's ArcGIS ArcMap program release 10.5 [12]. A county wide half-mile square polygon grid, or fishnet, data layer was created to further aggregate geo-coded intake data to density per half-mile and thus facilitate uniform comparisons of the distribution and areal density of the spatial intake data. Addresses of veterinary care and governmental offices were geocoded to create an animal resources spatial layer, which included the Alachua County shelter. Additional spatial data layers consisting of the US Census's socio-economic American Community Survey (ACS) Census 2011–2015 block groups, a county boundary, roadways, lakes, and rivers were acquired from the Florida Geographic Data Library [13]. Previously analyzed child maltreatment areal density data was provided courtesy of one co-author (NH) [6]. To standardize visual comparisons the maltreatment density data was resampled to the half-mile square fishnet grid. All spatial data layers were projected to a common planar coordinate system, the Florida Modified Albers (FMA) projection coordinate system. Animal intake density and child maltreatment density spatial data were simultaneously visualized for comparisons. A study area that contained high areal densities per half-mile common to both data sets was selected. The study area boundary GIS layer was created by selecting the 5 contiguous U.S. census block groups that contained the identified overlap. The census block group boundaries were then dissolved using ArcGIS to create a single study area boundary. Socio-economic summary statistics for the study area were generated.

2.2. *Field Observations*

The field-research team consisted of three graduate students (Jennifer W. Applebaum, Dorothy Berry, and Britan Ethridge) and three faculty members (Terry Spencer, Linda Behar-Horenstein, and Amber Emanuel) who in advance of field work practiced to develop consistency. Researchers conducted multiple site visits looking for: living conditions of pets and people, challenges to pet ownership, availability of veterinary resources, personal safety concerns, language barriers, and best places to post recruitment flyers as well as residents to be interviewed (See Appendix A: Field Observation Guide).

2.3. *Semi-Structured Interviews*

Researchers interviewed a convenience sample of 39 volunteers recruited from within the study zone by use of posted flyers, canvassing, and tabling at community-gatherings. Each interviewer followed a semi-structured interview guide (See Appendix B: Interview Guide). Interviews were recorded, transcribed by a professional transcription service, and then entered into NVivo 11 software (QSR International Pty Ltd, Doncaster, Australia) for open-coding. Each volunteer who completed the interview received compensation of a $10 VISA card. To ensure anonymity, interviewees created unique personal codes that consisted of the initials for their parents and digits for their birth date. Interview questions were designed to gather information about family demographics and other socioeconomic-risk factors that might affect the welfare of dogs and cats in the study area, such as: mental and physical health, housing insecurity, interpersonal violence, and communication disparities.

The questions were also intended to elicit attitudes toward domestic animals and ideas about the causes and solutions for the issue of animals entering the county shelter from the study area. Participants were White (22%), Black (53%), Hispanic and Other/Mixed Race, both (8%), 14% were retired, 33% unemployed, 23.5% employed, 23.5 % disabled, and 6% did not report this information.Twentynine percent had less than a high school education, 34% held a high school diploma or equivalent and 34% had some college or higher education. The remaining 6% did not report this information.

2.4. Analysis of Interview Transcripts

Two researchers (Terry Spencer and Linda Behar-Horenstein) applied grounded theory (GT) analytical techniques to inductively analyze the transcribed interviews in order to better understand the interviewee's perspectives, explain the phenomenon of interest in the interviewee's words, and to provide a framework for further study. They independently open-coded the data line-by-line in the transcripts then met to reach consensus about the initial coding and to agree on emergent themes [14,15]. The GT approach ensured that the researchers developed a deep understanding of the data by: looking and listening for cues about feeling and meaning; looking for how, when, and why people act; looking for what people do as well as what they say; and taking a critical stance toward the data, rather than the participants. Direct quotes were extracted from the coded-data to support the themes. This rigorous and systematic approach allowed the researchers to feel confident that what they report is representative of participants' perspectives.

3. Results

3.1. Animal Intake

The achieved cleaned, geocoded address result of 77.42% is 2.58% less than the ASPCA's recommended 80% cleaned data rate [16]. Intake mostly consisted of stray adult animals. Alachua County took in slightly more cats than dogs. Animals less than 6 months of age were classified as juveniles, but not all intake data included an age estimate. We did not use intake data that recorded the reproductive status of the animals because that data was not confirmed by a shelter veterinarian and the status sometimes changed during the animal's stay in the shelter (from not sterilized to sterilized) (See Appendix C).

3.2. GIS-Maps

Visualization of the clean, geocoded intake data revealed an area of high intake contained in 5 contiguous census block groups in Alachua County, Florida, which visually overlapped with a high-density of previously mapped Alachua County child maltreatment cases [6]. The outline of the 5 contiguous census block groups was used to determine the study area boundary and served as the focus for field investigations. Study area shelter records were aggregated to 59 unique study area addresses. The number of animals at each address ranged from minimum of 1 to a maximum of 8 per address. The average number of animals was 2.1 per address and the standard deviation was 1.8. The species maps revealed that a cluster of high animal-intake, particularly for dogs, overlapped with the child maltreatment data within the boundaries of the study area. A cluster of high-intake for cats occurred slightly south of the boundaries. (See Figures 1–5). The Alachua County shelter location in relation to the study area can be seen in Figure 5. The distance from the study area center to Alachua County animal Services location is approximately 19 km by road and approximately 14 km by straight line distance.

3.3. Field Observations

Pet policies varied widely among the interviewees because each apartment was privately owned by a landlord who set policy rather than a single property management group. Multiple veterinary clinics and a pet-friendly county park were near to the study area. A non-profit community resource center (SWAG) opened in 2012, was centered in the area and offered medical and dental services, a food bank, a community garden, computers, and a children's play area. County bus service was available, but ran infrequently on the weekends. The majority of bus stops were not protected from the weather. Pets were not allowed on public transportation unless they fit in carriers.

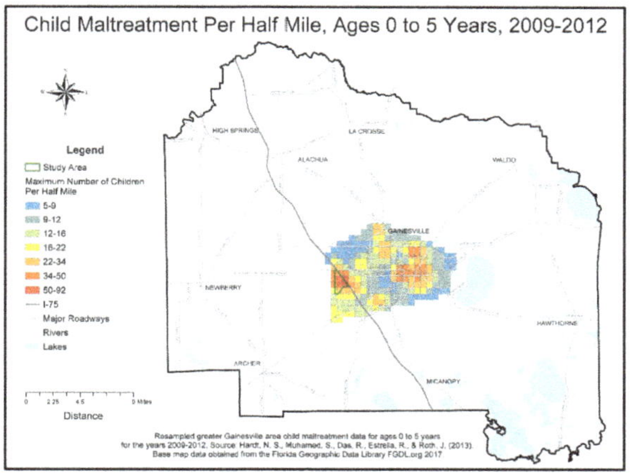

Figure 1. Child Maltreatment Density Map.

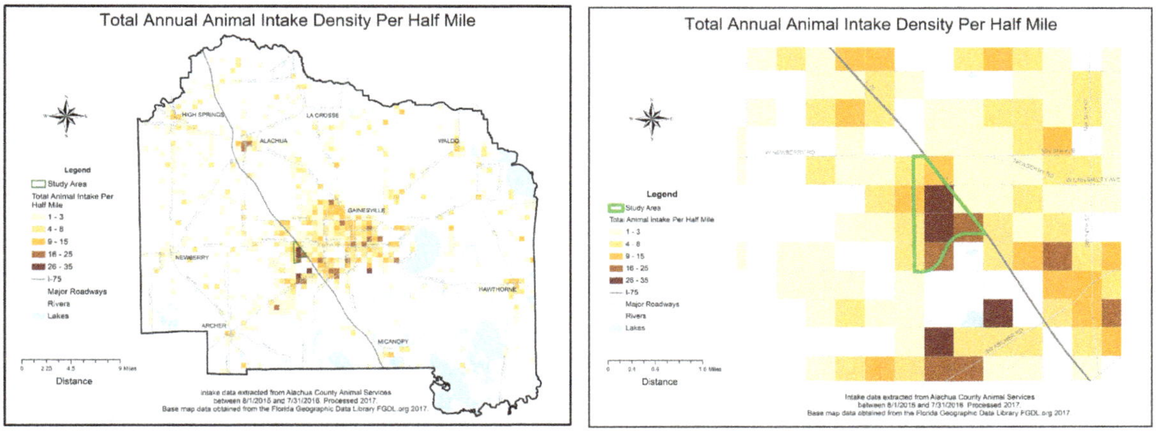

(**a**) (**b**)

Figure 2. (**a**): Total Annual Animal Intake Density Map. (**b**): Study Area Close Up.

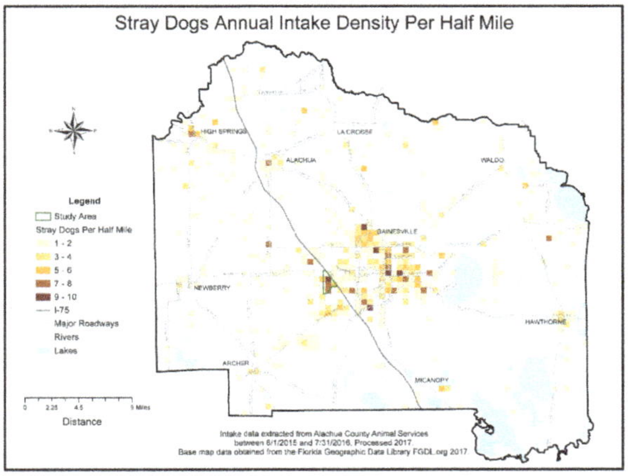

Figure 3. Stray Dog Annual Intake Density Map.

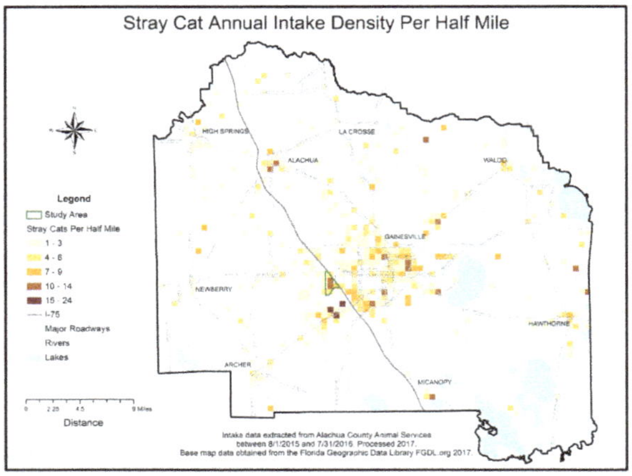

Figure 4. Stray Cat Annual Intake Density Map.

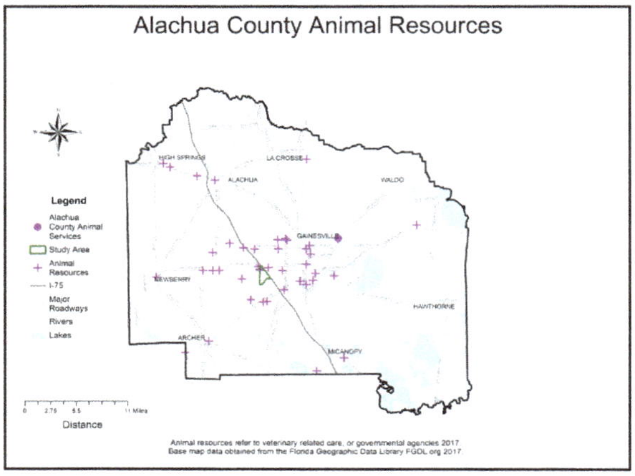

Figure 5. Alachua County Veterinary and Governmental Animal Resources.

Pet dogs were seen walking on leashes, off-leash under voice control, and tethered. One free-roaming dog and several free-roaming cats without visible collars or ear-tips were seen. The presence of pet-waste receptacles suggested that this was a pet-friendly neighborhood. Animal Control Officers reported concerns about crime in the study area, as well as ongoing issues with improper identification and confinement of dogs in the neighborhoods. One case of dog-fighting had previously been investigated in the study area. Many residences displayed signs stating, "No Trespassing" and "Beware of Dogs," which served as visual evidence of personal safety concerns present in the study area. Signs advertising pit-bull puppies for sale were also prominent. Evidence of recent evictions alluded to housing instability (See Figure 6).

3.4. Community and Participant Demographics

Block group census data was extracted from the 2011–2015 American Community Survey for the area of interest for this study, summed, and averaged (mean). The study area consists of 823 acres and a total population of 7921 people living in 3130 households. The household poverty level of the interviewees was determined by comparing reported family size to the 2017 federal poverty guidelines for the 48 contiguous states and the District of Columbia (See Table 1).

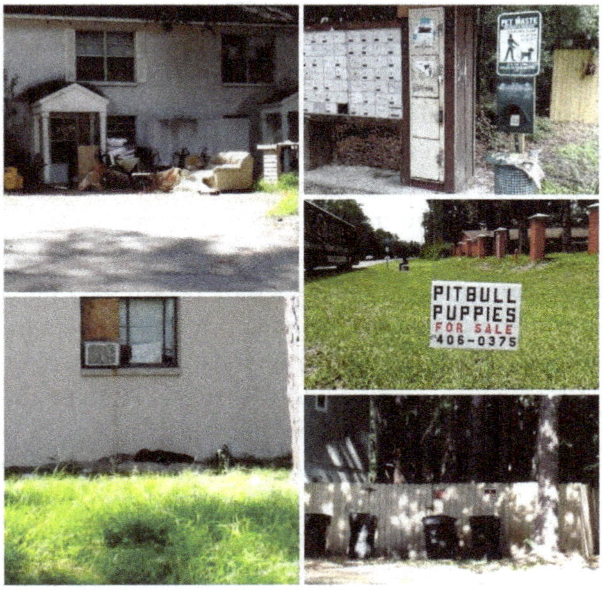

Figure 6. Images from Field Observations.

Table 1. Comparison of Demographics for Census Block Group and Interviewees.

Demographics	Census Data	Interviewees
Average household size (people)	3.87	2.5
Household income:		
Below poverty line	34%	80%
Above poverty line	66%	3%
Unknown	-	17%
Gender:		
Male	47%	50%
Female	53%	50%
Education:		
Less than high school	9%	29%
High school diploma or equivalent	29%	34%
Some college or higher ed	62%	34%
Race and ethnicity:		
White	48%	22%
Black	45%	53%
Hispanic	1%	8%
Other/Mixed race	7%	8%

3.5. Interview Findings

Only 35 of the 39 interviews were analyzed because one interview was interrupted before it could be completed and three recordings were not audible for transcription.

Fewer than half of the interviewees, 16 (46%) reported keeping pets in their homes. Seven were dog only homes (three kept single dogs, four kept two dogs). Five were cat only homes for one to three cats. Four homes kept both cats and dogs; including one with a single cat and single dog, two with two cats and one dog, and one home with one cat and two dogs. Pet dogs were described as either mixes or purebreds of various sizes. Cats were described as either indoor-only (one was walked outdoors on a leash) or allowed to freely roam outdoors. Nineteen (56%) of the interviewees reported not keeping pets in their homes. However, because they reported interacting with community and neighborhood pets, understanding the scope of their experiences with animals was considered important.

Many 22 (62.9%) interviewees mentioned community cats (unowned, free-roaming cats), although only 3 (8.6%) felt these cats were a nuisance. CAZ09 said "They don't seem abused or nothing" when describing the cats in the neighborhood "laying around that need somewhere to go." CEK16 said, "You see more cats than stray dogs" in the neighborhood. Several mentioned feeding the community cats, such as CLL12 who said, "I know it's a lot of cats on my street. Mostly the people that stay on my street, it's at least 3 or 4 houses that they all go to and they feed them. Like 8 or 10 cats."

The thirty-five interviewees reported frequent relocations within the past five years; 21 (60%) of respondents reported having moved from one to nine times within the past five years, and 11 (31.4%) of those who had recently moved reported currently keeping pets in their homes. We did not determine whether the pets had moved with the residents or had been obtained after the most recent move.

Some of the thirty-five respondents revealed previously experiencing controlling or threatening behavior by their partners toward themselves and their pets. Thirteen (37.1%) of the interviewees reported that they had previously felt controlled by their partners; 7 (20%) reported feeling afraid of their partners; 11 (31.4%) said they had been threatened by their partners; and 3 (8.6%) reported that their pets had been threatened, injured, or that fearing for their pet's welfare influenced their decision whether to leave or stay with a partner.

Thirty-four interviewees responded to questions about their physical and mental health. Ten (29.4%) rated their physical and mental health toward the fair to poor end of the rating scale. Twelve (35.3%) considered their physical and emotional health to be good to excellent. The remaining 12 (35.3%) considered their health to be mixed on the rating scale.

When directly asked about using dogs and cats as commercial commodities, (i.e., buying, selling, trading, breeding, betting, or gambling on animals), opinions varied. Most 23 (65.7%) disagreed with dog-fighting either because it was cruel or against the law. One interviewee (2.9%) felt it was all right to gamble on dogs, and a few 6 (17.1%) suspected illegal dog-fighting activities were occurring in their community but had not witnessed it. Some 10 (28.6%) thought it was all right to breed dogs and cats, 5 (14.3%) specifically agreed with selling puppies or kittens, and only 11 (31.4%) did not agree with breeding animals. Two (5.7%) suspected that dogs were being bred in the community to be used for dog fighting. Most 23 (75.7%) claimed they saw no evidence of breeding or selling of animals. MJM17 said "There's a lot of buying and selling" in the neighborhoods and there were no signs posted because the transactions occurred though "word of mouth and internet." CSN07 did not see a need for breeding and proffered that there are "so many that are in shelters already that they need to stop breeding their own." He suggested if people wanted a dog that they find a rescue dog. JJT19 commented that he had seen signs that pit bull puppies were for sale at the entrance of the neighborhood. He remarked that if there were dogs needed to be adopted "especially in that breed that you probably don't need to be breeding them."

Three themes emerged from the coded interview data related to concerns, attitudes, and disparities. The interviewees expressed concerns about animal welfare, personal safety, health, and money. They expressed how they regarded dogs and cats as household pets but also as useful commodities that could provide protection or produce money through breeding litters or betting on dog fights. Some viewed stray dogs and cats as pests in the neighborhoods. Others were indifferent to the animals in the neighborhood or resolved to accept the conditions they witnessed in the community. The disparities discussed by interviewees were associated with poverty (i.e., evictions, disabilities, jobs loss, financial insecurity, limited access to transportation and communications, risks for personal safety) (See Appendix D: Themes and Example Codes).

The interviewees explained why they thought animals ended up in shelters. Surprisingly, the cost of caring for pets was mentioned by just 9 (25.7%) of the interviewees. Merely 3 (8.6%) mentioned pet deposits as a deterrent to pet keeping and 4 (11.4%) mentioned size or breed restrictions as a problem. However, 14 (40%) of the interviewees suggested that abandonment of pets was the main explanation for animal homelessness in their neighborhood: 13 (37.1%) said pet keepers were unable to provide proper care and 10 (28.6%) suggested uncontrolled reproduction was an issue. Very few

1 (2.9%) suggested the problem was due to distance from veterinary care or sterilization services. DJJ16 explained it was "because a lot of people don't care for them or they move and can't take them where they're moving to." MWA22 said, "I guess the people that move and just leave their animals behind." SFF07 stated that, "They don't care about the dog no more." JJT19 suggested problems emanated from the properties being primarily rental and the high incidence of evictions. Researchers regard abandonment of pets as the act of knowingly leaving an animal at a location where they will not receive minimum care, while no longer taking responsibility for the care of the animal. This differs from voluntarily surrendering an owned animal to the shelter where they will receive care.

The interviewees offered a variety of suggestions for keeping pets out of shelters. The most commonly proposed solution was for more sterilization of dogs and cats in the community 9 (25.7%) followed by confining pets 7 (20%) and taking proper care of the pets 6 (17.1%). This was best expressed by CEK16 who suggested, "Feed them, take care of them, and give them shots." Other proposed solutions included educating residents about how to provide for the needs of pets 5 (14.3%), providing free veterinary care 4 (11.4%), and offering temporary shelter for pets 3 (8.5%). DAM18 suggested "temporary shelter for animals" was required when people experienced situational problems so that they could "get their animals back." MAM12 described pet homelessness as endemic to the community culture, "You know this neighborhood has changed so much over the past years. There are just a lot of irresponsible people in many ways that live here." CEK16 commented on the responsibilities associated with caring for pets, "Anybody who has a pet is going to become attached to the pet and they want to take care of their pet. And if they are not able to take care of the pet they have to give it up."

4. Discussion

This qualitative study described that poverty-related disparities increased the risk for pet abandonment within a geo-coded study area from which high numbers of stray dogs entered the local animal shelter. Field interviews with residents revealed their concerns about animal welfare, money, personal safety, and health. The respondents also revealed that their attitudes toward domestic animals varied from useful commodities (products) that could supplement income or guard property, to pests, or as household pets. This finding supports previous studies of attitudes toward animals as pets, pests, or profit [17]. Community-specific strategies for reducing local shelter intake of stray dogs should target the issue of pet abandonment by co-addressing veterinary needs of animals, socioeconomic disparities of residents, and attitudinal differences between residents and shelter professionals.

Limitations: Because more residents were available to interview when the SWAG Family Resource Center was open, the research team interviewed a more unemployed and older set of community residents who were using the services of the Resource Center than the census mean predicted. This convenience sample could have skewed the study findings. In addition, interviewees were informed at the time of recruitment that they would receive a $10 gift card in exchange for their time, which might have influenced the study sample. However, recruitment flyers posted in the study area advertised the $10 gift card, but did not yield any volunteers to be interviewed. The interview findings may only be representative of those community members who match the demographics of the residents we sampled. Thus findings of this study may not be generalizable to the larger community.

Our investigation revealed that residents were concerned about the welfare of homeless animals in their neighborhoods. Several previous studies have documented a decrease in animal homelessness after increasing access to spay-neuter surgeries within a city or a postal zip code [18–21]. Indeed, a nationwide emphasis on sterilization of pets has resulted in a corresponding decrease of animal intake to shelters in the United States over the past few decades to an estimated low of 7 million animals in 2015 from a high of 13 million in 1973 [20]. The current study area is located within a well-resourced veterinary community. Residents have access to nearby private veterinary clinics, veterinary college, veterinary clinic serving low-income clients, as well as a low-cost sterilization clinic for pets farther away on the northeast side of town. This community is definitely not a "veterinary desert" as previously documented by GIS mapping in Atlanta, GA, USA [22].

Despite ready access to veterinary services, our investigation revealed that few puppies, kittens, or community cats from this study area have benefitted from veterinary care. In addition to personal concerns about money, disparities with communications, transportation, and health might explain why some of the study-area residents failed to sterilize their pets. With intermittent phone and internet access, it can be difficult to schedule a veterinary appointment. Limited bus schedules and pet restrictions on public transportation likely present additional obstacles for accessing veterinary resources when they are not located within easy walking distance. Personal-health challenges might further limit a pet-owner's ability to obtain veterinary services without additional assistance. One solution for overcoming such disparities might be to regularly offer veterinary services, in addition to medical and dental services, at the SWAG community resource center located within the study area. Applying such a "one health" approach to the social determinants of community health could greatly improve animal welfare in this neighborhood.

The problem of abandoning unwanted pets, rather than surrendering them to the safety net of a local animal shelter, was identified during this investigation. This issue could be exacerbated by the number of socioeconomic disparities faced by community residents rather than by a lack of bonding with pets. In fact, the majority of interviewees responded favorably toward community cats and dogs rather than regarding them as pests. Unfortunately, the municipal animal shelter in this study charges a fee to surrender a pet and an additional fee to transport an unwanted pet to the county shelter. Such fees likely deter pet owners of this neighborhood from "doing the right thing" when faced with multiple and pressing socioeconomic challenges. Therefore, it might seem easier to turn a pet loose in the neighborhood where another neighbor might choose to care for it than to transport that unwanted pet across the county to the animal shelter during specific hours of operation and also pay a fee. A simpler solution would be to waive owner surrender and transport fees from this neighborhood.

Surprisingly, GIS maps of increased stray dog intake from this neighborhood visually corresponded with previously made GIS maps of increased cases of child maltreatment from the same study area. This potential correlation between homeless animals and child maltreatment hinted at a community facing multiple challenges and prompted the research team to further investigate the neighborhood. Certainly this observation deserves additional study to assess any relationship between adverse childhood events and animal homelessness. In the meantime, this community might benefit from encouraging animal control officers to work jointly with local social services personnel to support both people and their pets. For example, the municipal animal shelter could use volunteers to repair fences in order to better confine dogs in their owner's yards. The county could also provide pet-friendly emergency sheltering for families experiencing episodes of domestic violence or eviction. Such interagency cooperation might encourage pet retention rather than pet abandonment.

However, such solutions will likely not change the attitudes of those residents who view animals as commodities for protecting personal property and increasing personal income rather than as pets. This study reveals the need for a different solution that addresses the local economy of pets in order to reduce the urge to breed, sell, and gamble on dogs.

In 2011, the ASPCA collaborated with Portland, Oregon to provide free and reduced-cost veterinary services for geo-coded neighborhoods associated with high-intake to the local animal shelter and then assessed whether the intervention decreased shelter animal intake. This study documented a small reduction in the intake of surrendered cats but not a significant decrease in intake for surrendered dogs or any strays [23]. Perhaps these limited results were due to a mismatch between the proffered solutions and the true problems experienced by the residents of the study area. Our study indicates that community-specific findings, such as attitudinal differences and disparate socioeconomic conditions, should be addressed when designing strategies to reduce stray dog intake to the local animal shelter.

The institutional culture of animal shelters influences how the animal welfare community defines social problems, poses solutions, and serves its clients [24]. Increasing access to free or low-cost veterinary services is the primary solution offered by animal shelter professionals for solving the problem of unwanted domestic pets. However, as this study shows, not all who live with dogs and cats view these animals as pets nor is cost of pet care their primary concern. It is possible that animal welfare professionals are overlooking other solutions to the problem of pet homelessness because they lack familiarity with the lived-experiences of community residents or they apply a "moral certainty" to saving animals that divides clients into those who are "good" and "bad" pet caretakers [25]. For example, shelter professionals might view pet owners who choose to breed their pets as a "bad" owners, even if the money obtained by selling the litters helps the families put food on their tables. As poverty increases, shelter and security concerns arise for families, imperiling their ability to care for pets. As the urgency of socioeconomic disparities increases, pets become burdensome to at-risk families, which can result in their abandonment in an effort to preserve family resources.

5. Conclusions

In order to reduce the intake of homeless pets, shelter professionals should attempt to connect and empathize with their clients rather than judging them and using "institutional thinking" to solve problems. Giving a dog free vaccine, an identification tag, and a low-cost sterilization might be a solution to a veterinary-related problem, but it isn't the solution needed by a pet living in a household facing a socioeconomic issue such as eviction, domestic violence, or a family medical emergency. As this and previous studies allude, community-specific solutions are necessary to solve community-specific problems that involve both humans and animals. Although this study was specific to one neighborhood served by a municipal animal shelter in Florida, the findings are applicable to other shelters and communities. Talking directly with residents of communities responsible for high-intake to animal shelters should help us discover what residents need, so we can develop targeted solutions for addressing animal welfare issues.

Acknowledgments: This study, approved by the university's institutional review board (IRB02-1601287), was funded by a generous gift from the American Society for the Prevention of Cruelty to Animals (ASPCA). The funding included covering the costs to publish the manuscript in open access. Emily Weiss of the ASPCA originally suggested the idea of geo-mapping the shelter intake and correlating it with the community geo-data previously collected for child-maltreatment by Nancy Hardt. The views expressed in this manuscript represent those solely of the authors. In addition, two graduate students, Dorothy Berry and Britan Ethridge, assisted with the field observations and interviews.

Author Contributions: Linda Behar-Horenstein, Nancy Hardt, and Terry Spencer conceived and designed the project. Amber Emanuel, Nancy Hardt, and Natalie Isaza assisted with developing the field guides and interview guides. Jennifer W. Applebaum, Linda Behar-Horenstein, Emanuel, and Terry Spencer performed the field observations and interviews. Joe Aufmuth cleaned, geo-coded, and created the GIS maps from the data. Linda Behar-Horenstein and Terry Spencer analyzed the transcripts. Jennifer W. Applebaum, Joe Aufmuth, Linda Behar-Horenstein, and Terry Spencer wrote and revised the paper. Amber Emanuel, Nancy Hardt, and Natalie Isaza reviewed the first draft.

Conflicts of Interest: The authors declare no conflicts of interest.

Appendix A. Observation Field Notes Guide

Part 1: Initial Observations of GIS-Mapped Neighborhoods

1. Schedule an initial one-hour ride-along with an animal control officer or community service officer who is familiar with each GIS-mapped neighborhood to be observed.
2. Prior to the initial observation session, prepare to compile **FIELD NOTES** for each neighborhood that will include:

- the day of the week and date of the initial observations
- the scheduled start and stop times of the initial observations
- descriptive name for each observed neighborhood

- the boundaries for each neighborhood to be observed
- who will be present for the initial observations, including your name as the observer
- Background information from the accompanying animal control officer or community service officer that describes their familiarity with residents and domestic animals of each neighborhood (What is their professional knowledge about each neighborhood? What are the common types of service calls they make in each neighborhood? Do they know of any particular challenges or issues faced by the residents or dogs and cats of the neighborhoods? Why do they think so many dogs and cats enter the county animal shelter from these locations? What do they think is needed to prevent dogs and cats from these neighborhoods ending up at the county animal shelter?)

3. During the initial observation session in each neighborhood, and with the ride-along officer's assistance, record or document in the FIELD NOTES evidence of/that:

   - dogs and cats being kept as household or personal **pets**
   - the neighborhood is **pet-friendly**
   - dogs and cats being cared for as community animals or **loosely-owned** animals
   - dogs and cats are viewed as **pests** or problematic in the community
   - dogs and cats are viewed as **products** in the community (for breeding, gambling, fighting, selling, trading, etc.)
   - specific neighborhood **culture or language** exists
   - specific neighborhood **socioeconomic status** exists
   - access to basic **community services** for people
   - access to pet care and **veterinary care services**
   - residents of the neighborhood feel **safe**
   - special **challenges/issues** facing this community

   The evidence you collect should note or document the presence or absence of such things as (sort of like a scavenger hunt):

   - Freely-roaming animals
   - Pets on leashes
   - Pet collars (and note types of collars, such as heavy chains, pinch, harness, etc.)
   - Tethered pets
   - Pet-feeding stations
   - Pet-water bowls
   - Pet-food bowls
   - Pet housing
   - Fenced yards
   - Dog parks
   - Animals feeding from waste containers
   - Lost pet flyers
   - Posted signs restricting animals
   - Advertisements for "pet-friendly" housing or businesses
   - Posted policies restricting breeds of dogs
   - Public transportation
   - Community centers
   - Libraries
   - Schools

- Churches
- Grocery stores
- Recreation areas
- Veterinary care centers
- Foreclosures
- Real estate for sale or for rent
- Whether people are active outdoors and estimate their ages
- Language used on posted signs
- Predominant housing type (apts/mobiles/single family/tents/dorms) and density of residences.
- Whether housing has bars on windows and doors
- "Community watch" or community protest signs
- Predominant types of businesses in neighborhood
- Other pertinent items, such as abandoned houses, construction, etc.

4. After each observation session, take time to reflect on everything you observed or documented. What information do you still need? What questions remain? What concerns you about the relationships between people and pets in this neighborhood? Add a summary of your impressions of each neighborhood to the Field Notes.
5. Compile the Field Notes, your summary, and debrief with the research team.

Part 2: Follow-up Observations of GIS-Mapped Neighborhoods

1. Schedule follow-up observation sessions to each neighborhood with a partner who can drive while you compile Field Notes. The purpose of the follow-up sessions is to validate your initial observations on different days of the week and at different times of day.
   a. Plan to make a total of 4 observations in each neighborhood, including the initial observation session, which occurs on a: weekday during normal business hours; weekday evening; weekend day; and weekend evening.
   b. Plan any additional observations only if necessary to answer questions that arise during the sessions or to clarify data you were unsure of from previous sessions.
   c. Use the same format for your Field Notes as during the initial observation. Your follow-up sessions should each be one-hour in duration and serve to supplement the data you collect from prior observations.
   d. Reflect and write your summary notes after each session.
2. Compile the Field Notes, your summaries, and debrief with the research team.

Appendix B. Interview Guide

PART I: Obtain Informed Consent and Create Personal ID Code

A. Read page one of the *Informed Consent Document* to each volunteer, or ask him/her to read page one of the *Informed Consent Document*.
B. Have volunteer sign, date, and provide their mailing address on page two of the *Informed Consent Document* if they agree to participate.
C. Collect page two of the *Informed Consent Document* and give page one to the volunteer.
D. Ask the volunteer to create a personal identification code and remind them to note that code on their copy of the *Informed Consent Document* so they will have it in the future if they wish later to withdraw from the study. Volunteers create a person ID code using the following system:

a. First letter of mother's first name
b. First letter of father's first name
c. Middle initial of volunteer
d. Day of their birth (two digits)

EXAMPLE: my mother's name is Rosie and my father's name is Milton. My middle initial is G for Gale. I was born on the second day of the month. My personal code is: RMG02.

PART II: Demographics (closed-ended questions)

Inform volunteer that you will be recording the interview. Turn on the recorder and test that it is recording properly. Then begin the interview by asking the following questions:

Q1. What is your personal ID code for participating in this interview?
Q2. Please describe the location of this neighborhood and tell me whether you live or work here. *(EXAMPLE: This is Haile Plantation. I work here.)*
Q3. If you live here, please describe the type of housing you live in: *(EXAMPLE: is it an apartment, mobile home, condo, tent, dorm, assisted living facility, single-family home,* etc.)

NOTE: if they do not live in this neighborhood, skip to Q4.

Q3A. How many people share your residence? *(CHOOSE ONE: I live alone; I live with _______others)*
Q3B. Do any dogs or cats share your residence? *(CHOOSE ONE: No, Yes ________ dogs and _______cats)*
Q3C. How many times have you moved (relocated your residence) in the past 5 years? *(CHOOSE ONE: 0 times, 1–2 times, 3–4 times, 4–5 times, more than 5 times)*

Q4. If you work here, please describe what you do: (*EXAMPLE: shopkeeper, clerk, law-enforcement officer,* etc.)

Q4A. Do any dogs or cats live at your place of work? *(CHOOSE ONE: No, Yes _______ dogs and _______ cats.)*

Q5. How long have you lived or worked in this neighborhood? *(CHOOSE ONE: Less than1 year, 1–5 years, 6–10 years, 11 or more years)*
Q6. What is your sex? *(CHOOSE ONE: Male, Female, Other, or I prefer not to say.)*
Q7. What is your age? *(CHOOSE ONE: 18–25 years, 26–30 years, 31–40 years, 41–50 years, 51–60 years, 61–70 years, 71–80 years, 81–90 years, 91 years or older)*
Q8. What is the highest level of education you completed? *(CHOOSE ONE: I did not finish high school, GED or high school diploma, vocational certificate or some college, undergraduate degree, graduate degree (such as a master's or PhD), professional degree (such as an MD or DVM), post-doc)*
Q9. How do you prefer to describe your racial and/or ethnic group? *(EXAMPLES: Hispanic White, Non-Hispanic Black, Asian,* etc.*)*
Q10. What is your primary spoken or written language? *(EXAMPLES: American English, Spanish, French, Chinese, Russian,* etc.*)*
Q11. What is the approximate annual income for you and your immediate family? *(CHOOSE ONE: less than $10K, $11–20K, $21–30K, $31–40K, $41–50K, $51–60K, $61–70K, $71–80K, $81–90K, $91–100K, More than $100K.)*

Q11A. How many people are dependent on this income? *(CHOOSE ONE: only me, 2, 3, 4, 5, 6, 7, 8, 9, 10,* etc.*)*

Q12. How often do you have access to a computer and the internet for personal use? *(CHOOSE ONE: Always, Never, only Sometimes–if chooses Sometimes, ask for an explanation for why)*
Q13. How often do you have access to a smartphone (cell phone) for personal use? *(CHOOSE ONE: Always, Never, Sometimes—if chooses sometimes, ask for an explanation for why)*

PART III: Health and Relationships (closed-ended questions)

Q14. How do you rate your physical health? *(CHOOSE ONE: excellent, good, fair, poor)*
Q15. How do you rate your emotional health? *(CHOOSE ONE: excellent, good, fair, poor)*
Q16. Have you ever had a partner try to control you? *(CHOOSE ONE: yes, maybe, no)*
Q17. Have you ever had a partner of whom you were afraid? *(CHOOSE ONE: yes, maybe, no)*
Q18. Have you ever had a partner who threatened you? *(CHOOSE ONE: yes, maybe, no)*
Q19. Have you ever had a partner who threatened to hurt or kill your pet? *(CHOOSE ONE: yes, maybe, no)*
Q20. Have you ever had a partner who intentionally hurt or injured your pet? *(CHOOSE ONE: yes, maybe, no)*
Q21. Has concern about your pet's welfare ever affected your decision-making about whether to stay or leave your partner? *(CHOOSE ONE: yes, maybe, no)*

PART IV: Attitudes toward Animals, Attachment to Pets, Accessibility of Veterinary Resources (open-ended questions)

Q22. Tell me about any dogs or cats that you keep or care for.
Q23. Tell me about any other dogs or cats that live within your neighborhood.
Q24. Tell me about any issues or problems you have with the dogs and cats in your neighborhood.
Q25. Tell me how you feel about breeding of dogs and/or cats in this neighborhood.
Q26. Tell me how you feel about betting (gambling) on dogs and/or cats in this neighborhood,
Q27. Tell me how you feel about buying/selling/trading of dogs and/or cats in this neighborhood.
Q28. How would you feel about a neighborhood dog or cat being euthanized (put down) at the county animal shelter?
Q29. Why do you think so many dogs and cats from this neighborhood end up at the county animal shelter?
Q30. What do you think would help keep dogs and cats from this neighborhood out of the county animal shelter?

PART V: Confirmation of Findings

Q31. Thank the volunteer for participating in this study and ask if you can contact them again to confirm the finding. (CHOOSE ONE: Yes or No)

Q31a. If the answer is NO, this completes the interview.
Q31b. If the answer is YES, ask them how to contact them when you are ready. NOT COLLECT THEIR NAME, you will refer to them using their personal ID code when you contact them—Record the phone number or other method they want you to use to contact them.)

Appendix C. Annual Animal Intake for Alachua County, Florida

	Count By Species and Age			Percent of 2901 Animals		
Intake Age	Canine	Feline	total	Canine	Feline	total
Adult	1121	875	1996	38.64%	30.16%	68.80%
Juvenile	77	291	368	2.65%	10.03%	12.69%
Unknown Age	210	327	537	7.24%	11.27%	18.51%
	Count By Species and Type			**Percent of 2901 Animals**		
Intake Type	Canine	Feline	total	Canine	Feline	total
Stray	955	1279	2234	32.92%	44.09%	77.01%
Surrendered or Returned	409	211	620	14.10%	7.27%	21.37%
Confiscated	40	3	43	1.38%	0.10%	1.48%
Other Disposition	4	0	4	0.14%	0.00%	0.14%
Total	1408	1493	2901	48.53%	51.47%	100.00%

Appendix D. Themes and Example Codes

THEMES	Codes	Examples
CONCERNS	Animal Welfare	**RCV12:** I think we can't take care of them, don't breed them. You see them [indiscernible] you see now these guys out here, under the sun, and I'm like it breaks my heart to see a dog like that. He sits there all day chained to a tree. **EEL05:** Sometimes they don't have shelter at night. They are not getting fed on a regular basis. Just whatever they can scrounge. **MBE12:** I know of a lady trying to help sell little Chihuahuas. I drove them to Missouri and was bottle feeding this one all of the way over. How cruel, taking it away from its mother early age. To each its own, but—I'm powerless. **DAM18:** Cats are in their apartment and dogs are in their apartments. Until they put them out on a chain outside. Well, they do let the pit bulls outside. Well, they're pit bulls, and when you pass by, they start barking. Chaining them out there—I heard they ate a few cats when they got off that chain.
	Personal Safety	**MWA22:** We ain't in a very good neighborhood. **JJD12:** I've seen them [dogs in the neighborhood] but they're not stray. Yeah they're owned by others, yes ma'am. Yes, yes definitely [on leashes]. All but one. All but one they never have on a leash and I'm terrified of that dog. I'm just afraid of big dogs. I think it's a pit. It's like, um, like a pit bulldog. **CLL12:** I'm scared of anyone [dogs] that's not mine. But for the most part I've never had a bad experience. **MJM17:** Because people don't want to take care of their own animals. They just want to let them do whatever they want and that's why dogs have to be—dogs and cats get put down because they want to attack people because they do what they want when they want to and when you try to change their habit you can't. There are little kids around here.
	Money	**CLL12:** My neighbors have two dogs and it's not the same situation. A small dog is like 6 months and really, really skinny. I try to feed sometimes but I can't afford to feed it. **JJD30:** We did have to [pay] $200 [pet fee for cat in apartment]. That's expensive for a cat. Like my husband back in the day, he used to work for (indiscernible). So [landlord] allowed us to pay him so much a month. We're on limited income and have to pay the light bill and everything else, but he's working with us. **OHM29:** The main problem when you have to pay the vet. The test is really expensive. I think maybe the community have to have options for veterinary care that's free. Because one time I bring the pet to UF vet, they treat everything. It's expensive, $2000. My son requested a loan from the college because he has scholarships. It doesn't pay for [Indiscernible] he doesn't want to lose the dog [Indiscernible] he wants to treat. It's really expensive. Because my son is a student. I make eight hundred a month and we live alone.
	Health	**EJM18:** I would [have a cat] if I could. My daughter, she's pregnant. The doctor told her (indiscernible). She can't be around them. That's the only reason I won't bring the cat in. I wish I could open the door and bring them in. **MOE20:** It's my daughter's dog. My daughter, I told you she has got the heart transplant. They all fell in love with her. She fell in love with the golden doodle that used to come to the hospital visiting. So, they all got together with the breeder, and the breeder gave her the golden doodle, and throughout the—all the doodle owners, they all pay for everything for my daughter for the dog. They pay for the vet, groom, feed. They take care of it all. Once it gets to be a year old, it's going to be put on as a service dog. They are paying for everything. It's something for my daughter. But right now, it's my dog. It bites too much right now. **DAM18:** I had two dogs and they took them to [low-cost vet clinic] and she took because I had hip replacement and she took one of my puppies away because I was sick and can't take care of the puppy. And I still could have took care of the puppy. No. I wasn't, [OK with that] because you know it was my dog and she kept convincing me that I—and I had to go to surgery. I don't have any family except my boyfriend. It was hard—I should have took her back.

THEMES	Codes	Examples
ATTITUDES	Animals are pets	**JJD30:** If you have got an animal, take care of it. Keep it inside. Like me, my cat, that's my girl. I take care of her. She's a member of my family. Everybody in the family loves her. If she wants to go out and use the bathroom, she will sit there. So, I keep my spray bottle because there's a calico and a black and white one. The other day they literally tore her collar off. I just take a little water bottle and spray. I wouldn't hurt nobody's animal, but I'm going to protect mine at the same time. **FJL31:** I have one dog. Her name is Lee-Lou. That in Mandarin Cantonese means perfect. And she is. She was abused as a puppy. It's actually my brother's dog but he passed away a year ago and I promised him I would keep her forever. She's 10 years older than I am and I'm 67 and she'll healthy as a horse just like I am. We go walk about 6 to 8 times a day. I give her at least a mile to a mile and a quarter every day and I try to get 3 miles myself. **CAZ09:** I had a dog when I was like 11 or 12 and it got ran over. That really hurt me. [Indiscernible] because I get attached to them. I love them but I can't get one because something happen to them that's how I know other people feel about animals when they get hurt. I don't get none of them.
	Animals are commodities/ products	**CGR11:** I don't think people should be able to breed. Certain people, especially if you want to know their background. Some of them be breeding dogs, having them fight. You know, I don't know. They breed them, fight them. **MRJ14:** My neighbor has one, and he's pretty good. Real friendly. And sometimes when they go out of town, we keep him. Pit bull but really trained and loves kids. He's not no mean dog. He won't bite unless you try to, you know, go outside, don't know you, or you try to go in there or something like that. Yeah, really protective. More protection and part of the family. **CEK16:** Dogs do have litters. They can't keep all of the litter so they sell the dogs. **DWM26:** But I was telling [Indiscernible] our daughter bets on dogs but I ain't never seen it. **SRT02:** Yeah, it was a business. And were they purebred dogs like they have papers or—I don't know that much. I know he has a website and that's his main income in Gainesville. Yeah, I know him and some other guys that do dog shows and stuff like that. I know like some people have those stands or whatever to make dogs breed and stuff like that. But I don't know. I never went inside and seen his operation.
	Animals are pests	**MJM17:** There's a lot of dogs and cats around here but the thing about it is when people don't want to have their dogs around no more they let them run loose. That's a big deal because basically if you wanted a pet that's your pet. Not other people's. Other people trying to take care of your pets that you have that you let loose. A lot of them have owners but the owners let them run loose anytime they want until they want to come home. **MOE20:** The cats. They just roam. My cats never see outside. But other people, they've got the outside cats and they always run around in my yard. **JWG09:** Barking at 4 in the morning. **RCV12:** I just don't like them hanging around because of thieves and stuff. Crapping in the yard.
	Acceptance/ Pragmatic	**MWA22:** Just let the, the animal just stay around. They ain't causing no trouble. Feed off the land. **SFF07:** If you're doing it [breeding] for the right reason, yeah, but if you do it for the wrong reason, no. Right reasons to protect your family. If you got paperwork on the dogs and you're doing it legal, yeah. It's okay. **DWM26:** I would buy a dog. I would buy a pit dog. **MRJ14:** Because betting on it you get caught you're going to jail. Fighters dog the same thing, going to jail. I'm against that.
	Angry/Fearful	**CLL12:** Hard to answer. I'm like an anti-person. [Indiscernible] I don't know anyone so... **JCL21:** Because see I have to stay in my house. When I my house I get in my vehicle and I leave out of here. Come back home and go to my house again. I don't really—I'm not really from this side of town. I was raised on the Southeast side of town. So that's where all my family is so when I leave my house—Yeah. I don't really know what's going on out here. I have an American Rock and Pit mixed with Jack Russel. They bite. I keep them in my house. [Laughing] **JJD30:** Because me, I stay in the house. I go in, come out and get grandbabies and go to the grocery store. I stay by myself. **MRJ14:** Well, over there where I live, I don't see that. Not where I live at. No. Well, I don't go on that side. I just keep to myself.
DISPARITIES	Economic insecurity	**JFE21:** They can't take care of them. Can't take care of themselves. **MBE12:** Same way a lot of people end up over here. The poverty situation.

THEMES	Codes	Examples
	Housing insecurity	**EJM18:** Yeah. It was the home in Bronson. We had the job situation. Someone stole my car, so I couldn't get to work. So, the money wasn't coming in, so they foreclosed and we had to move. We gave the cats to the shelter. **JJT19:** There is a lot of evictions around here. I own three units here. I don't allow pets. Yes, we had one tenant who temporarily had his son's pit bull, and the next thing we knew, there was a whole litter of puppies. They wrecked the whole back fence and chewed holes in the walls. **MOE20:** Well, what most of the new landlords are doing around here, they are making it pet free. No pits. There's a pit bull right now—people are moving, and they have got to find a home for it. It's a five-month-old puppy. They can't find a place for it.
	Transportation challenges	**MBE12:** Now, I need a vehicle to put myself out of the neighborhood and change this. I have a driver's license, which is—and I'm 63 though, finding a job, good luck. We just got buses on Sunday. What a God send. Thank you for the little blessing. They feel like they are throwing us a big bone. Look at the bus stops. If it rains, we are still standing in the rain. Go to any other neighborhoods and they have nice covered—don't get me started. **OHM29:** Some people maybe don't bring the pets in for the vaccine. [Indiscernible] I see some dogs. It's really long trip, painful.
	Communication challenges	**MRJ14:** I don't have neither one of them [smart phone or internet]. I have a government phone. Just plain phone. **SJR16:** Yeah, you got my cell number [to call back]. It's disconnected, unless I pay.
	Personal health challenges	**DCA08:** No. I can't [work]. I just got—I fell and broke my ribs. I just got out of rehab. That's why I'm here **FJL31:** Unfortunately, I was born with congenital cataracts. I'm almost blind in my left eye because it's hemorrhaged 5 times. **MRJ14:** I had three jobs. Not now. I'm disabled now.

References

1. Esri. *Community Analyst: Reports Reference Guide*; ESRI: Redlands, CA, USA, 2012.
2. Patronek, G.J. Mapping and measuring disparities in welfare for cats across neighborhoods in a large us city. *Am. J. Vet. Res.* **2010**, *71*, 161–168. [CrossRef] [PubMed]
3. Patronek, G.J. Use of geospatial neighborhood control locations for epidemiological analysis of community-level pet adoption patterns. *Am. J. Vet. Res.* **2010**, *71*, 1321–1330. [CrossRef] [PubMed]
4. Aguilar, G.D.; Farnworth, M.J. Stray cats in Auckland, New Zealand: Discovering geographic information for exploratory spatial analysis. *Appl. Geogr.* **2012**, *34*, 230–238. [CrossRef]
5. Miller, G.S.; Slater, M.R.; Weiss, E. Effects of a geographically-targeted intervention and creative outreach to reduce shelter intake in Portland, Oregon. *Open J. Anim. Sci.* **2014**, *4*, 165–174. [CrossRef]
6. Hardt, N.S.; Muhamed, S.; Das, R.; Estrella, R.; Roth, J. Neighborhood-level hot spot maps to inform delivery of primary care and allocation of social resources. *Perm. J.* **2013**, *17*, 4–9. [CrossRef] [PubMed]
7. Dolan, E.D.; Scotto, J.; Slater, M.; Weiss, E. Risk factors for dog relinquishment to a Los Angeles municipal animal shelter. *Animals (Basel)* **2015**, *5*, 1311–1328. [CrossRef] [PubMed]
8. Weiss, E.; Slater, M.; Garrison, L.; Drain, N.; Dolan, E.; Scarlett, J.M.; Zawistowski, S.L. Large dog relinquishment to two municipal facilities in New York city and Washington, DC: Identifying targets for intervention. *Animals (Basel)* **2014**, *4*, 409–433. [CrossRef] [PubMed]
9. Searchable Database to Compare Community Lifesaving. Available online: http://www.maddiesfund.org/searchable-database.htm (accessed 20 June 2017).
10. Taylor, B.; Francis, K. *Qualitative Research in the Health Sciences: Methodologies, Methods and Processes*; Routledge: New York, NY, USA, 2013.
11. Charmaz, K. *Constructing Grounded Theory*; SAGE: Los Angeles, CA, USA, 2014.
12. Maps throughout this Article were Created Using ArcGIS®Software by Esri. ArcGIS®and ArcMap™ are the Intellectual Property of Esri and are Used Herein under License. Copyright © Esri. All Rights Reserved. For More Information about Esri®Software. Available online: www.esri.com (accessed on 20 June 2017).
13. Florida Geographic Data Library, University of Florida, Gainesville, Florida. Available online: http://www.FGDL.org (accessed 20 June 2017).
14. Creswell, J.W. *Educational Research: Planning, Conducting, and Evaluating Quantitative and Qualitative Research*, 4th ed.; Pearson: Boston, MA, USA, 2012.

15. Saldaña, J. *The Coding Manual for Qualitative Researchers*, 2nd ed.; SAGE: Los Angeles, CA, USA, 2013.
16. X Maps Spot. Available online: http://www.aspcapro.org/resource/saving-lives-research-data/x-maps-spot-gis-program (accessed 4 June 2017).
17. Taylor, N.; Signal, T.D. Pet, pest, profit: Isolating differences in attitudes towards the treatment of animals. *Anthrozoos* **2009**, *22*, 129–135. [CrossRef]
18. Frank, J.M.; Carlisle-Frank, P.L. Analysis of programs to reduce overpopulation of companion animals: Do adoption and low-cost spay/neuter programs merely cause substitution of sources? *Ecol. Econ.* **2007**, *62*, 740–746. [CrossRef]
19. Kass, P.H.; Johnson, K.L.; Weng, H.Y. Evaluation of animal control measures on pet demographics in Santa Clara County, California, 1993–2006. *PeerJ* **2013**, *1*, e18. [CrossRef] [PubMed]
20. Scarlett, J.; Johnston, N. Impact of a subsidized spay neuter clinic on impoundments and euthanasia in a community shelter and on service and complaint calls to animal control. *J. Appl. Anim. Welf. Sci.* **2012**, *15*, 53–69. [CrossRef] [PubMed]
21. White, S.C.; Jefferson, E.; Levy, J.K. Impact of publicly sponsored neutering programs on animal population dynamics at animal shelters: The New Hampshire and Austin experiences. *J. Appl. Anim. Welf. Sci.* **2010**, *13*, 191–212. [CrossRef] [PubMed]
22. Basic Animal Data Matrix. Available online: https://www.shelteranimalscount.org/docs/default-source/DataResources/sac_basicdatamatrix.pdf (accessed on 20 June 2017).
23. Pets by the Numbers: U.S. Pet Ownership, Community Cat and Shelter Population Estimates. Available online: http://www.humanesociety.org/issues/pet_overpopulation/facts/pet_ownership_statistics.html (accessed on 7 March 2017).
24. Irvine, L. The problem of unwanted pets: A case study in how institutions think about clients' needs. *Soc. Probl.* **2003**, *50*, 550–566. [CrossRef]
25. Taylor, N. In it for the nonhuman animals: Animal welfare, moral certainty, and disagreements. *Soc. Anim.* **2004**, *12*, 317–339. [CrossRef]

6

Estimating the Availability of Potential Homes for Unwanted Horses in the United States

Emily Weiss *, Emily D. Dolan, Heather Mohan-Gibbons, Shannon Gramann and Margaret R. Slater

Research and Development, Community Outreach, American Society for the Prevention of Cruelty to Animals (ASPCA ®), New York, NY 10128, USA; emily.dolan@aspca.org (E.D.D.); heather.mohan-gibbons@aspca.org (H.M.-G.); shannon.gramann@aspca.org (S.G.); margaret.slater@aspca.org (M.R.S.)

* Correspondence: Emily.weiss@aspca.org

Simple Summary: There are approximately 200,000 unwanted horses annually in the United States. Many are shipped to slaughter, enter rescue facilities, or are held on federal lands. This study aimed to estimate a potential number of available homes for unwanted horses in order to examine broadly the viability of pursuing re-homing policies as an option for the thousands of unwanted horses in the U.S. The results of this survey suggest there could be an estimated 1.2 million homes who have both the perceived resources and desire to house an unwanted horse. This number exceeds the approximately 200,000 unwanted horses living each year in the United States. These data suggest that efforts to reduce unwanted horses could involve matching such horses with adoptive homes and enhancing opportunities to keep horses in the homes they already have.

Abstract: There are approximately 200,000 unwanted horses annually in the United States. This study aimed to better understand the potential homes for horses that need to be re-homed. Using an independent survey company through an Omnibus telephone (land and cell) survey, we interviewed a nationally projectable sample of 3036 adults (using both landline and cellular phone numbers) to learn of their interest and capacity to adopt a horse. Potential adopters with interest in horses with medical and/or behavioral problems and self-assessed perceived capacity to adopt, constituted 0.92% of the total sample. Extrapolating the results of this survey using U.S. Census data, suggests there could be an estimated 1.25 million households who have both the self-reported and perceived resources and desire to house an unwanted horse. This number exceeds the estimated number of unwanted horses living each year in the United States. This study points to opportunities and need to increase communication and support between individuals and organizations that have unwanted horses to facilitate re-homing with people in their community willing to adopt them.

Keywords: horses; slaughter; rescue; adoption

---

1. Introduction

Estimates of the total number of horses in the U.S. vary widely but the number most often cited, 9.2 M, comes from a 2005 economic impact study commissioned by the American Horse Council (AHC) [1]. There are many thousands of unwanted horses annually in the United States. Unwanted is defined by the Unwanted Horse Coalition [1] (a program of the AHC) as "horses which are no longer wanted by their current owner because they are old, injured, sick, unmanageable, fail to meet their owner's expectations (e.g., performance, color or breeding), or their owner can no longer afford them". The reasons horses are unwanted are varied. An estimated 6000–10,000 horses are housed in horse rescues at any given time [2,3]. One U.S. study found that the most common reasons horses were relinquished to rescue organizations were health (54%), lack of suitability for desired purpose (28%), and behavioral problems of the horses (28%) [2]. Owner-related factors most commonly reported were financial hardship (52%) physical illness or death of the owner (27%), and lack of time for the horse (16%) [2]. Horses who were relinquished were most commonly thoroughbreds (22%) and quarter horses (19%) and 51% were geldings, 7.5% colts/stallions and 42% mares. A wide range of ages were reported with a mean of 12 years old. In another national U.S. study of horses seized in cruelty, neglect or abandonment investigations, the most common reasons leading to the investigation were: owner ignorance, economic hardship, and lack of responsibility [4]. Many unwanted, but otherwise re-homable, horses are among the estimated 82,000 to 150,000 horses that are shipped annually to Mexico or Canada for slaughter [5–8]. Among the horses shipped to slaughter between 2002 and 2005, the demographics were similar to the U.S. horse population, indicating that the option of slaughter was applied across the spectrum of horse ownership [9]. There are also more than 100,000 horses being held long-term on open lands by the Bureau of Land Management (both on and off-range) [10].

A number of options exist for unwanted horses, including relinquishment to rescue organizations, donation to universities or law enforcement agencies, sale to or adoption by new owners (re-homed) or euthanasia [9]. For some of these outcomes, the horses must meet specific criteria; for others, the owner must be aware of the option, there must be space available in the program for the horse, or the owner must be able to afford euthanasia and disposal. Increasing the ability of existing horse owners to re-home their horses to private households is one potential way to reduce the number of unwanted horses and improve their welfare and longevity. Horses typically have multiple owners [11], can live up to 30 years [7,11], and are expensive to keep [12,13] making life-long housing difficult to ensure. In order to determine if re-homing is a viable option, it is important to know if there are enough homes to accommodate the number of unwanted horses. To our knowledge, there is no current evidence to inform this question.

This study used a national survey to gather information about the number of potential homes for these horses in a "horse-interested population" (defined as currently owning a horse, having owned a horse in the past 5 years, or interested in owning a horse in the near future). This survey examined whether people would be willing to adopt unwanted horses, what characteristics were required of horses to be considered "adoptable" in the respondent's opinion, and whether potential adopters thought they had adequate resources to keep a horse. From this survey, an estimate for the number of potential homes for horses in the United States was extrapolated in order to broadly examine the viability of pursuing re-homing policies as an option for the thousands of unwanted horses in the U.S.

2. Materials and Methods

2.1. Survey

A telephone-based survey of the general adult population was conducted by Edge Research using CARAVAN ® ORC International. Telephone calls were made between 24 September 2015 and 11 October 2015. The CARAVAN ® Omnibus telephone survey is a nationally projectable study conducted among a probability sample of U.S. residents, 18 years of age and older. See Appendix A for ORC International's complete methodology. The horse survey questions were included in a larger bank of questions asked during the interview. This study was conducted using two probability samples: randomly selected landline telephone numbers and randomly selected mobile (cellular) telephone numbers. The combined sample consists of 3036 adults (18 years old and older) living in the continental United States. Of the 3036 interviews, 1536 were from the landline sample and 1500 from the cell phone sample. The survey had a response rate of 20% and a completion rate of 86%.

The survey sample size was selected to serve two purposes. First, we wanted to ensure that the estimate of potential adoptive households for horses in the U.S. (the primary outcome) was sufficiently precise (i.e., had a narrow enough confidence interval) to be useful in practice. For hypothetical scenarios where either 0.5%, 1.0%, or 1.5% of households were interested in and capable of adopting a horse, an overall sample size of 3000 would ensure confidence interval coverage of no more than ±0.5%. In other words, the 95% confidence interval would cover less than 0.5% in either direction from the point estimate for all three scenarios. Secondly, we continued the survey to ensure that at least 500 respondents were *horse-interested*. This ensured that the confidence interval coverage would be no more than ±5% for any proportions calculated from this subgroup.

2.2. Survey Questions

To estimate the number of individuals with available homes, respondents were asked: "Which, if any, types of animals do you personally own?"; "Have you ever owned a horse?"; "How long ago did you last own a horse?"; "How interested are you in obtaining a horse at some point in the future?" with possible responses "very interested", "somewhat interested", and "not interested"; and for three scenarios (1. A horse that no longer has an owner; 2. A horse that has medical or behavioral challenges; 3. A horse that might be abandoned if a new owner is not found): " ... how interested you would be in adopting a horse in those circumstances" with possible responses "very interested", somewhat interested", and "not interested"; and "Do you currently have the space and resources necessary to house and care for a newly-adopted horse on your own property or at a local barn or stabling facility?" with possible responses "yes", "no", or "don't know".

2.3. Final Sample

Among the total sample of individuals reached, we first identified those who were *horse-interested* as defined by reporting currently having or having had a horse (in the last 5 years) and/or being interested in getting a horse in the future. Among these, we further defined a target subgroup of *potential adopters* who have the interest in and perceived resources and capacity to take a re-homed horse based on two additional criteria. To qualify for this subgroup, respondents must first have reported *strong interest* in adoption under all 3 scenarios of interest (a horse facing abandonment, a horse without an owner, and a horse with medical or behavioral problems). We used the criterion of strong interest to ensure that casual or circumstantial interest was not included in the definition. Secondly, the respondent must also have endorsed currently perceiving that they have the space and resources to care for a horse. A further, smaller, subgroup was *experienced potential adopters*, which included only those *potential adopters* who also have previous experience owning a horse in the last 5 years, indicating that they had an understanding of the demands needed to care for a horse.

2.4. Analysis

Characteristics of respondents were described using frequencies and percentages. We used proportions from our sample to make inferences about the corresponding population level proportions using standard statistical methods [14]. For key results, exact 95% confidence intervals were calculated using the standard Cloppe-Pearson method in our software [15]. The confidence intervals generated can be interpreted as a plausible range for the proportion of the U.S. population that would provide the same answers. This method was selected because it provides more conservative estimates (i.e., wider confidence intervals) for small proportions (i.e., accounts for uncertainty in the population level estimates that could be related to relying on a random sample). Numbers of U.S. households and U.S. adults meeting our definitions of *horse-interested* and *potential adopters* were estimated by multiplying proportion estimates and lower and upper confidence intervals with corresponding population totals [16]. Demographics of *potential adopters* were compared to other *horse-interested* respondents using the chi-square test. For adults in the U.S., 2015 total population of 321,418,820 was multiplied by the percentage of adults in the U.S. in 2015 (77.1%, with the percentage of the population <18 = 22.9%) [16]. This equaled ~247,813,910 U.S. adults. For total U.S. households, the estimated number in 2016 was 135,697,926 [16]. Analyses were run using Stata/IC 13.1 (StataCorp LP, College Station, TX, USA).

2.5. Ethical Statement

All respondents were informed that their responses would be kept confidential. The only identifying information collected was their first name; recorded for quality control purposes. Institutional review was not sought because the data were collected as part of a larger, national opinion survey and this type of research is considered to be exempt from review by IRBs.

3. Results

3.1. Demographics

Data describing the characteristics of the national sample and the *horse-interested* sample are in Table 1.

Table 1. The unweighted frequencies and percentages for the nationally representative sample of 3036 respondents and the 500 who were *horse-interested*.

Respondent Characteristics	National Sample, n = 3036		*Horse-Interested* Sample, n = 500	
	Frequency	Percent	Frequency	Percent
Gender				
Male	1529	50.4	262	52.4
Female	1507	49.6	238	47.6
Total	3036	100.0	500	100.0
Age range				
18–29	473	15.5	126	25.2
30–49	678	22.3	148	29.6
50–64	888	29.3	146	29.2
65 or older	961	31.7	77	15.4
Refused/No Response	36	1.2	3	0.6
Total	3036	100.0	500	100.0
Total household income (before tax, 2014)				
Under U.S.$25,000	585	19.3	107	21.4
$25,000 but less than $50,000	735	24.2	144	28.8
$50,000 but less than $100,000	628	20.7	91	18.2
$100,000 or more	519	17.1	80	16.0
Don't know/Refused/NR	569	18.7	78	15.6
Total	3036	100.0	500	100.0
Region				
North East	551	18.2	75	15.0
Midwest	675	22.2	97	19.4
South	1136	37.4	208	41.6
West	674	22.2	120	24.0
Total	3036	100.0	500	100.0

3.2. Sample Breakdown Based on Responses

The breakdown of respondents, in a flow diagram, is shown in Figure 1.

Among the 3036 individuals contacted, seventeen percent (95% CI 15–18%; n = 500) met our criteria for *horse-interested* by reporting that they either currently own a horse, want to own a horse in the near future, or have owned a horse within the past 5 years. These respondents then reported their interest in horse ownership based upon three common scenarios, shown in Figure 2.

Nine percent (45; 95% CI 7–12%) of the *horse-interested* sample reporting being "very interested" in obtaining a horse under all three scenarios. Further, 46% of the *horse-interested* sample (230; 95% CI 42–51%) reported having the resources (which we termed perceived capacity) to house and care for a newly-adopted horse.

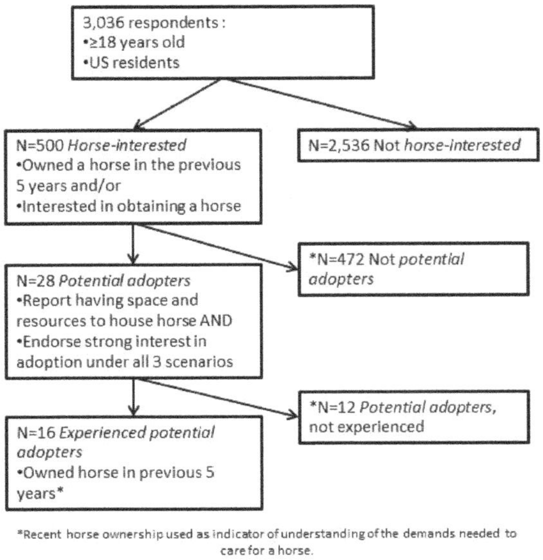

Figure 1. Proportion of respondents in each category.

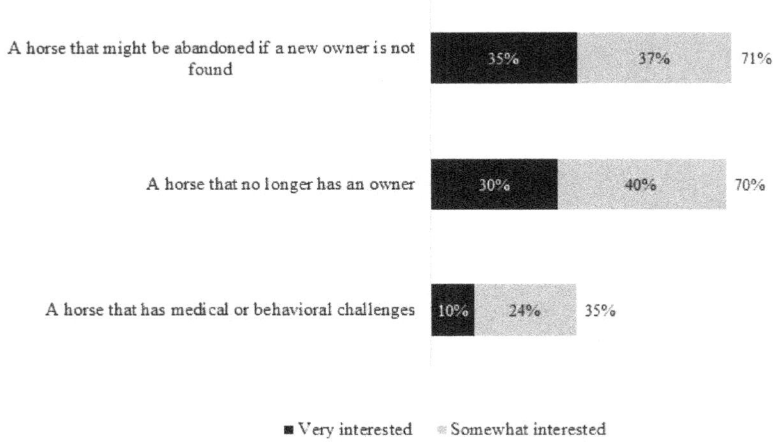

Figure 2. Reported interest by 500 *horse-interested* U.S. residents in adopting unwanted horses under various scenarios.

Among the *horse-interested* sample, 5.6% (28; 95% CI 4–8%) were classified as *potential adopters* based on reporting both strong interest in each of the three adoption scenarios as well as having the perceived capacity to house and care for a horse. These 28 *potential adopter* respondents represent 0.92% (CI = 0.6–1.3%) of the total sample. There were no significant or important differences between *potential adopters* and other horse interested responders (gender $p = 0.5$; age $p = 0.5$; income $p = 0.7$). Among *potential adopters*, 53% were women, 32% were 18–29 years old, 36% 30–49 years old, 25% 50–64 years old and 7% age 65 and up. For income, 30% of *potential adopters*' income was <U.S. $25,000, 26% was $25,000 to <$50,000, 29% from $50,000 to <100,000 and 26% had incomes of $100,000 and above.

3.3. Population Estimates

Applying the percentage of *potential adopters* to the number of U.S. households yielded an estimated 1.25 million households (0.0092 × 135,697,926; CI = 0.83–1.80 million) that had strong interest and perceived they could house a horse. Multiplying 0.92% times the U.S. population of 247,813,901 adults for a less conservative estimate, we estimated that approximately 2.28 million people

(CI = 1.52–3.30 million) in the U.S. would have strong interest in obtaining a horse and perceived they could house a horse.

Twelve respondents who met the criteria through their interest in future horse ownership had not previously owned a horse. Excluding those from the 28, 16 or 0.53% (CI = 0.3–0.9%) *experienced potential adopters* reported strong interest, perceived capacity, and currently or had ever owned a horse. Applying this more conservative criteria, an estimated 0.72 million U.S. households (CI = 0.41–1.22 million), with first-hand knowledge of the challenges of owning a horse, could house a horse.

4. Discussion

There are many reasons why horses may be abandoned and in need of re-homing. One central reason is affordability; however, owners reported a variety of more specific reasons including old age, injuries, horse behavior, factors in the owner's household such as divorce or lack of time and the horse not meeting expectations [2,8]. Regardless of the underlying industry reasons for the numbers of horses needing to be re-homed, finding previously untapped homes is critical. The current study found an estimated 1.25 million households with the interest and self-assessed, perceived capacity to adopt a horse in the United States, including those with medical or behavioral problems. Accounting for the uncertainty that comes with applying a sample proportion to the entire population, the true count could reasonably be expected to lie between 0.83 and 1.80 million households. This estimate was based on only those surveyed who reported strong interest in all three categories of unwanted horses: a horse that might be abandoned; a horse that is homeless; a horse with medical or behavioral issues. Excluding qualified respondents who had never owned a horse, 0.72 million households are estimated to have the perceived capacity and interest in owning a horse. These numbers of interested households who perceive themselves as qualified, suggests that there are more available homes than previously thought to accommodate unwanted horses in the U.S.

The number of respondents in our survey was large and this probability sample was representative of the population of the United States, indicating that our estimates are reflective of national interest in horse ownership. When using a less conservative approach by estimating based on individuals rather than households, 2.28 million individuals reported self-assessed perceived capacity to house horses. We looked at individual estimates because we assumed respondents to be from different households and because each individual with interest might be engaged in horse related support programs. However, we acknowledge that it is more conservative to consider households as the basis for estimating new homes for horses.

What is not clear is how long the number of potential homes would exceed the number of horses that need those homes. It is theorized that as horses move amongst homes, the availability of new homes would likely remain stable. Further, although the lifespan of a horse is not short, each year, homes with horses have horses that die or are euthanized, opening some spaces for new horses. It could be, however, that the number of homes would eventually become saturated. If that were true, this estimate remains crucial for three primary reasons: (1) even just a short time of reducing the prevalence of unwanted horses in the United States would free available resources for expanded access to safety net services and resources in order to reduce the incidence of newly unwanted horses; (2) the estimate provides a rich and substantial target for recently launched innovative industry programs such as Time To Ride [17], which aims to grow the horse industry though engaging new audiences, and The Right Horse [18], which aims to promote horse adoption in innovative ways; and (3) moving even a small number of additional horses into homes would allow for a decreased prevalence of unwanted horses.

In addition to increasing the likelihood of finding more homes for horses, there may be opportunities to keep horses in their current homes with more community support. Given limited capacity and funding of rescue organizations [2,8], safety net and expanded programs can improve the industries' reach. Equine rescue organizations in the U.S. have an estimated capacity of 13,400 horses per year [2]. Given that 53% of those horses arrived in poor health or body condition and that 23% of owners reported economic or financial hardship as a contributor to relinquishment [2], programs which better support current owners and their horses with food and/or accessible and affordable veterinary care have promise in potentially preventing re-homing. Another survey of horse owners [19] found that 47% noted they believed cost of care was a problem and a contributing factor to unwanted horses. It is possible that providing assistance to horse owners in times of crisis could keep some horses in their current homes, preventing the re-homing in the first place. These shifts, as well as support for end-of-life, including providing affordable, accessible, humane euthanasia [20], would allow resources to focus on moving horses from conditions of neglect and cruelty toward higher welfare living arrangements.

It is possible that some of those people who noted having both strong interest and perceived capacity would not actually adopt when given the opportunity. Further, the present study only considered the respondents perceptions and self-assessment of whether they have adequate resources to care for a horse and not any objective measures of the adequacy of the available resources. It is acknowledged that reported intent could overestimate actual behavior. However, the responders' ability to house and care for a horse may or may not be related to previous horse ownership. Previous owners may know what is needed to appropriately house a horse or may continue to provide care which some might view as less than optimal. New horse owners could spend time becoming well informed or simply dive into horse ownership. We strongly encourage additional information sharing with current and new horse owners as an additional method of improving horse welfare. The data presented, however, suggest preliminarily that there may be many homes for unwanted horses that have not yet been accessed. Surprisingly, people showed interest in adopting horses that needed extra support, as 35% of respondents reported being "very or somewhat interested" in adopting a horse that was medically or behaviorally unsound. It is possible that some responses were due to social desirability bias, where the respondents gave the socially correct or pleasing answer. That would tend to increase the likelihood of respondents saying that would be interested in adopting a horse. In this survey, the presence of multiple other questions and the neutral organization administering the survey, as well as the lack of face to face interaction would partially help to mitigate this bias [21,22].

Because the characteristics of horses likely contribute to their ability to be re-homed [23], research into increasing access to services such as medical, and behavior support would provide important information in expanding the link between unwanted horses and potential adopters. This would also be important to ensure that adopters of horses with medical or behavior issues receive the counseling and support needed to appropriately care for those horses.

The current results that there are currently untapped potential adopters seems to contradict previously published results that shelters and rescues are overwhelmed [2]. In general, this contradiction points to a potential gap in communication or understanding between horse organizations that have horses available for adoption and people interested in adopting. It is possible that horse organizations have not embraced the importance of finding new homes as an important part of allowing them to save more horses. Additionally, finding new and creative ways to connect potential homes with current owners who need to re-home is likely to be critically important [22]. Future research that focuses on the opportunities to increase adoptions from horse organizations would provide valuable information to the body of unwanted horse literature. A study of people who have re-homed their horses is currently being conducted to complement the findings presented here and to update findings in Holcomb et al. [20].

5. Conclusions

This study found an estimated 1.24 million households interested in and potentially able to adopt a horse and 2.26 million potential horse advocates for horses in need. While this estimate may not reflect an immediate, objectively suitable set of adopters, these numbers of people who report their willingness and perceived ability exceed the known estimates of unwanted horses and suggest a substantial and underutilized resource. This study points to opportunities to increase communication and support between individuals and organizations that have unwanted horses to facilitate re-homing with people in their community willing to adopt them.

Acknowledgments: The authors thank the following for their time and expertise: Carolyn Schnurr, Nancy Perry, Bert Troughton, Justine Dang, Vic Spain, and Maya Gupta. We also thank Edge Research, Inc. (Arlington, VA, USA) for administering the survey and compiling the data.

Author Contributions: Emily Weiss, Heather Mohan-Gibbons, Shannon Gramann and Margaret R. Slater conceived and designed the experiments; Emily D. Dolan and Margaret R. Slater analyzed the data; Emily Weiss, Emily D. Dolan, Heather Mohan-Gibbons, Shannon Gramann and Margaret R. Slater wrote the paper. Edge Research was contracted to conduct the survey and compile the data.

Conflicts of Interest: The authors declare no conflict of interest.

Appendix A

ORC International's Random Digit Dial (RDD) telephone sample is generated using a list-assisted methodology. The standard GENESYS RDD methodology produces a strict single stage, EPSEM (Equal Opportunity of Selection Method) sample of residential telephone numbers. The cell phone sample, also RDD, has been supplied by SSI, Inc. using their proprietary Cell/WINS technology. The cell phone sample is generated from cell phone 1000 series blocks (a "block" is defined by the first seven digits of the phone number) with all permutations within each block included. The sampling interval is then calculated by dividing the universe of all possible numbers by the number of records desired, thus specifying the size of the frame subdivisions. Within each of the subsets, one number is selected at random giving all numbers an equal probability of selection.

Surveys are collected by trained and supervised U.S. based interviewers using ORC International's computer assisted telephone interviewing (CATI) system. Final data are adjusted to consider the two sample frames and then weighted by age, gender, region, race/ethnicity and education to be proportionally representative of the U.S. adult population. Weighting adjustments are used to reduce the potential for biases that may be present due to incomplete frame coverage and survey nonresponse-both inherent in all telephone surveys. Each respondent is given a score based on their reported information and then either weighted up (if they are underrepresented) or weighted down (if they are overrepresented).

References

1. Deloitte Consulting, LLP. *2005 The Economic Impact of the Horse Industry on the United States*; National Report: Louisville, KY, USA, 2005.
2. Holcomb, K.E.; Stull, C.L.; Kass, P.H. Unwanted horses: The role of nonprofit equine rescue and sanctuary organizations. *J. Anim. Sci.* **2010**, *88*, 4142–4150. [CrossRef] [PubMed]
3. Lenz, T.R. The unwanted horse in the United States: An overview of the issue. *J. Equine Vet. Sci.* **2009**, *29*, 253–258. [CrossRef]
4. Stull, C.L.; Holcomb, K.E. Role of U.S. animal control agencies in equine neglect, cruelty, and abandonment investigations. *J. Anim. Sci.* **2010**, *92*, 2342–2349. [CrossRef] [PubMed]
5. Taylor, M.; Sieverkropp, E. The impacts of U.S. horse slaughter plant closures on a western regional horse market. *J. Agric. Resour. Econ.* **2013**, *38*, 48–63.
6. Stull, C.L. The journey to slaughter for North American horses. *Anim. Front.* **2012**, *2*, 68–71. [CrossRef]
7. American Veterinary Medical Association. Unwanted Horses and Horse Slaughter FAQ. Available online: https://www.avma.org/KB/Resources/FAQs/Pages/Frequently-asked-questions-about-unwanted-horses-and-horse-slaughter.aspx (accessed on 28 February 2017).
8. Unwanted Horse Coalition/The American Horse Council. 2009 Unwanted Horses Survey. Available online: http://www.unwantedhorsecoalition.org/wp-content/uploads/2015/09/unwanted-horse-survey.pdf (accessed on 14 July 2017).
9. Lenz, T.R. The unwanted horse—A major welfare issue. In *Equine Welfare*; McIlwraith, C.W., Rollin, B.E., Eds.; Blackwell Publishing Ltd.: West Sussex, UK, 2011; pp. 425–441, ISBN-10: 1405187638.

10. U.S. Department of the Interior, Bureau of Land Management. Wild Horse and Burro Quick Facts. Available online: https://www.blm.gov/programs/wild-horse-and-burro/about-the-program/program-data (accessed on 3 April 2017).
11. Animal Welfare Council. The Life Cycle and Recycle of Horses. Available online: http://animalwelfarecouncil.com/wp-content/uploads/2012/02/UHLessons-3-THREE-Aug-6-2012.pdf (accessed on 28 February 2017).
12. Schueler, K.A. The Perceptions of the Unwanted Horse Population in Illinois. Master's Thesis, Illinois State University, Normal, IL, USA, 2015.
13. North, M.S.; Bailey, D.; Ward, R.A. The potential impact of a proposed ban on the sale of U.S. horses for slaughter and human consumption. *J. Agribus.* **2005**, *23*, 1–17.
14. Ott, R.L.; Longnecker, M.T. Chapter 5: Inferences about population central values. In *An Introduction to Statistical Methods and Data Analysis*, 7th ed.; Cengage Learning: Boston, MA, USA, 2016; pp. 232–299.
15. Clopper, C.J.; Pearson, E.S. The use of confidence or fiducial limits illustrated in the case of the binomial. *Biometrika* **1934**, *26*, 404–413. [CrossRef]
16. U.S. Census. U.S. Census Quick Facts. Available online: https://www.census.gov/quickfacts (accessed on 8 June 2017).
17. American Horse Council. Time to Ride. Available online: https://www.timetoride.com/ (accessed on 28 February 2017).
18. The Right horse Initiative. The Right Horse. Available online: https://www.therighthorse.org/ (accessed on 28 February 2017).
19. Stowe, J.C. Results from 2012 AHP Equine Industry Survey. American Horse Publications, 2012. Available online: http://www.americanhorsepubs.org/equine-survey/2015-equine-survey/ (accessed on 28 February 2017).
20. Holcomb, K.E.; Stull, C.L.; Kass, P.H. Characteristics of relinquishing and adoptive owners of horses associated with U.S. nonprofit equine rescue organizations. *J. Appl. Anim. Welf. Sci.* **2012**, *15*, 21–31. [CrossRef] [PubMed]
21. Chang, L.; Krosnick, J.A. National surveys via RDD telephone interviewing versus the internet: Comparing sample representativeness and response quality. *Public Opin. Quart.* **2009**, *73*, 641–678. [CrossRef]
22. King, J.F.; Bruner, G.C. Social desirability bias: a neglected aspect of validity testing. *Psychol. Mark.* **2000**, *17*, 79–103. [CrossRef]
23. Stowe, C.J.; Kibbler, M.L. Characteristics of adopted thoroughbred racehorses in second careers. *J. Anim. Sci.* **2016**, *19*, 81–89. [CrossRef] [PubMed]

7

Ranging Behaviour of Commercial Free-Range Broiler Chickens 2: Individual Variation

Peta S. Taylor [1,*], Paul H. Hemsworth [1], Peter J. Groves [2], Sabine G. Gebhardt-Henrich [3] and Jean-Loup Rault [1]

1 Animal Welfare Science Centre, Faculty of Veterinary and Agricultural Sciences, University of Melbourne, Parkville, VIC 3010, Australia; phh@unimelb.edu.au (P.H.H.); raultj@unimelb.edu.au (J.-L.R.)
2 Poultry Research Foundation, School of Veterinary Science, Faculty of Science, The University of Sydney, Camden, NSW 2570, Australia; peter.groves@sydney.edu.au
3 Research Centre for Proper Housing: Poultry and Rabbits (ZTHZ), Division of Animal Welfare, University of Bern, CH-3052 Zollikofen, Switzerland; sabine.gebhardt@vetsuisse.unibe.ch
* Correspondence: ptaylo37@une.edu.au

Simple Summary: Although the consumption of free-range chicken meat has increased, little is known about the ranging behaviour of meat chickens on commercial farms. Studies suggest range use is low and not all chickens access the range when given the opportunity. Whether ranging behaviour differs between individuals within a flock remains largely unknown and may have consequences for animal welfare and management. We monitored individual chicken ranging behaviour from four mixed sex flocks on a commercial farm across two seasons. Not all chickens accessed the range. We identified groups of chickens that differed in ranging behaviour (classified by frequency of range visits): chickens that accessed the range only once, low frequency ranging chickens and high frequency ranging chickens, the latter accounting for one-third to one half of all range visits. Sex was not predictive of whether a chicken would access the range or the number of range visits, but males spent more time on the range in winter. We found evidence that free-range chicken ranging varies between individuals within the same flock on a commercial farm. Whether such variation in ranging behaviour relates to variation in chicken welfare remains to be investigated.

Abstract: Little is known about broiler chicken ranging behaviour. Previous studies have monitored ranging behaviour at flock level but whether individual ranging behaviour varies within a flock is unknown. Using Radio Frequency Identification technology, we tracked 1200 individual ROSS 308 broiler chickens across four mixed sex flocks in two seasons on one commercial farm. Ranging behaviour was tracked from first day of range access (21 days of age) until 35 days of age in winter flocks and 44 days of age in summer flocks. We identified groups of chickens that differed in frequency of range visits: chickens that never accessed the range (13 to 67% of tagged chickens), low ranging chickens (15 to 44% of tagged chickens) that accounted for <15% of all range visits and included chickens that used the range only once (6 to 12% of tagged chickens), and high ranging chickens (3 to 9% of tagged chickens) that accounted for 33 to 50% of all range visits. Males spent longer on the range than females in winter ($p < 0.05$). Identifying the causes of inter-individual variation in ranging behaviour may help optimise ranging opportunities in free-range systems and is important to elucidate the potential welfare implications of ranging.

Keywords: poultry; pasture; outdoor; range; meat chicken; welfare; Radio Frequency Identification (RFID)

1. Introduction

Broiler chicken ranging behaviour remains poorly understood, particularly on free-range commercial farms. A greater understanding of ranging behaviour can assist to ensure optimal opportunities to range through the provision of adequate environment and management practices, and possibly by selecting pertinent chicken characteristics.

The majority of studies on broiler chicken ranging behaviour to date report variability in range use at flock level [1–3], which has been attributed to environmental conditions such as resources on the range (such as artificial and natural shelters, hay bales, perches and panels [4–8]) and weather variables (including outdoor temperature and Ultra Violet index [4,7,9]). Yet, very little is known about variation between individual broiler chickens within a flock. Genetics and rearing environments have been shown to alter ranging behaviour within a flock [6,10] but relationships with individual ranging behaviour is unknown.

Heterogeneous ranging behaviour may result in variation in individual welfare and reduce uniformity in flocks. There are various beliefs that accessing an outdoor range will impact an animal's welfare state; some consumers believe that accessing an outdoor range will have positive effects on broiler chicken welfare (e.g., increased expression of natural behaviours) and other groups (such as some farmers and veterinarians) are concerned with negative welfare consequences of range access, such as increased health risks due to increased exposure to parasites and extreme weather conditions [11,12]. However, there is very little scientific evidence of the impact of range access on broiler chicken welfare. Chicken welfare assessments are often reported as flock averages, however if variation in ranging behaviour exists then chickens within the same flock may have different welfare implications from ranging depending on the degree of variation. If welfare is compromised with increased range use, productivity (growth) may also be affected and result in reduced flock uniformity; an additional challenge for free-range flock management.

In order to assess whether heterogeneous ranging behaviour exists in commercial broiler chicken flocks, we monitored individual broiler chicken ranging behaviour to determine the variation in ranging behaviour between individuals within commercial free-range flocks.

2. Materials and Methods

All animals used in this study were approved by the University of Melbourne Animal Ethics Committee (approval number 1413428.3). A full description of the methodology is provided in part one of this paper series "Commercial free-range broiler chicken ranging behaviour 1: factors related to flock variability"; however, it is briefly outlined below.

2.1. Study Site and Animals

Four flocks (A–D) of ROSS 308 broiler chickens were studied across two seasonal replicates on one commercial farm during the Austral winter (flocks A and B) and summer (flocks C and D). All sheds had chicks from the same hatchery, same feed, same manager, and comparable management practices. Seasonal replicates occurred within the same sheds (Shed one: 40.5 m × 9.3 m, housing approximately 6000 chickens, flocks A and C; Shed two 50.5 m × 12.3 m, housing approximately 10,000 chickens, flocks B and D). Flocks had access to adjacent range areas (54.1 × 13.9 m and 77.9 × 16.4 m adjacent to the shed wall and 13.6 × 9.3 m and 27.5 × 12.3 m at the back of the shed, for shed one and two respectively) accessible through manually operated 1.3 × 0.4 m doors described hereafter as "pop-holes" and spaced 5.65 m apart, with six pop-holes for shed one and seven pop-holes for shed two. Feed and water were provided ad libitum inside the shed, but never in range areas.

2.2. Tracking Individual Range Use

Individual range use was tracked by the Gantner Pigeon Radio Frequency Identification (RFID) System (2015 Gantner Pigeon Systems GmbH, Benzing, Schruns, Austria), with a bespoke program,

Chicken Tracker. Chickens (n = 300/flock) were randomly selected and fitted with a silicone leg band that automatically loosened with leg growth (Shanghai Ever Trend Enterprise, Shanghai, China). Each leg band contained a unique ID microchip (Ø4.0/34.0 mm Hitag S 2048 bits, 125 kHz) that registered as the chickens walked over the antenna. Antennas were attached to both sides of each pop-hole (i.e., indoor and outdoor) to determine the direction of movements by each tagged chicken; allowing calculation of the frequency and duration of range visits for each individual.

Chickens were tracked from the first day that range access was permitted (21 days of age) until a few days before partial depopulation (30–33 days of age) in winter flocks due to logistical reasons. However, chickens in summer flocks were tracked until complete depopulation for slaughter (43–45 days of age). Sex and weight of individuals (flock A: n = 83, flock B: n = 97, flock C: n = 280, flock D: n = 290) were recorded at the end of the study when leg bands were removed.

2.3. Statistical Analysis

RFID data were cleaned with SAS™ (v. 9.3, SAS Institute Inc., Cary, NC, USA) using a modified macro [13]. All range visits <10 s were treated as false positives and removed from analysis.

Descriptive data are presented for each flock.

Statistical analysis was performed with SPSS statistical software (v. 22, IBM Corp, Armonk, NY, USA).

Latency to access the range data did not meet the criteria of normality; therefore, Spearman's rho correlation coefficients were used to examine relationships between latency to first access the range and total frequency and duration of range visits and duration per range visit, relationships between total time spent on the range, total number of range visits and average time spent on the range per range visit. Chi square analysis was used to determine if there was a difference in the number of females and males that accessed the range. Flock could not be included as a random variable in non-parametric Spearman correlations or chi square analysis, but each correlation and chi square analysis was initially performed on each flock and there were no differences in direction or significance values between flocks within season. Hence, flocks were pooled and are presented within season. Ranging data were log transformed and subsequently met the criteria of normality and homogeneity of variance; hence, General Linear Mixed models were used to determine the effect of sex on the total frequency and duration of range visits, average time spent on the range per visit and the number of days an individual accessed the range, with flock and individual nested within flock as random factors, in addition to running the model both with and without final body weight as covariate. General Linear Mixed models were used to compare the average time spent on the range per range visit between chickens that accessed the range only once and chickens that accessed the range more than once, with flock and individual nested within flock included as random factors and weight as a covariate. Results are presented as raw means $\pm$ SE unless otherwise noted.

3. Results

3.1. Range Availability

Winter flocks had fewer opportunities (days and hours per day) to access the range than summer flocks; data were presented in part one of this paper series "Commercial free-range broiler chicken ranging behaviour 1: factors related to flock variability". Briefly, in winter, management provided the flocks with access to the range for a mean of 5.6 $\pm$ 0.4 h a day, for 70% of the days prior to partial depopulation. In summer, management provided the flocks with access to the range for a mean of 10.4 $\pm$ 0.6 h a day, for 75% and 76.2% of the days prior to complete depopulation, in flocks C and D respectively.

3.2. Inter-Individual Variation in Ranging Behaviour

There was individual variation in ranging frequency and duration within all flocks (Figures 1 and 2). The mean number of daily visits made by an individual varied between 0–11.8 and 0–12.7 visits in winter and summer flocks, respectively. The mean time an individual spent on the range daily varied between 0–76.6% and 0–65.7% of the available ranging time, equivalent to 0–4.3 h and 0–6.8 h, in winter flocks and summer flocks respectively.

The total number of range visits made by an individual varied between 0–71 and 0–167 visits over the course of the study in winter flocks and summer flocks respectively. The total duration an individual spent on the range over the course of the study varied between 0–23.0% and 0–40.2% of available overall ranging time, equivalent to 0 to 8.7 h and 0 to 40.7 h, in winter flocks and summer flocks respectively.

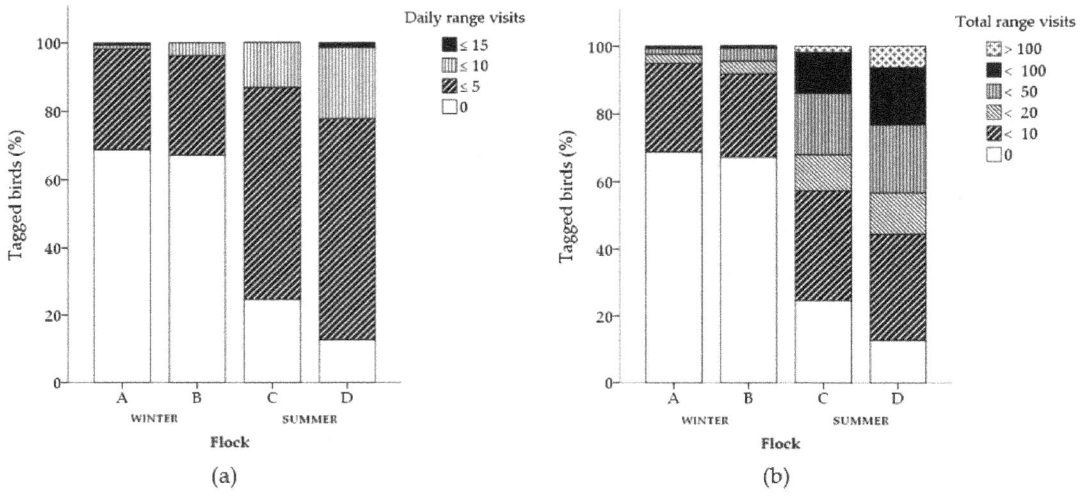

Figure 1. Frequency of range visits for individual chickens within each flock (winter flocks: A and B; summer flocks: C and D). Patterns within stacked bars represent the number of chickens (% successfully tracked) in each ranging frequency category, daily mean (**a**) and total number of visits throughout the study (**b**) for each flock.

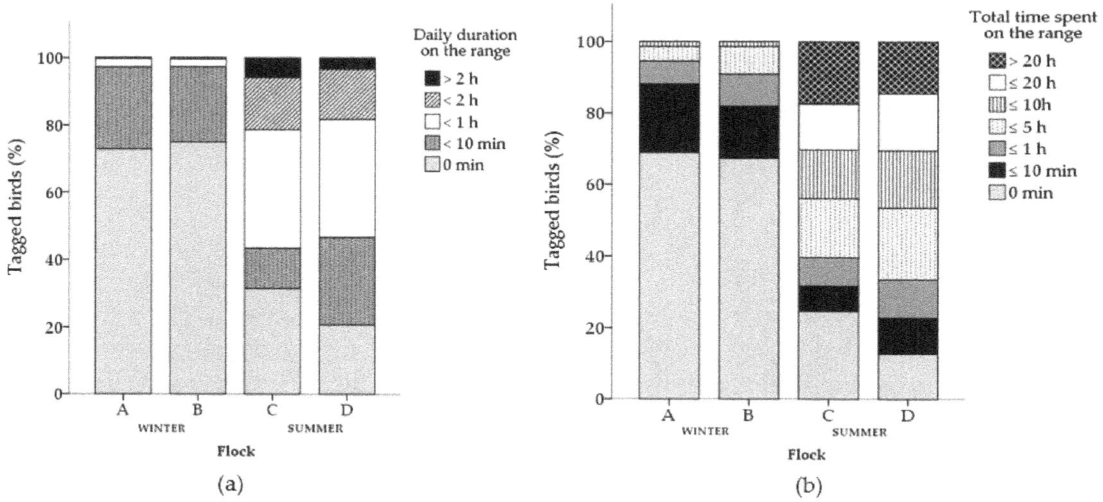

Figure 2. Duration of range visits for individual chickens within each flock (winter flocks: A and B; summer flocks: C and D). Patterns within stacked bars represent the number of chickens (% successfully tracked) in each ranging duration category, daily mean (**a**) and total time spent on the range throughout the study (**b**) for each flock.

3.3. Latency to Access the Range

The number of days before an individual first accessed the range after range access was first provided (hereafter referred to as "latency to access the range") was negatively correlated with an individual's total number of range visits (winter: $r_{(188)} = -0.41$, $p < 0.001$; summer: $r_{(460)} = -0.44$, $p < 0.001$) and total duration of range visits (winter: $r_{(188)} = -0.34$, $p < 0.001$; summer: $r_{(460)} = -0.35$, $p < 0.001$), but not the mean duration per visit. Latency to access the range was also negatively correlated with the number of days an individual accessed the range in summer flocks ($r_{(460)} = -0.33$, $p < 0.01$), but not in winter flocks.

When individual ranging data were corrected for number of available ranging days remaining after the range was first accessed, to assess ranging patterns after first range access, latency to access the range was still negatively correlated with range use in both seasons (frequency: winter—$r_{(143)} = -0.24$, $p < 0.01$; summer—$r_{(450)} = -0.34$, $p < 0.00$; duration: winter—$r_{(143)} = -0.20$, $p < 0.05$; summer—$r_{(450)} = -0.31$, $p < 0.001$).

3.4. High Frequency Ranging Chickens

A total of 1434 range visits were recorded in winter flocks (flock A: 573 visits; flock B: 861 visits) and 14,008 range visits in summer flocks (flock C: 5644 visits; flock D: 8364 visits). The top 10% of ranging chickens, based on frequency of range visits, accounted for approximately half of the range visits in winter flocks (flock A: 9 chickens accounted for 57% total range visits; flock B: 10 chickens accounted for 47% total range visits) and one-third of range visits in summer flocks (flock C: 21 chickens accounted for 34% of range visits; flock D: 25 chickens accounted for 33% total range visits). The top 50% of ranging chickens accounted for 89–91% of all range visits, irrespective of season (Figure 3). Thus, the bottom 50% of ranked ranging chickens accounted for <15% of the total range visits (winter: flock A—13.3% of total visits, flock B—13.5% of total visits; summer: flock C—4.8% of total visits, flock D—5.9% of total visits).

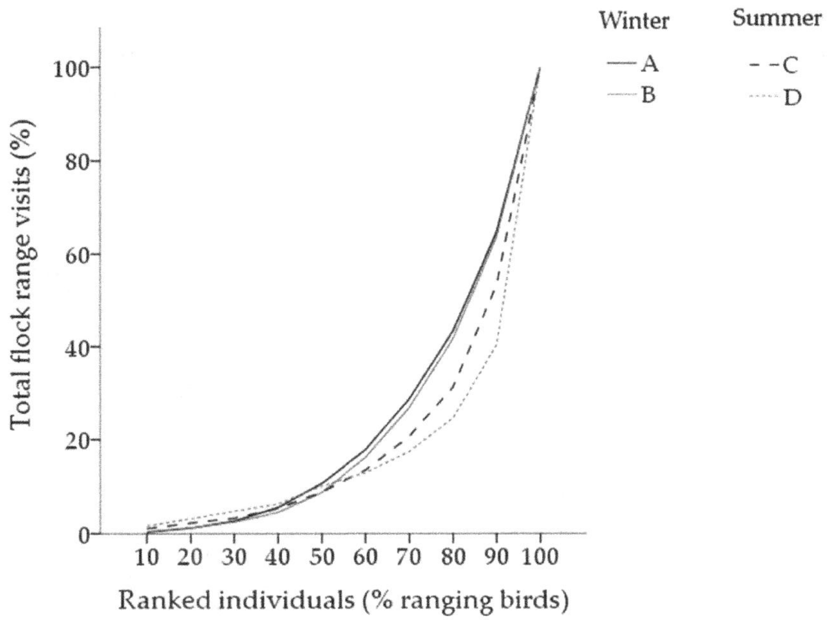

Figure 3. The proportion of range visits (% total flock range visits) that was attributed to ranked individuals. Ranging chickens were ranked on the total number of range visits and are displayed from lowest to highest percentage of ranging chickens in each flock; solid lines represent winter flocks (flocks A and B), dotted lines represent summer flocks (flocks C and D). Chickens that did not access the range are not included.

The top 10% of ranging chickens, based on the total time spent on the range, accounted for more than half of the total flock time spent on the range in winter flocks (flock A: 70%, flock B: 54%) and more than one-third of the total flock time spent on the range in summer flocks (flock C: 38%, flock D: 37%). The average time spent on the range per visit did not differ between birds ranked in the top 10%, top 11–49% or bottom 50% in either season (winter: $F_{(2,184)} = 0.4$, $p > 0.05$; summer: $F_{(2,456)} = 0.3$, $p > 0.05$).

3.5. Chickens that Accessed the Range Once

There was a relatively consistent proportion of chickens, across all flocks, that accessed the range only once throughout the study (flock A: 12%, flock B: 10%, flock C: 6%, flock D: 8%). The total number of one-time ranging chickens on a particular day was positively correlated with the number of chickens on the range daily in summer flocks ($r_{(30)} = 0.49$, $p < 0.05$), but not in winter flocks. Conversely, the total number of chickens that ranged only once was positively correlated with age in winter flocks ($r_{(11)} = 0.63$, $p < 0.05$), but not in summer flocks.

Chickens that accessed the range only once spent longer on the range during that visit than the average time per visit by chickens that accessed the range more than once in summer flocks (one-time ranging chickens: 31.3 ± 12.4 min/visit, more than once ranging chickens: 22.4 ± 4.3 min/visit; $F_{(1,457)} = 11.5$, $p < 0.01$), but there was no difference in winter flocks (one-time ranging chickens: 9.9 ± 4.7 min/visit, more than one-time ranging chickens: 6.6 ± 0.9 min/visit; $F_{(1,65)} = 3.46$, $p > 0.05$).

3.6. Individual Ranging Variation and Relationships with Sex

The proportion of females and males that accessed the range did not differ (winter flocks: female ranging chickens—52.9%; male ranging chickens—47.1%; $\chi^2_{(1,180)} = 1.43$, $p > 0.05$; summer flocks: female ranging chickens—57.3%; male ranging chickens—42.7%; $\chi^2_{(1,437)} = 0.05$, $p > 0.05$). The number of days that males and females accessed the range did not differ (winter flocks: females—3.7 ± 0.4 days; males—3.3 ± 0.3 days; $\chi^2_{(1,65)} = 0.46$, $p > 0.05$; summer flocks: females—7.6 ± 0.3 days, males—7.0 ± 0.3 days; $\chi^2_{(1,437)} = 2.09$, $p > 0.05$).

The overall frequency of range visits did not differ between males and females in both seasons (winter flocks: female—18.8 ± 3.2 visits, male—12.8 ± 2.6 visits; summer flocks: female—35.3 ± 2.3 visits; male—24.3 ± 2.0 visits; all $p > 0.05$). Noteworthy, when weight was not included in the analysis females (lighter in weight than males) accessed the range more frequently than males in summer flocks ($F_{(1,442)} = 8.18$, $p < 0.01$) but not winter ($F_{(1,66)} = 1.26$, $p > 0.05$).

Males spent longer on the range overall than females in winter flocks (females: 2.0 ± 0.4 h, males: 2.3 ± 0.4 h; $F_{(1,66)} = 3.92$, $p = 0.052$) but not summer (females: 11.8 ± 0.9 h, males: 9.5 ± 0.8 h; $F_{(1,442)} = 1.14$, $p = 0.29$). Males spent longer on the range per range visit than females in winter (females: 12.7 ± 7.1 min, males: 17.3 ± 5.7 min; $F_{(1,65)} = 5.8$, $p < 0.05$) but not summer (females: 20.3 ± 1.2 min, males: 27.1 ± 2.5 min; $F_{(1,442)} = 0.47$, $p > 0.05$).

4. Discussion

Our results showed that ranging behaviour varied greatly between individuals within flocks from the same hatchery, genetic lines, feed composition and availability, management regime, stock people and environmental and range conditions. In all flocks, not all chickens accessed the range and the number of visits and time spent on the range varied greatly between individuals. Although the data clearly identifies a continuum of ranging variation, we have categorized chickens in this paper for simplicity and acknowledge that such categories are arbitrary. We observed chickens that accessed the range only once, high frequency ranging chickens that accounted for one-third to half of all of the range visits (depending on season) and low frequency ranging chickens ranked in the bottom 50% of all tagged chickens that ranged but accounted for less than 15% of all range visits throughout the study.

The variation in ranging behaviour may reflect differences in the motivation to access the range. The high frequency ranging chickens accessed the range more frequently but also for a longer period of time overall and sooner after range access was first provided. High frequency ranging chickens have also been reported in commercial laying hens [13,14] and may be of particular interest to industry and consumers. Consumers that support free-range products often feel betrayed with reports of low range use in commercial flocks, leading to controversy and revised labelling regulations of free-range egg products in Australia for instance [15]. Determining the characteristics that result in high frequency range use may permit early life environmental interventions or breeding programs to encourage range use. A thorough understanding of such interventions is critical and the appropriate application will depend on the characteristics involved and the outcomes on the welfare of the chicken.

Variation in ranging behaviour between individuals may also reflect individual experiences on the range. The most interesting group in this regard is the chickens that accessed the range only once throughout the study (6–12% of tracked chickens). Although the sample size of this group was low within each flock, it was relatively similar between flocks. One-time ranging chickens were not necessarily "accidental" range users, because the duration of range visits was greater for chickens that accessed the range once compared to those that accessed the range more than once in summer flocks. Perhaps the first range visit was a frightening experience for these chickens, which may have discouraged the chicken from going out again. Indeed, there are reports of links between exposure to a range environment and fearfulness in broiler chickens [16] and the number of days an individual visits the range and fearfulness in laying hens [17–19]. However, the direct relationship between fearfulness and individual broiler chicken ranging behaviour is unknown. It would be interesting to investigate the ranging experience of these particular chickens that accessed the range only once, such as the individual's location, behaviour and environmental stimuli on the range during this single visit.

Our results demonstrated that it is important to disentangle the effects of sex and weight on ranging behaviour of broiler chickens. Females and males did not differ in their ranging frequency when weight was included in the analysis, as lighter chickens accessed the range more frequently. This suggests that weight should always be included when comparing sex effects on ranging behaviour in broiler chickens, given the marked sexual-dimorphic growth of broiler chickens. Our findings differ from Chapuis, Baudron, Germain, Pouget, Blanc, Juin and Guemene [2] who monitored a slower growing strain of broiler chicken and conversely found that males made up a higher percentage of the top ranging chickens (60%) and females made up the majority of the lowest ranging chickens (70%). Although Chapuis, Baudron, Germain, Pouget, Blanc, Juin and Guemene [2] did not control for weight, sexually dimorphic contrasts are greater in slow growing broiler than fast growing strains [20] and it is likely that the difference in findings between the two studies would be exacerbated if growth was controlled for in their study. We hypothesize that our and Chapuis, Baudron, Germain, Pouget, Blanc, Juin and Guemene [2] findings may reflect temperament differences between sexes of chickens. Independent of weight, we found that males spent more time on the range overall and per visit than females, in winter flocks. Hence, sex characteristics other than weight may be associated with ranging behaviour such as those reported in strains of laying chickens, including fearfulness, exploratory behaviour or social behaviour [21,22].

We found a high level of variation in range use between seasons and within flocks. These findings highlight the importance of monitoring individual chickens when investigating relationships between range access and welfare. For example, if we were to measure a welfare indicator on our winter flocks, the likelihood of obtaining a measure from an animal that accessed the range at least once would have been only 33%, and a low 10% chance that the chicken would have accessed the range frequently. Clearly, there is a need to determine individual ranging patterns to understand the welfare implications of range use. In addition, the welfare implications of range restriction (during periods of extreme weather conditions or prior to depopulation, a typical commercial practice) on the behaviour and welfare of chickens that differ in their ranging behaviour and motivation remains to be elucidated.

This study provides details of individual variation in the ranging behaviour of broiler chickens on a commercial free-range farm. However, this study was only conducted on one broiler strain and one farm, and the external validity of the findings to other broiler strains, geographical areas, and farms with different flock sizes and range design is unknown. Further investigation is needed to determine the causal factors for this variation, since variation was observed between individuals in the same flock, with the same breeding and hatching history, same shed and range design and similar management practices. This knowledge could lead to science-based improvements in ranging opportunities of commercial free-range broiler chickens.

5. Conclusions

Ranging behaviour varied between individuals within the same commercial flocks, revealing chickens that never accessed the range, chickens that accessed the range only once, low frequency ranging chickens, and high frequency ranging chickens, with the latter accounting for a third to a half of all range visits. Males spent more time on the range than females in winter flocks, but frequency of range visits was related to weight rather than sex in summer flocks.

These findings suggest that individual characteristics and/or early life experience partly determine ranging behaviour in commercial conditions, which subsequently results in heterogeneous flock ranging behaviour. The causes for this inter-individual variation in ranging behaviour within flocks should be investigated to ensure that chickens in free-range systems are best suited to such housing conditions and thus facilitate optimal ranging behaviour on commercial farms.

Acknowledgments: This work was partly funded by Rural Industries Research and Development Corporation (RIRDC), Chicken Meat. The authors would like to thank industry participants and staff and students the Animal Welfare Science Centre for their help with the experimental work and Michael Toscano from the Research Centre for Proper Housing: Poultry and Rabbits (ZTHZ), Division of Animal Welfare, University of Bern, Switzerland for his assistance with RFID equipment.

Author Contributions: Jean-Loup Rault and Paul H. Hemsworth obtained the funding. Peta S. Taylor, Jean-Loup Rault, Paul H. Hemsworth and Peter J. Groves conceived and designed the experiment; Peta S. Taylor performed the experiment, analysed data and drafted the manuscript; Sabine G. Gebhardt-Henrich contributed macros and assisted with RFID data and equipment. All authors contributed to writing the manuscript.

Conflicts of Interest: The authors declare no conflicts of interest. The founding sponsors had no role in the design of the study; in the collection, analyses, or interpretation of data; in the writing of the manuscript, and in the decision to publish the results.

References

1. Durali, T.; Groves, P.; Cowieson, A.; Singh, M. Evaluating range usage of commercial free range broilers and its effect on birds performance using radio frequency identification (RFID) techology. *Aust. Poult. Sci. Symp.* **2014**, *25*, 103–106.
2. Chapuis, H.; Baudron, J.; Germain, K.; Pouget, R.; Blanc, L.; Juin, H.; Guemene, D. Characterization of organic free range broiler exploratory behaviour obtained through RFID technology. In Proceedings of the 9ème Journées de la Recherche Avicole, Tours, France, 29–30 March 2011.
3. Taylor, P.S.; Groves, P.; Hemsworth, P.H.; Rault, J.L. Patterns of range access of individual broiler chickens in commercial free-range flocks. In Proceedings of the 50th Congress of the International Society of Applied Ethology, Edinburugh, UK, 12–15 July 2016.
4. Dawkins, M.S.; Cook, P.A.; Whittingham, M.J.; Mansell, K.A.; Harper, A.E. What makes free-range broiler chickens range? In situ measurement of habitat preference. *Anim. Behav.* **2003**, *66*, 151–160. [CrossRef]
5. Fanatico, A.C.; Mench, J.A.; Archer, G.S.; Liang, Y.; Gunsaulis, V.B.B.; Owens, C.M.; Donoghue, A.M. Effect of outdoor structural enrichments on the performance, use of range area, and behavior of organic meat chickens. *Poult. Sci.* **2016**, *95*, 1980–1988. [CrossRef] [PubMed]
6. Gordon, S.H.; Forbes, M.J. Management factors affecting the use of pasture by table chickens in extensive production systems. In Proceedings of the UK Organic Research 2002 Conference, University of Wales Aberystwyth, Wales, UK, 26–28 March 2002; pp. 269–272.
7. Rodriguez-Aurrekoetxea, A.; Leone, E.H.; Estevez, I. Environmental complexity and use of space in slow growing free range chickens. *Appl. Anim. Behav. Sci.* **2014**, *161*, 86–94. [CrossRef]
8. Rivera-Ferre, M.G.; Lantinga, E.A.; Kwakkel, R.P. Herbage intake and use of outdoor area by organic broilers: Effects of vegetation type and shelter addition. *NJAS Wagen. J. Life Sci.* **2007**, *54*, 279–291. [CrossRef]

9. Jones, T.; Feber, R.; Hemery, G.; Cook, P.; James, K.; Lamberth, C.; Dawkins, M. Welfare and environmental benefits of integrating commercially viable free-range broiler chickens into newly planted woodland: A UK case study. *Agric. Syst.* **2007**, *94*, 177–188. [CrossRef]
10. Nielsen, B.L.; Thomsen, M.G.; Sorensen, P.; Young, J.F. Feed and strain effects on the use of outdoor areas by broilers. *Br. Poult. Sci.* **2003**, *44*, 161–169. [CrossRef] [PubMed]
11. Howell, T.J.; Rohlf, V.I.; Coleman, G.J.; Rault, J.L. Online chats to assess stakeholder perceptions of meat chicken intensification and welfare. *Animals* **2016**, *6*, 67. [CrossRef] [PubMed]
12. De Jonge, J.; van Trijp, H.C. The impact of broiler production system practices on consumer perceptions of animal welfare. *Poult. Sci.* **2013**, *92*, 3080–3095. [CrossRef] [PubMed]
13. Gebhardt-Henrich, S.G.; Toscano, M.J.; Frohlich, E.K.F. Use of outdoor ranges by laying hens in different sized flocks. *Appl. Anim. Behav. Sci.* **2014**, *155*, 74–81. [CrossRef]
14. Larsen, H.; Cronin, G.M.; Gebhardt-Henrich, S.G.; Smith, C.L.; Hemsworth, P.H.; Rault, J.-L. Individual ranging behaviour patterns in commercial free-range layers as observed through RFID tracking. *Animals* **2017**, *7*, 21. [CrossRef] [PubMed]
15. Free Range Egg Labelling. http://www.treasury.gov.au/ConsultationsandReviews/Consultations/2015/Free-range-egg-labelling (accessed on 10 June 2017).
16. Zhao, Z.G.; Li, J.H.; Li, X.; Bao, J. Effects of housing systems on behaviour, performance and welfare of fast-growing broilers. *Asian Australas. J. Anim. Sci.* **2014**, *27*, 140–146. [CrossRef] [PubMed]
17. Hernandez, C.; Lee, C.; Ferguson, D.; Dyall, T.; Belson, S.; Lea, J.; Hinch, G. Personality traits of high, low, and non-users of a free range area in laying hens. In Proceedings of the 48th Congress of the International Society for Applied Ethology, Vitoria-Gasteiz, Spain, 29 July–2 August 2014; Estevez, I., Manteca, X., Marin, R., Averos, X., Eds.; 2014; p. 89.
18. Hartcher, K.M.; Hickey, K.A.; Hemsworth, P.H.; Cronin, G.M.; Wilkinson, S.J.; Singh, M. Relationships between range access as monitored by radio frequency identification technology, fearfulness, and plumage damage in free-range laying hens. *Animal* **2016**, *10*, 847–853. [CrossRef] [PubMed]
19. Campbell, D.L.M.; Hinch, G.N.; Downing, J.A.; Lee, C. Fear and coping styles of outdoor-preferring, moderate-outdoor and indoor-preferring free-range laying hens. *Appl. Anim. Behav. Sci.* **2016**, *185*, 73–77. [CrossRef]
20. Fanatico, A.C.; Pillai, P.B.; Cavitt, L.C.; Owens, C.M.; Emmert, J.L. Evaluation of slower-growing broiler genotypes grown with and without outdoor access: Growth performance and carcass yield. *Poult. Sci.* **2005**, *84*, 1321–1327. [CrossRef] [PubMed]
21. Vallortigara, G.; Cailotto, M.; Zanforlin, M. Sex differences in social reinstatement motivation of the domestic chick (gallus gallus) revealed by runway tests with social and nonsocial reinforcement. *J. Comp. Psychol.* **1990**, *104*, 361. [CrossRef] [PubMed]
22. Jones, R.B. Sex and strain differences in the open-field responses of the domestic chick. *Appl. Anim. Ethol.* **1977**, *3*, 255–261. [CrossRef]

8

Welfare Status of Working Horses and Owners′ Perceptions of Their Animals

Daniela Luna [1], Rodrigo A. Vásquez [2], Manuel Rojas [3] and Tamara A. Tadich [4,*]

1 Programa Doctorado en Ciencias Silvoagropecuarias y Veterinarias, Universidad de Chile, Santa Rosa 11315, La Pintana, Santiago 8820000, Chile; danluna@veterinaria.uchile.cl
2 Instituto de Ecología y Biodiversidad, Departamento de Ciencias Ecológicas, Facultad de Ciencias, Universidad de Chile, Las Palmeras 3425, Ñuñoa, Santiago 7800003, Chile; rvasquez@uchile.cl
3 Departamento de Ingenieria Industrial, Facultad de Ciencias Físicas y Matemáticas, Universidad de Chile, Beauchef 851, Santiago 8370456, Chile; manuelrojas@uchile.cl
4 Departamento de Fomento de la Producción Animal, Facultad de Ciencias Veterinarias y Pecuarias, Universidad de Chile, Santa Rosa 11735, La Pintana, Santiago 8820000, Chile
* Correspondence: tamaratadich@u.uchile.cl

Simple Summary: Appropriate strategies aimed at improving the welfare of working horses should contemplate the assessment of welfare status, as well as the evaluation of the human–animal relationship within each geo-cultural context. We assessed and compared the welfare status of working horses in two administrative regions of Chile and explored the nature of the owner–horse relationship from the perspective of the owner. The overall prevalence of health problems and negative behavior responses was low. However, significant differences between regions exist in the presence of lesions and the person responsible for managing horseshoeing. Two differing views were found regarding the owners' perception of their horse: predominantly affective or instrumental. Despite the instrumental perception predominantly residing in one region, the affective perception was widely shared by owners in each region. The findings suggest that Chilean working horses have a, generally, good welfare and that the development of an affective owner–horse relationship is possible. Additionally, the results suggest that affective and instrumental perceptions of these animals can coexist.

Abstract: Appropriate interventions to improve working equine welfare should be proposed according to scientific evidence that arises from different geo-cultural contexts. This study aims to assess and compare the welfare status of working horses in two administrative regions of Chile and to determine how owners perceive their horses. Horses' welfare status was assessed through direct indicators (direct observation and clinical examination) and indirect indicators (an interview with the owner). Owners′ perceptions of their horses were determined through a discourse analysis of their statements. In total, 100 horses and 100 owners were assessed. Results showed a low prevalence of health problems and negative behavior responses among horses in the two regions evaluated. Significant associations were found between inadequate body condition and the absence of deworming, and between hoof abnormalities and a low frequency of shoeing. Between regions, significant differences were found in the presence of lesions and the person responsible for horseshoeing. In regards to the owners′ appreciations, two differing perceptions of working horses were found: a predominantly affective perception and a perception of the animal as a working instrument. Although the instrumental perception was more frequent in the Araucania region, the affective perception was widely shared by both owner populations. The results reveal a good welfare status in working horses and suggest that both affective and instrumental perceptions of these animals can coexist.

Keywords: equine welfare; working horses; urban draught horses; semantic analysis; human-animal relationship

1. Introduction

In developing countries such as Chile [1], working animals provide an essential resource of power for millions of people who live in poverty [2,3]. In the case of working equids, there is increasing evidence of their socioeconomic contribution to human livelihood through their direct and indirect impact in generating income for thousands of households worldwide [3,4]. It has been reported that the welfare state of these equids is usually poor and impacts directly on their health, mental state and working capacity [5–7]. This may seriously compromise the well-being of these animals and the families they work for. For this reason, the World Organization for Animal Health (OIE) recently decided to develop the first welfare standards for working equids used for traction, transport and income generation [8]. However, it is important to highlight that welfare problems, husbandry practices, and the role that these working animals play can vary between countries, through time, and even within the same community or locality [5,6]. Consequently, appropriate intervention strategies to improve the welfare of these animals should be proposed and implemented according to, at least, two criteria: the main welfare problems found in the different geo-cultural contexts, and the assessment of the quality of the human–animal relationship [5].

The welfare status of working horses in Chile has been previously assessed [9–11], but the influence of geographic and cultural differences on these animals' welfare is unknown. Differences within a country could modify the risk factors associated with the animals' welfare. For example, heat stress and dehydration are conditions that negatively affect the welfare of equids in countries with arid climates, such as India or Pakistan, where ambient temperatures of up to 48 °C are found [5]. However, these problems have not been observed in countries that have internally varying climatic and geographic conditions, such as Chile. For example, the Metropolitana de Santiago region has the largest urban population in the country and is characterized by a warm, temperate climate with a prolonged dry season [12]. In comparison, the Araucania region is one of the poorest regions and has the highest percentage of rural population. It is characterized by four types of climate, predominantly a rainy, temperate climate [13]. Moreover, this region has the largest number of individuals belonging to the Mapuche group, an indigenous, ethnic population that preserves their ancient traditions [14].

Strategies oriented to improve the welfare of animals, their owners and caretakers, also require an appropriate understanding of the human–animal relationship and the multiple factors that modulate it [15]. To form these strategies, the motivational considerations (bases) that underlie attitudes towards animals must be identified [16], primarily the emotional or instrumental ways in which people relate to animals [17,18]. Attitudes toward animals are often the focus of human–animal interaction studies [19–22] and, more recently, of welfare studies [23]. However, most research on human–horse relationships and the perceptions of horses has been centered in the equestrian world (horses that are kept primarily for recreational riding or competition) [18,24–26] and there is little information on working equids despite the important implications that this knowledge holds for improving these animals' welfare. Moreover, welfare studies on working horses have not addressed the importance of owner's attitudes and perceptions on their interactions with their animals [4–7,9].

Most studies that focus on the perception of horses have been limited to the use of traditional qualitative discourse analysis to determine the diverse perceptions and conceptions that people have in relation to these animals [18,25–27]. None of these studies have explored the representation of the meaning of words from other perspectives, such as Latent Semantic Analysis (LSA). LSA is a mathematical tool that has been proposed by psychology researchers as a method for extracting and representing the meaning of words [28] obtained in written texts, interviews or free text surveys [29,30]. In contrast to traditional discourse analysis, LSA is an interesting tool for inferring much deeper relationships between words and results in better predictions of human judgments [28]. Determining, through LSA, how owners or caregivers perceive their animals could reflect the nature of the owner–horse interaction and the animals' role in their lives. It could also infer the owner's motivation for improving the well-being of their horse.

Taking this into account, the aims of this study were, firstly, to assess and compare the welfare status of working horses in two different administrative regions of the country: one of which is closer to an ethnic-rural background (Araucania region) than the other (Metropolitana de Santiago region), and to determine whether any associations between direct and indirect welfare indicators exist. Secondly, we aimed to determine owners′ perceptions of their horses in both regions.

2. Materials and Methods

This study was carried out in peri-urban neighborhoods in two regions of Chile: the Metropolitana de Santiago region (33°26′16″S 70°39′01″O) and the Araucania region (38°54′00″S 72°40′00″O), between March of 2015 and January of 2016. The welfare assessment protocol was approved by the Animal Use and Care Ethical Committee of the School of Veterinary Sciences of the University of Chile (N° 06-2015). Owner participation was voluntary after signing an informed consent agreeing to participate in the study under the understanding that no economic benefit was involved. The owners were informed of all the aims of the study.

Due to a lack of information on the number of working horses in Chile, a convenience sample size was used. To localize owners and their horses, an electronic consultation with local municipalities was held to geographically locate the cities in Chile where a significant number of these horses were currently working. The researchers then visited the residences of the owners to invite them to participate in this study.

2.1. Welfare Assessment Protocol for Working Equines

A total of 100 urban draught horses (48 from the Metropolitana de Santiago region and 52 from the Araucania region) were assessed using a welfare assessment protocol for working equids, based on previously published literature [4,5,9,31]. When the owner had more than one horse, one horse per owner was randomly selected for the analysis. This protocol included a set of direct indicators, such as health parameters and behavioral observations, for assessing the general welfare status of horses (Tables 1 and 2). In addition, indirect indicators, such as resource-based measures (the provision of food and water, management practices, etc.) were included (Table 3). Additional information about the general characteristics of the horses, such as age, sex, conformation type and estimated live weight, were also recorded. These indicators were evaluated at the owners' households. First, the behavioral assessments were performed through direct observation. The health parameters were then assessed through clinical examination (direct indicators). Finally, each horse owner answered a specific questionnaire to obtain general characteristics of their horse, and the main resources and management that they received (indirect indicators).

Table 1. Description of the direct health parameters applied.

Welfare Indicators	Categorization	Description
Skin lesion	Present/absent	Wounds of any size and severity were recorded according to their location. Lesions at labial commissures of the mouth were also included.
Body condition score	Adequate/inadequate	Assessed on a five-point scale from 1 (emaciated) to 5 (obese) including half scores [6,32]. Scores of 3, 3.5 and 4 were considered adequate.
Hoof health	Adequate/inadequate	Quality, shape and conformation of hoofs were assessed. The hooves were considered adequate if these were round and smooth, had no cracks or sections missing, and did not show defects of the hoof capsule [31,33].
Coat and skin condition	Adequate/inadequate	The coat and skin condition was recorded adequate if the hair coat was uniform, with a general healthy aspect (shiny), without dryness or dirt (mud or feces) or presence of ectoparasites of any species (in hair or skin) [5,33].
Gait abnormalities	Present/absent	Assessed by observation of the horse while walking in a straight line for approximately 20 meters. The observer assessed presence of lameness, uneven stride, reluctance to put weight on one or more limbs, uneven head-nodding or hip movement [31].

Table 2. Description of the direct behavioral observations applied.

Indicator	Categorization	Description
General attitude	Alert/apathetic or depressed	The horse was observed (only by observer) from a distance of 3 to 5 meters for 60 seconds [4]. The horse's response was categorized as: Alert: when the animal was attentive and responds to the different stimuli of the environment (eyes wide open, active movement of the ears, head, tail and/or skin to keep away flies) [4]. Apathetic or depressed when it showed decreased responses to the environmental stimuli (head lowered, eyes half closed, complete or partial cessation of tail and skin movements to avoid insects, reduced ear movement) [4,31]. Apathetic and depressed were combined in the current study based on criteria of Burn et al. [31].
Approximation test	Indifference/friendliness/avoidance/aggressiveness	The observer approached at an angle of approximately 20° to the sagittal plane of the animal's body and stopped at a distance of 30 cm from the head of the horse [4]. The observer recorded the horse's response at the moment that he stopped. The owner was instructed to perform exactly the same procedure and then the observer recorded the animal's response [4]. Responses were recorded as: Indifference: Immobile and relaxed without attempts to approach or move away from the observer/owner, depressed or relaxed body position and facial expression (with or without the ears moving, relaxed lips, possibly eyes half closed) [4]. Friendliness: Movement of the head toward the observer/owner, with relaxed face and normally open eyes, ears turned forward, no wrinkling around the mouth or nostrils [4]. Avoidance: The horse is immobile with a tense body position and facial expression (head up, eyes wide open and lips held tight) or the animal turning the head or attempts to move away from observer/owner [4]. Aggressiveness: The horse attempts to kick or bite, eyes fully opened and head oriented toward observer/owner, nostrils are dilated with or without wrinkles around the mouth, may paw or stomp the ground [4].
Walk down side	Indifference/friendliness/avoidance/aggressiveness	The observer walked alongside the horse toward its rear and back again, maintaining a distance of 30 cm from its body, then the observer recorded the horse's response [4]. The owner was instructed to perform the same procedure. The horse's response was categorized exactly as in the approximation test [4].
Chin contact	Accepts/avoids	The observer slowly placed their hand under the animal's chin and assessing if the horse accepted or avoided the contact [31]. The owner was instructed to perform the same procedure. The horse's response was categorized as accepts or avoids the contact [31].
Allows to pick up a limb	Accepts/avoids	The observer assessed if the horse resisted or not the lifting up of their left front limb. The owner was instructed to perform the same procedure.

Table 3. Description of the indirect indicators (resources and management) applied.

Welfare Indicators	Categorization	Description
1. Feeding practices		
Frequency of feeding	Once a day/twice a day/three or more per day	The owner was asked how many times per day he/she supplied water to their horse.
Water availability	*Ad libitum/not ad libitum*	The owner was asked if their horse had water available *ad libitum* when not working.
2. Working practices		
Frequency of use per day	Days per week	The owner was asked about how many days per week he/she uses the horse for work
Frequency of use per week	Hours per day	The owner was asked how many hours per day he/she uses the horse for work.
Work type	Type of load	The owner was asked about the activities in which he/she uses the horse.
3. Shoeing practices		
Frequency of shoeing	Every 15/between 16–30/>30 days	The owner was asked about the frequency that his/her horse is shod.
Responsible person	Farrier/owner	The owner was asked about the main person responsible of the shoeing of the horse.
4. Preventive management		
Deworming	Never/<6 month/>6 month	The owner was asked when was the last time his/her horse was dewormed. The response was categorized as never; less than 6 months ago; or more than 6 months ago.
5. Veterinary consultation	Never/<1 year/>1 year	The owner was asked about the last time his/her horse was examined by a veterinarian. The response was categorized as never (if the horse has never been examined by a veterinarian); less than a year ago; or over a year ago.

2.1.1. Direct Welfare Indicators

An assessment of horses' health status and behavioral parameters was performed by an observer (a veterinarian) using a standardized protocol (Tables 1 and 2). The presence and location of skin lesions were recorded, including lesions at the labial commissures of the mouth. A Body Condition Score (BCS) was recorded for each horse using a standard scoring scale from 1 (emaciated) to 5 (obese) [32], including half scores [6]. The assessment of hoof shape, conformation and quality was based on the criteria of Popescu et al. [33] and Burn et al. [31] (see Table 1). Hair coat and skin condition, including the presence of external ectoparasites, were assessed based on the criteria of Pritchard et al. [5]. Gait abnormalities (lameness) were assessed and recorded by observing a horse walking in a straight line for approximately 20 m. The presence of lesions, hoof abnormalities, coat and skin conditions, and gait abnormalities were recorded as either present or absent, and adequate or inadequate, in terms of whether an indicator was altered or within the normal range. Body condition was recorded as a score and either adequate or inadequate (scores of 3, 3.5 and 4 were considered adequate, whereas scores of 1, 1.5, 2, 2.5, 4.5 and 5 were considered inadequate). The health parameters were also assessed in the same process as described previously.

Observations of horses′ general attitudes (alert, apathetic or depressed), horses′ responses to the approach of both the observer and the owner, horses′ responses to the observer and the owner walking down the animal's side ("walk-by"), and horse's responses to chin contact by the observer and the owner were made similarly, in accordance to the welfare assessment protocol as described by Popescu and Diugan [4] and Burn et al. [31] (see Table 2). In the approach and "walk-by tests", each horse's response toward the observer and the owner was categorized as indifference, friendliness, avoidance or aggressiveness, based on the criteria of Popescu and Diugan [4] (see Table 2). The horse's response in the "chin contact test" was categorized as avoidance or acceptance based on the criteria of Burn et al. [31] (Table 2). Finally, the "picking up a limb test" was also included in this study. Each horse's response towards the owner and the observer, when they picked up a limb, was recorded as either avoidance or acceptance. This was done in order to assess the response towards a common handling routine (cleaning of the hoof and shoeing). The behavioral tests were made first by the observer and then by the owner. Each horse's response was observed and recorded by the observer. The only procedure not made by the owner was that of general attitude (Table 2).

2.1.2. Indirect Welfare Indicators

A total of 100 urban working horse owners (n = 48 from the Metropolitana de Santiago region and n = 52 from the Araucania region), most of them men (n = 90) ranging in age from 17 to 83 years (average = 43; SD = 15.03), participated in this study. Each owner was interviewed using a standardized, structured questionnaire, which included a combination of open and closed questions to register information about feeding practices (frequency of feeding, water availability), working practices (work type and frequency), horse shoeing practices (frequency), preventive managements (deworming), and veterinary consultation (Table 3).

2.1.3. Horses′ General Characteristics

(a) Age, determined by the horse's history as recounted by the owner and confirmed by the inspection of teeth; (b) Sex, recorded by observing the external genitalia; (c) Anamorphosic Index (AI), calculated to establish whether the horse had a speed or draught type morphology, based on the equation and criteria described by Cassai [34]: $(HG)^2/HW$, where HG is the heart girth and HW is the height to the withers, expressed in meters. The equine is considered to be of draught type if the AI is greater than 2.12, and a speed type if it is lower. (d) Estimated live weight, using the modified equation for Chilean horses described by Meyer [35]: $HG^2 \times EIL/11.462$, where HG is the heart girth and EIL the shoulder-*tuber ischii* length, expressed in centimeters.

2.2. Owners′ Perception of Their Horses

After recording the direct and indirect welfare indicators, owners were asked to answer the following open question: "What does your working horse mean or represent for you?". Owners with more than one horse were asked to answer for the totality of their horses. This broad question gave owners the opportunity to express, in their own words, their subjective perception and point of view about how they conceive and conceptualize their working horse. Such an approximation has been used in other studies, such as Birke [26], Birke et al. [25] and Shuurman [18]. The owners′ responses were recorded and transcribed verbatim by the researcher.

2.3. Statistical Analysis

The data from 100 horses and 100 owners was incorporated and stored in an Excel spreadsheet (Microsoft Office Excel® 2013) and then exported to SPSS (IBM version 22.0.0.0 for Windows, Armonk, NY, USA) for further statistical and graphical analysis. Descriptive statistics (means, standard deviation and percentages) were used to summarize the information on the general characteristics and welfare state of horses for each location.

The Wilcoxon rank sum test and the Student's t-test were applied to determine significant differences between regions for the variables: age, estimated live weight, AI, BCS, feeding frequency, work and shoeing frequency. The association between the frequency of each indicator and location, as well as interactions between animal-based and resource-based information, were examined using the Chi-squared test and Fisher's exact test. A statistical significance level of $p < 0.05$ was established.

In order to determine the differences in the conceptions and interpretations of working horses embedded in the statements of owners, a linguistic analysis of each owner's discourse, through text analytics (TA), was applied. These analyses included cluster analysis, correlations of terms (Spearman's correlation) and Latent Semantic Analysis (LSA).

Latent Semantic Analysis is a natural language processing technique developed by Landauer and Dumais [36]. The technique allows for a mathematical representation of the relationship of meaning between words and sentences contained in a written text. Initially developed for library indexing, it is now applied to the analysis of quantitative literature reviews, textual data in computer-mediated communication, interviews, and management of knowledge repositories [29]. This is all possible through the statistical techniques that identify relationships between sentences in a collection of documents, thus generating a specialized domain of evaluation that allows an analysis to be performed.

The LSA was used to identify the degree of proximity of words within the semantic space evaluated. In this study, the LSA was applied in conjunction with cosine measurement to determine the contextual meaning and latent relationships behind the words of owners, specifically to determine concepts associated with the term "horse" in each region.

In the determination and implementation of the different semantic algorithms of this study, the following tools and software-type product-license and versions were used: (a) System Operative Linux Debian-Ubuntu, GPL-GNU (General Public License v3.0), kernel 3.13.0-35-generic; (a: Linux open source, New York, USA) (b) R-CRAN, Cluster of library of analysis and modeling, v3.1.1; (c) IDE RSTUDIO, Integrated Developmental Environment-GNU, v.098.1028; (d) Package tm, Library R-CRAN-GNU, v0.6-2; (e) Package lsa, Library R-CRAN-GNU, v0.6-2; (f) Package ggplot2, Library R-CRAN-GNU, v1.0.0; (g) Package igraph, Library R-CRAN-GNU, v0.7.0. (b–g: The R Project, open source, The R Foundation, Vienna, Austria)

The Unix-based operating system, and particularly Linux, added to the tools used which are supported under the GLP-GNU licenses-General Public License V3.0-features that allow them to be used to study, share and copy, as well as modifying the software. This environment allows an open implementation and the ability to make the modifications and adjustments necessary to achieve the integration of different tools.

3. Results

3.1. General Characteristics

The average age of the study population of working horses was 8.7 years (range = 2–25; SD = 4.5 years). Sixty-one percent of horses were mares, 29% geldings, and 10% stallions (Table 4). Most horses (70%) had a speed type conformation according to their anamorphosic index and the average estimated live weight was 413 kg (range = 185–632; SD = 82 kg). General characteristics of horses within each region are shown in Table 4. Between regions, no significant differences were found in relation to age (Wilcoxon rank sum test, $p = 0.572$), estimated live weight (Student's t–test, $p = 0.315$), and the percentage of horses with draught conformation ($X^2 = 2.21$; $p = 0.190$). However, there was a higher frequency of geldings in the Araucania region, compared to the Metropolitana de Santiago region (Fisher's exact test, $p = 0.001$), which, in turn, presented a higher frequency of stallions (Fisher's exact test, $p = 0.006$).

Table 4. General characteristics of urban working horses ($n = 100$) assessed from the Metropolitana de Santiago ($n = 48$) and Araucania ($n = 52$) regions in Chile. Results are expressed as average, standard deviation (SD), range, percentage (%) and number (n).

Descriptor	Metropolitana de Santiago ($n = 48$)	Araucanía ($n = 52$)	Total ($n = 100$)
Average age (years (SD))	8.1 (3.7) [a]	9.2 (5.1) [a]	8.7 (4.5)
Age range	2–15	2.5–25	2–25
Estimated live weight average (kg (SD))	388 (81.4) [a]	436 (76.1) [a]	413 (82)
Anamorphosic index adequacy for draught activities (% (n))	23 (11) [a]	37 (19) [a]	30 (30)
Geldings (% (n))	10 (5) [a]	46 (24) [b]	29 (29)
Stallions (% (n))	19 (9) [a]	2 (1) [b]	10 (10)
Mares (% (n))	71 (34) [a]	52 (27) [a]	61 (61)

[a, b] Different letters denote significant differences ($p < 0.05$) between administrative regions.

3.2. *Working Horse Welfare Assessment: Direct and Indirect Indicators*

The summarized results of health indicators and behavioral parameters assessed in working horses in each region are shown in Tables 5 and 6, respectively. Significant differences between regions are also indicated.

The evaluation of health indicators in the total population showed that most horses (83%) had an adequate body condition score (average = 3.3; SD = 0.56; range = 2–5). No significant differences (Wilcoxon rank sum test, $p = 0.08$) were found between regions in relation to the BCS of the horses. The main welfare problems found were hoof abnormalities (53%) and the presence of skin lesions (47%), which were mostly simple excoriations located on harness-related areas. Other less frequent problems found were inadequate skin or coat condition (14%), limb-associated abnormalities such as gait abnormality and lameness (13%), and lesions at the labial commissures of the mouth (3%). Significant differences between regions were found only in the presence of lesions: horses from the Metropolitana de Santiago region had a significantly higher ($X^2 = 4.7$; $p = 0.03$) frequency of skin lesions, primarily in the head and neck area (Fisher's exact test, $p = 0.0003$).

Table 5. Descriptive statistics of health indicators of 100 draught horses assessed from the Metropolitana de Santiago ($n = 48$) and Araucania regions ($n = 52$) in Chile, expressed in number (n) and percentage (%) within each region. Significant differences between regions are also shown.

Indicators	Metropolitana de Santiago n (%)	Araucania n (%)	Total n (%)	p-Value
Inadequate body condition score	7 (15)	10 (19)	17 (17)	0.53
Presence of body lesions (skin)	30 (63)	17 (33)	47 (47)	<0.05
Lesions at the labial commissures	3 (6)	0	3 (3)	0.10
Head/neck	17 (35)	3 (6)	20 (20)	<0.001
Breast/shoulder	9 (19)	6 (12)	15 (15)	0.31
Thorax/abdomen	13 (27)	11 (21)	24 (24)	0.48
Hindquarters/tail base	9 (19)	4 (8)	13 (13)	0.13
Forelegs/hindlegs	10 (21)	4 (8)	14 (14)	0.08
Abnormal coat and skin	8 (17)	6 (12)	14 (14)	0.46
Abnormalities of hoof	25 (52)	28 (54)	53 (53)	0.85
Abnormal gait/lameness	9 (19)	4 (8)	13 (13)	0.13

Table 6. Descriptive statistics of behavioral indicators of 100 draught horses assessed from the Metropolitana de Santiago ($n = 48$) and Araucania regions ($n = 52$) in Chile, expressed in number (n) and percentage (%) within each region. Significant differences between regions are also shown.

Indicators	Metropolitana de Santiago n (%)	Araucania n (%)	Total n (%)	p-Value
General attitude				
Alert	46 (96)	51 (98)	97 (97)	0.60
Apathetic/depressed	2 (4)	1 (2)	3 (3)	0.60
Response to observer approach				
Indifference	11 (23)	1 (2)	12 (12)	<0.001
Friendly	30 (63)	44 (85)	74 (74)	<0.01
Avoidance	3 (6)	6 (12)	9 (9)	0.49
Aggression	4 (8)	1 (2)	5 (5)	0.19
Response to owner approach				
Indifference	8 (17)	2 (4)	10 (10)	<0.05
Friendly	32 (67)	46 (88)	78 (78)	<0.01
Avoidance	5 (10)	3 (6)	8 (8)	0.47
Aggression	3 (6)	1 (2)	4 (4)	0.34
Response to observer walking down side				
Indifference	16 (33)	5 (10)	21 (21)	<0.01
Friendly	26 (54)	38 (73)	64 (64)	<0.05
Avoidance	2 (4)	6 (12)	8 (8)	0.27
Aggression	4 (8)	3 (6)	7 (7)	0.70
Response to owner walking down side				
Indifference	9 (19)	2 (4)	11 (11)	<0.05
Friendly	32 (67)	44 (85)	76 (76)	<0.05
Avoidance	3 (6)	5 (10)	8 (8)	0.71
Aggression	4 (8)	1 (2)	5 (5)	0.19
Response to observer making chin contact				
Acceptance	36 (75)	41 (79)	77 (77)	0.64
Avoidance	12 (25)	11 (21)	23 (23)	0.64
Response to owner making chin contact				
Acceptance	39 (81)	43 (83)	82 (82)	0.85
Avoidance	9 (19)	9 (17)	18 (18)	0.85
Response to observer picking up a limb				
Acceptance	44 (92)	49 (94)	93 (93)	0.70
Avoidance	4 (8)	3 (6)	7 (7)	0.70
Response to owner picking up a limb				
Acceptance	45 (94)	50 (96)	95 (95)	0.66
Avoidance	3 (6)	2 (4)	5 (5)	0.66

The prevalence of behavioral responses displayed by the horses toward their owner and the observer for the five behavioral tests applied are shown in Table 6. Most horses had an alert attitude (97%) and presented positive (friendly) responses towards both the owner and the observer. No significant differences ($p > 0.05$) were found in a horse's responses towards their owner and the observer in the four behavioral tests. Regarding the differences between regions, there was a significantly higher ($p < 0.01$) frequency of horses with friendly responses toward both the observer and the owner in the approach test and the walk-by test in the Araucania region in comparison to the Metropolitana de Santiago region (see Table 6). Aggressiveness, avoidance and indifference were the least frequent reactions in all tests, with no significant differences between regions ($p > 0.05$) (Table 6).

No significant difference (Wilcoxon rank sum test, $p = 0.15$) was found between the two regions regarding feeding frequency. The availability of water had no significant association (Fisher's exact test, $p = 0.18$) with the horses' regions of origin. Most horses (83%) were fed twice or three times per day and almost all of them (90%) had access to drinking water throughout the entire day, as declared by the owner. Regarding equine healthcare, approximately half of the horses (49%) had been dewormed within the last 6 months. However, 17% of them had never received an antihelmintic drug and 42% of horses had never been examined by a veterinarian. In relation to shoeing practices, all owners declared that their horses were periodically shod, every 15, 16–30 or more than 30 days (14%, 33% and 53% respectively). Within this sample, 48% were shod by the owner and 52% by a farrier.

However, only 75% of horses were shod at the moment of inspection. Within the horses that were not shod (25%), 52% of them had a poor hoof quality. A significantly higher number of horses (75%) (X^2 = 31.64; p = 0.000) were shod by owners in the Araucania region, while in the Metropolitana de Santiago region most horses (81%) were shod by a farrier. No significant difference (Wilcoxon rank sum test, p = 0.51) was found in the frequency of shoeing. Most owners reported that their horses were used for transporting their families (52%) and for carrying diverse types of loads. The diversity of products transported varied depending on the region where they were located. In the Araucania region, horses were mainly used for carrying agricultural products (58%) and wood for construction and fuel (42%). In the Metropolitana de Santiago region, horses were used for transporting potting soil (27%), agricultural products for sale in markets (27%), and rubble (8%). The average frequency of use was 3.25 days per week (SD = 1.65; range = 1–7) at an average of 3.87 hours per day (SD = 2.03; range = 1–12). No significant differences in the frequency of hours per day of work (Wilcoxon rank sum test, p = 0.58) and of days per week (Wilcoxon rank sum test, p = 0.11) were found between regions.

3.3. Interactions between Animal-Based Information and Indirect Indicators

The presence of skin lesions was not significantly associated with any of the following variables: poor body condition score (Fisher's exact test, p = 0.06), inadequate conformation for draught activities (X^2 = 0.37; p = 0.54), work frequency greater than 5 days/week (Fisher's exact test, p = 0.13) or more than 6 hours/day (Fisher's exact test, p = 0.10), nor individuals older than 15 years of age (Fisher's exact test, p = 1.00). However, there was a tendency for horses to present lesions on the skin when in a poor body condition.

Inadequate body condition (found in 17% of the horses) was significantly associated (X^2 = 4.86; p = 0.02) with the absence of deworming. However, no association (Fisher's exact test, p = 0.14) was found between poor body condition and feeding less than two times per day, work frequency more than 5 days per week (Fisher's exact test, p = 0.69) or more than 6 hours per day (Fisher's exact test, p = 0.73), nor with individuals older than 15 years of age (Fisher's exact test, p = 1.00).

Hoof abnormalities were significantly associated (X^2 = 5.63; p = 0.01) with a low frequency of shoeing (over 30 days). In this study, no significant association (X^2 = 0.05; p = 0.82) was found between an appropriate hoof condition and a history of the horse being shod by a farrier. There was no association between the presence of lameness and hoof abnormalities (Fisher's exact test, p = 0.36).

Most horses had an alert attitude (97%), which was significantly associated with unlimited access to drinking water during the day, while they were not working. (Fisher's exact test, p = 0.03).

3.4. Perceptions of Horses

Linguistic (semantic) analyses carried out showed that horse owners' discourses were characterized by a high degree of subjectivity, with a high recurrence of terms such as "partner", "like", "friend", "my life", "all for me" or "family member" (e.g., "He is a family member, as a son"). Moreover, this subjectivity can be reflected in that all horses had a name given by the owner (e.g., Shakira or Gringo). In addition, 60% of the owners (63% from the Metropolitana de Santiago and 58% from the Araucania region) used possessive adjectives, such as "my horse is . . . " or "they are my . . . ", in their declarations.

Cluster analysis was used to identify related concepts or words according to their proximity in the owners' answers. As shown in Table 7, both groups of owners tended to conceptualize their horses as either a friend, a family member, part of the home and/or a working tool. Furthermore, they recognized psychological attributes in their horses, such as the capacity for understanding or loyalty, and stated that they like having a horse. Some differences were observed between regions; five out of nine of the word clusters from the Metropolitana de Santiago region identified concepts that reflected the willingness of owners to care for their animals (e.g., "feed them", "take care of them") and indicated that horses were perceived mainly as a source of recreation for their owners (e.g., "hobby", "toy", "pets", "favorite"). On the other hand, most word clusters (seven) from Araucania region identified a wider range of concepts that reflected the use of animals as the owners' main source of subsistence (e.g., "transport", "food", "we eat", "source", "bread", "plough").

A Semantic Correlation Analysis among the most frequently used words by horse owners of the Metropolitana de Santiago region revealed a significant and very high correlation ($r = 1.00$; $p < 0.05$) between the following terms: "family" and "house", "favorite" and "pet", "distracts" and "take care of them", "feed them" and "pet", and "favorite" and "feed them". Terms such as "feed them", "take care of them", "distracts", "pet" and "favorite" were highly correlated with the term "like" ($r = 0.7$; $p < 0.05$) and the terms "livelihood" and "tool" were moderately correlated ($r = 0.57$; $p < 0.05$).

In the Araucania region, analysis of semantic correlation showed a very high and significant correlation ($r = 1.00$; $p < 0.05$) between the terms: "provide" and "food", "transport" and "like", "apart" and "brother", "transport" and "house", and "bread" and "feeding". A high correlation ($r = 0.7$; $p < 0.05$) was found between: "source" and "feeding", "tool" and "food", "income" and " plough", "family" and "apart", "source" and "we eat", "brother" and "family", and "unique" and "understands".

Acording to the LSA cosine value, the terms used by the owners of the Metropolitana de Santiago region, such as "loyal" (cosine value 1.00) and "favorite" (cosine value 0.005), showed the closest proximity to the term "horse". In the Araucania region, the words with the strongest proximity to the word "horses" were "family" (cosine value 0.99), "brother" (cosine value 0.98) and "friend" (cosine value 0.30).

Table 7. Word clusters originated from working horse owners' answers to the open question "What does your working horse mean or represent for you?", according to the location of origin, either Metropolitana de Santiago region ($n = 48$) or Araucania region ($n = 52$).

Clusters	Metropolitana de Santiago Region	Araucania Region
cluster 1	friend, animals, horses	son, horse, feeding
cluster 2	tools, animals, feed them	house, like, transport
cluster 3	hobby, toy, time	friend, rescuer, feeding
cluster 4	loyal, horse, feed them	food, tool, bread
cluster 5	feed them, like, pets	unique, understands, life
cluster 6	horse, feel, feed them	we eat, eat, source
cluster 7	tool, animal, livelihood	foods, provides, feeding
cluster 8	like, feed them, take care of them	friend, apart, horse
cluster 9	friend, favorite, home	source, plough, feeding

4. Discussion

Working equids contribute directly and indirectly to the livelihoods of the poorest communities around the world [4,32]. More specifically, they contribute to income-generating activities through the transport of goods, people, water, agricultural products, and building material. They also provide draught power for agriculture, among other activities [5,8,9,37,38].

In the present study, most working horses assessed had a good welfare status. This is contrary to reports from other studies that describe the welfare status of these animals as often poor [5,6,38–40]. This may be the result of labor demands by their owners and their relationship with people and their environment [41]. Additionally, welfare status may be affected by individualities (i.e., personality).

In this study, most owners showed a preference for using mares (61%), differing from previous studies where geldings were more frequent [9,10]. This could be due to owners' considering the reproduction of their animals as an extra source of income through the sale of the offspring or as an asset, by keeping the newborns as replacement. Lanas et al. [11] reported a similar preference for mares in peri-urban areas of central and southern Chile.

Age is an important factor that may affect the welfare of working equids. McLeod [42] suggests that horses work better between 4–12 years of age. This is because horses reach their zootechnical maturity at approximately 4 years of age. In comparison, after 12 years of age, work, efficiency progressively decreases [42,43]. The average age of horses in this study was 8.7 years, which is similar to findings for other areas of Chile [9–11,42]. In addition, the majority of the horses (68%) included in this study were within the optimum age range, which as Sáez et al. [10] pointed out, has positive implications for the welfare of these animals.

A small percentage of horses examined in this study (17%) had an inadequate BCS, similar to the findings of Tadich et al. [9] in urban draught horses in southern Chile and contrary to those reported by Pritchard et al. [5], where the majority of working horses (70%) in Middle East and Central Asia had a BCS of 2 or less. It has been described that a low BCS can be caused by multiple factors, such as malnutrition, overwork, parasitism and diseases [7]. However, in this study, the only factor significantly associated with a poor BCS was the absence of deworming practices, found in 17% of all horses. Therefore, educational programs for owners focusing on good husbandry and healthcare practices should be implemented, with the purpose of improving owners′ understanding on how these practices can potentially affect the health and well-being of their animals.

Lameness, hoof abnormalities and poor hoof care are problems widely reported in working equines [6,9], mostly in urban or peri-urban areas [44]. In this study, hoof abnormalities were the most frequent problem, with no differences between the administrative regions studied. The majority of horses included in this study (53%) were shod infrequently, at intervals longer than 30 days. Additionally, there was a significant association between hoof abnormalities and shoeing intervals higher than 30 days. This low frequency of shoeing could be attributed to the lack of knowledge of owners about hoof balance and care and the lack of availability and accessibility of the service, taking into account the economic constraints of this population [3,45]. Furthermore, there was no association between adequate hoof conformation and horses being shod by a farrier, which may indicate deficiencies in farrier knowledge and skills, probably due to a lack of farrier training courses in the country [9]. Thus, farrier training courses on hoof care, aimed at local farriers and owners, should be implemented to increase horse owner awareness of the negative consequences that hoof abnormalities can bring for their animals′ welfare. This could be organized by local government agencies or non-governmental organizations in Chile.

Skin lesions have been reported as one of the most common afflictions found in working equids worldwide [5,6,9,10,38]. In this study, skin lesions were the second most frequently observed problem, especially in the Metropolitana de Santiago region. Most of these were simple excoriations, located mainly on harness related areas and in concordance with the patterns reported by other studies [9,44]. In our study, skin lesions tended to be associated with poor body condition, whereas age (individuals over 15 years of age) was not significantly associated with the presence of lesions, contrary to the findings of Burn et al. [6]. Likewise, the presence of skin lesions was not associated with an inadequate conformation of horses for draught activities (based on their AI and work frequency). Thus, other factors, probably extrinsic to the animal, could be associated with the occurrence of skin lesions, for example, ill-fitting harnesses, poor maintenance of equipment, or overload, all of which have been previously related to skin lesions in working equids [4,9].

Behavioral tests are considered an essential component in the welfare assessment of working equids. These tests indicate, to some extent, how the animal interacts with its environment [5] and can help to understand the quality of the human–equid relationship [7]. The high prevalence of an alert attitude and of positive (friendly) responses towards both the owner and the observer, in all tests in the two administrative regions evaluated, indicates the existence of a good human–animal relationship between owners and their animals and an appropriate handling of horses by owners. These results differ from previous studies, in which indifference, aggressiveness and avoidance were the most prevalent behaviors observed in working equids [4,5]. Several studies in the livestock industry have provided evidence on the relationship between the attitudes of a stockperson towards the animals and their behavioral responses and welfare [23,46]. For example, Hemsworth et al. [47] found that positive attitudes toward animals were associated with more positive interactions and that these positive interactions were negatively associated with the animals′ fear towards humans. Therefore, it could be possible that the owners assessed in this study have more positive attitudes towards their animals, which may be reflected in the high prevalence of the positive responses found. The discourse analysis of statements made by owners, focusing on how they conceptualize and describe their horses, showed similarities and differences, depending on the geographical location of their owners. Owners of both regions generally tended to see their animals as subjective individuals, capable of occupying a place in their lives as either a member of the family or as a friend. This indicates an affective relationship between the owners and their horses. Owners also conferred their animals' mental abilities similar to those of humans (e.g., "He is the only one who understands me"), which can be interpreted as a clear tendency to anthropomorphize the behavior of their animals [48]. However, the recognition of these animals as a work tool (instrumental role) was also present in the discourse of owners, especially in owners from the Araucania region (e.g., "He is the one that provides my everyday bread, the food, my work tool").

Generally, the emotionality (affection for animals) and instrumentality (e.g., economic or utilitarian considerations) with which animals are perceived are often seen as opposite to each other [17,18]. In this sense, the emotional aspect of the human–animal relationship has often been linked to the keeping of companion animals [49], whereas keeping farm animals has been mainly associated to financial or utilitarian purposes, especially in the poorest communities worldwide. However, the results from this study suggest that affective and instrumental perceptions can coexist. Coincidentally, Schuurman [18] showed that conceptions of horses, particularly of equine welfare, consist of a combination of emotional and instrumental relations. Other authors have also reported that instrumental and affective components within livestock production can, and do, co-exist. In other words, animals can be thought of as simultaneously a friend and source of food [50,51].

How people relate to animals cannot be isolated from the cultural contexts in which they are embedded [51]. In the Metropolitana de Santiago region, owners′ descriptions of their horses indicated predominantly affective-type ascriptions. Based on a correlations analysis, the results indicate that the owners of this region showed a certain predilection for this species and tend to perceive their horse, primarily, as a pet, even in some cases as a friend. Moreover, they recognized certain needs of their animals and demonstrated a clear intention in taking care of them (e.g., "They are my favorite pets, I like to feed them well"), most likely to keep their animals as healthy as possible. This could be interpreted as the owners having an empathetic understanding of their horses as individuals (in some cases, they are considered companion animals or pets) that feel and have needs. In the Araucania region, owners′ discourse emphasized the perception of their horse as a family member. This indicates, based on the LSA results, that there may be an affective owner–horse relationship, due to the owners′ perception of kinship and closeness with their animals. The owner′s discourse also highlighted the instrumental role that their horse plays for the family′s subsistence, exalting a feeling of gratitude towards the animal′s labor. One possible explanation for this ambivalence could be that owners from the Araucania region are closer to an ethnic-rural background than those from the Metropolitana de Santiago region. The Mapuche ethnicity, which is concentrated in the Araucania region [14], has been characterized as a culture that historically conceived the horse (a species introduced by the Spanish) as an animal of burden and transport [52]. Therefore, it might be expected that these owners have acquired a utilitarian vision of these animals through family and cultural inheritance. On the other hand, owners from the Metropolitana de Santiago region do not see the horse exclusively as a working instrument but also as a companion animal. In fact, results of the LSA suggest that there is predilection towards horses, rather than a sense of labor in keeping these animals, and that owners from this region also recognize human psychological characteristics in their animals.

It is important to note that most owners stated that they enjoyed keeping these animals, which could indicate that many of them chose this occupation voluntarily, rather than simply continuing a family activity. These perceptions, added to the good welfare status found in most of the horses evaluated in the study, illustrate that working horse owners in Chile, generally, do not mistreat or exploit their animals.

This study may be limited by owners′ bias when asked about perceptions regarding their horses. Survey respondents may have tried to anticipate what they thought researchers wished to hear. This is especially likely to occur within this population, considering that working horses are constantly in the public eye [53–55]. Given that an owner's appreciation of his/her horse(s) was determined through only one question, future studies could include the use of validated attachment scales and attitudes towards animals.

5. Conclusions

The low prevalence of health problems and negative behavior responses found in horses of the two administrative regions evaluated indicate a good welfare status for the majority of horses in this study. The main problems observed in the assessed animals were hoof abnormalities and skin lesions. A possible explanation for these observations is a lack of understanding of basic husbandry practices, which may be improved through the implementation of educational programs for owners and local farriers. There were significant differences between the two regions in the presence of lesions on horses and the person responsible for shoeing management. This study provides preliminary information about owners' relationships with their animals and both affective and instrumental perceptions of working horses were observed. While the instrumental perception was more common in the Araucania region, the affective perception was widely shared by both groups of owners. This indicates that working horses are considered part of the family or a friend and suggests an affective relationship of owners with their animals.

Acknowledgments: This study was supported by CONICYT National Doctorate Fellowship (NO. 21130091), and FONDECYT 11121467 and FONDECYT 1161136. RAV acknowledges support from ICM-P05-002, PFB-23-CONICYT and FONDECYT 1140548. We especially thank all working horse owners and their families for their contribution to the study and Benjamin Uberti for revising the English of the manuscript.

Author Contributions: Daniela Luna, Tamara A. Tadich and Rodrigo A. Vásquez conceived and designed the experiments; Daniela Luna performed the experiments; Daniela Luna, Tamara A. Tadich and Manuel Rojas analyzed the data; Tamara A. Tadich contributed reagents/materials/analysis tools; Daniela Luna wrote the paper, which was revised by all authors.

Conflicts of Interest: The authors declare no conflict of interest.

References

1. World Economic Situation and Prospect 2017. Country Classifications Data Sources, Country Classifications and Aggregation Methodology, United Nation. 2017. Available online: https://www.un.org/development/desa/dpad/wp-content/uploads/sites/45/publication/2017wesp_full_en.pdf (accessed on 27 June 2017).
2. The Role, Impact and Welfare of Working (Traction and Transport) Animals. Animal Production and Health Report. No. 5. Food and Agriculture Organization, FAO: Rome, Italy, 2014. Available online: http://www.fao.org/3/a-i3381e.pdf (accessed on 21 February 2017).
3. Tadich, T.A.; Stuardo-Escobar, L.H. Strategies for improving the welfare of working equids in the Americas: A Chilean example. *Rev. Sci. Tech. Off. Int. Epiz.* **2014**, *33*, 203–211. [CrossRef]
4. Popescu, S.; Diugan, E.A. The relationship between behavioral and other welfare indicators of working horses. *J. Equine Vet. Sci.* **2013**, *33*, 1–12. [CrossRef]
5. Pritchard, J.C.; Lindberg, A.C.; Main, D.C.J.; Whay, H.R. Assessment of the welfare of working horses, mules and donkeys, using health and behaviour parameters. *Prev. Vet. Med.* **2005**, *69*, 265–283. [CrossRef] [PubMed]
6. Burn, C.C.; Dennison, T.L.; Whay, H.R. Environmental and demographic risk factors for poor welfare in working horses, donkeys and mules in developing countries. *Vet. J.* **2010**, *186*, 385–392. [CrossRef] [PubMed]
7. Ali, A.B.; El Sayed, M.A.; Matoock, M.Y.; Fouad, M.A.; Heleski, C.R. A welfare assessment scoring system for working equids—A method for identifying at risk populations and for monitoring progress of welfare enhancement strategies (trialed in Egypt). *Appl. Anim. Behav. Sci.* **2016**, *176*, 52–62. [CrossRef]
8. Chapter 7.12, Welfare of Working Equids. World Organization for Animal Health, OIE, 2016. Available online: http://www.oie.int/fileadmin/Home/eng/Health_standards/tahc/current/chapitre_aw_working_equids.pdf (accessed on 21 February 2017).
9. Tadich, T.; Escobar, A.; Pearson, R.A. Husbandry and welfare aspects of urban draught horses in the south of Chile. *Arch. Med. Vet.* **2008**, *40*, 267–273. [CrossRef]

10. Sáez, M.; Escobar, A.; Tadich, T. Morphological characteristics and most frequent health constraints of urban draught horses attending a free healthcare programme in the south of Chile: A retrospective study (1997–2009). *Livestock Res. Rural Dev.* **2013**, *25*, 91.
11. Lanas, R.; Luna, D.; Tadich, T. The link between animal welfare of urban draught horses and livelihoods of their owners: The case of Chile. In Proceedings of the 24th Annual Conference of the International Society for Anthrozoology, Saratoga Springs, New York, NY, USA, 7–9 July 2015.
12. Región Metropolitana Información Regional, Oficina de Estudios y Políticas Agrarias, ODEPA, 2015. Available online: http://www.odepa.cl/wp-content/files_mf/1437485361Metropolitanajulio.pdf (accessed on 5 February 2017) . (In Spanish)
13. División Político Administrativa y Censal, Región de la Araucania. Instituto Nacional de Estadística, INE, 2007. Available online: http://www.inearaucania.cl/archivos/files/pdf/DivisionPoliticoAdministrativa/araucania.pdf (accessed on 5 February 2017). (In Spanish)
14. Estadísticas sociales de los pueblos indígenas en chile-censo. Instituto Nacional de Estadísticas, INE, 2002. Available online: http://www.ine.cl/canales/chile_estadistico/estadisticas_sociales_culturales/etnias/pdf/estadisticas_indigenas_2002_11_09_09.pdf (accessed on 21 February 2017). (In Spanish)
15. Waiblinger, S.; Boivin, X.; Pedersen, V.; Tosi, M.V.; Janczak, A.M.; Visser, E.K.; Jones, R.B. Assessing the human-animal relationship in farmed species: A critical review. *Appl. Anim. Behav. Sci.* **2006**, *101*, 185–242. [CrossRef]
16. Hills, A.M. The motivational bases of attitudes toward animals. *Soc. Anim.* **1993**, *1*, 111–128. [CrossRef]
17. Serpell, J.A. Factors influencing human attitudes to animals and their welfare. *Anim. Welfare* **2004**, *13*, 145–151.
18. Schuurman, N. Conceptions of equine welfare in Finnish horse magazines. *Soc. Anim.* **2015**, *23*, 250–268. [CrossRef]
19. Waiblinger, S.; Menke, C.; Coleman, G. The relationship between attitudes, personal characteristics and behaviour of stockpeople and subsequent behaviour and production of dairy cows. *Appl. Anim. Behav. Sci.* **2002**, *79*, 195–219. [CrossRef]
20. Signal, T.D.; Taylor, N. Attitudes to animals: Demographics within a community sample. *Soc. Anim.* **2006**, *14*, 147–157. [CrossRef]
21. Ellingsen, K.; Zanella, A.J.; Bjerkås, E.; Indrebø, A. The relationship between empathy, perception of pain and attitudes toward pets among Norwegian dog owners. *Anthrozoös* **2010**, *23*, 231–243. [CrossRef]
22. Muri, K.; Tufte, P.A.; Skjerve, E.; Valle, P.S. Human-animal relationships in the Norwegian dairy goat industry: Attitudes and empathy towards goats (Part I). *Anim. Welfare* **2012**, *21*, 535–545. [CrossRef]
23. Kielland, C.; Skjerve, E.; Østerås, O.; Zanella, A.J. Dairy farmer attitudes and empathy toward animals are associated with animal welfare indicators. *J. Dairy Sci.* **2010**, *93*, 2998–3006. [CrossRef] [PubMed]
24. Fureix, C.; Pagès, M.; Bon, R.; Lassalle, J.M.; Kuntz, P.; Gonzalez, G. A preliminary study of the effects of handling type on horses' emotional reactivity and the human-horse relationship. *Behav. Processes* **2009**, *82*, 202–210. [CrossRef] [PubMed]
25. Birke, L.; Hockenhull, J.; Creighton, E. The horse's tale: Narratives of caring for/about horses. *Soc. Anim.* **2010**, *18*, 331–347. [CrossRef]
26. Birke, L. Talking about Horses: Control and Freedom in the World of "Natural Horsemanship". *Soc. Anim.* **2008**, *16*, 107–126. [CrossRef]
27. Birke, L. "Learning to speak horse": The culture of "Natural Horsemanship". *Soc. Anim.* **2007**, *15*, 217–239. [CrossRef]
28. Landauer, T.K.; Foltz, P.W.; Laham, D. An introduction to latent semantic analysis. *Discourse Process.* **1998**, *25*, 259–284. [CrossRef]
29. Evangelopoulos, N.; Zhang, X.; Prybutok, V.R. Latent semantic analysis: five methodological recommendations. *Eur. J. Inf. Syst.* **2012**, *21*, 70–86. [CrossRef]
30. Dam, G.; Kaufmann, S. Computer assessment of interview data using latent semantic analysis. *Behav. Res. Methods* **2008**, *40*, 8–20. [CrossRef] [PubMed]
31. Burn, C.C.; Dennison, T.L.; Whay, H.R. Relationship between behaviour and health in working horses, donkeys, and mules in developing countries. *Appl. Anim. Behav. Sci.* **2010**, *126*, 109–118. [CrossRef]
32. Carroll, C.L.; Huntington, P.J. Body condition scoring and weight estimation of horses. *Equine Vet. J.* **1988**, *20*, 41–45. [CrossRef] [PubMed]

33. Popescu, S.; Diugan, E.A.; Spinu, M. The interrelations of good welfare indicators assessed in working horses and their relationships with the type of work. *Res. Vet. Sci.* **2014**, *96*, 406–414. [CrossRef] [PubMed]
34. Cassai, G. El caballo de Labranza. *Revista El Campesino* **1944**, *96*, 7–10. (In Spanish)
35. Meyer, K. A study of the condition of working horses in Chile. Master's Thesis, University College of North Wales, Bangor, Gwynedd, UK, 1992.
36. Landauer, T.K.; Dumais, S.T. A solution to Plato's problem: The latent semantic analysis theory of acquisition, induction, and representation of knowledge. *Psychol. Rev.* **1997**, *104*, 211–240. [CrossRef]
37. Blakeway, S. The multi-dimensional donkey in landscapes of donkey-human interaction. *Rel. Beyond Anthropocentrism* **2014**, *2*, 59–77. [CrossRef]
38. de Aluja, A.S. The welfare of working equids in Mexico. *Appl. Anim. Behav. Sci.* **1998**, *59*, 19–29. [CrossRef]
39. Biffa, D.; Woldemeskel, M. Causes and factors associated with occurrence of external injuries in working equines in Ethiopia. *Intern. J. Appl. Res. Vet. Med.* **2006**, *4*, 1–7.
40. Swann, W.J. Improving the welfare of working equine animals in developing countries. *Appl. Anim. Behav. Sci.* **2006**, *100*, 148–151. [CrossRef]
41. Geiger, M.; Hovorka, A. J. Using physical and emotional parameters to assess donkey welfare in Botswana. *Vet. Rec. Open.* **2015**, *2*, e000062. [CrossRef] [PubMed]
42. Mac-Leod, C. Estudio de los equinos carretoneros atendidos en un policlínico en Valdivia, caracterizando aspectos de hipometría, patologías, alimentación, cascos y herrajes. Memoria de título, Escuela de Medicina Veterinaria, Universidad Austral de Chile: Valdivia, Chile, 1999. (In Spanish)
43. Beltrán, J.M. *Ganado Caballar*, 1st ed.; Salvat Editores SA: Barcelona, Spain, 1954. (In Spanish)
44. Morgan, R. The epidemiology of lameness in working donkeys in Addis Ababa and the central Oromia region of Ethiopia: a comparative study of urban and rural donkey populations. In Proceedings of the Fifth International Colloquium on Working Equines. The future for working equines, Addis Ababa, Ethiopia, 30 October–2 November 2006; pp. 99–106.
45. Upjohn, M.M.; Pfeiffer, D.U.; Verheyen, K.L.P. Helping working equidae and their owners in developing countries: Monitoring and evaluation of evidence-based interventions. *Vet. J.* **2014**, *199*, 210–216. [CrossRef] [PubMed]
46. Hemsworth, P.H. Human-animal interactions in livestock production. *Appl. Anim. Behav. Sci.* **2003**, *81*, 185–198. [CrossRef]
47. Hemsworth, P.H.; Coleman, G.J.; Barnett, J.L.; Borg, S. Relationships between human-animal interactions and productivity of commercial dairy cows. *J. Anim. Sci.* **2000**, *78*, 2821–2831. [CrossRef] [PubMed]
48. Serpell, J.A. Anthropomorphism and anthropomorphic selection-beyond the "Cute Response". *Soc. Anim.* **2003**, *11*, 83–100. [CrossRef]
49. Voith, V.L. Attachment of people to companion animals. *Vet. Clin. N. Am. Small Anim. Pract.* **1985**, *15*, 289–295. [CrossRef]
50. Holloway, L. Pets and protein: Placing domestic livestock on hobby-farms in England and Wales. *J. Rural Stud.* **2001**, *17*, 293–307. [CrossRef]
51. Wilkie, R. Sentient commodities and productive paradoxes: The ambiguous nature of human-livestock relations in Northeast Scotland. *J. Rural Stud.* **2005**, *21*, 213–230. [CrossRef]
52. Montero, G. Ladran Sancho II. El caballo en el mundo ceremonial indígena. XII Jornadas Interescuelas/Departamentos de Historia. Departamento de Historia, Facultad de Humanidades y Centro Regional Universitario Bariloche. Universidad Nacional del Comahue, San Carlos de Bariloche, 2009. Available online: http://cdsa.aacademica.org/000-008/1380 (accessed on 21 February 2017) . (In Spanish)
53. Emol Nacional. Available online: http://www.emol.com/noticias/Nacional/2016/02/04/786822/Vina-del-Mar-Municipio-se-querello-por-maltrato-animal-en-coches-victoria.html (accessed on 29 June 2017).
54. CNN International Edition. Available online: http://edition.cnn.com/2014/01/23/opinion/bershadker-ban-horse-drawn-carriages/index.html (accessed on 29 June 2017). (In Spanish)
55. New York Times. Available online: https://www.nytimes.com/2014/01/19/nyregion/who-speaks-for-the-horses-in-battle-over-carriages.html (accessed on 29 June 2017).

9

Veterinary and Equine Science Students' Interpretation of Horse Behaviour

Gabriella Gronqvist [1], Chris Rogers [1], Erica Gee [1], Audrey Martinez [2] and Charlotte Bolwell [1,*]

[1] Massey Equine, Institute of Veterinary, Animal and Biomedical Sciences, Massey University, Private Bag 11-222, Palmerston North 4442, New Zealand; g.gronqvist@massey.ac.nz (G.G.); c.w.rogers@massey.ac.nz (C.R.); e.k.gee@massey.ac.nz (E.G.)

[2] École Nationale Vétérinaire de Toulouse 23 Chemin des Capelles, BP 87614, 31076 Toulouse CEDEX 3, France; a.martinez_14@envt.fr

* Correspondence: c.bolwell@massey.ac.nz

Simple Summary: We assessed first-year veterinary science and veterinary technology and undergraduate equine science students interpretation of expressive horse behaviours. Previous experience with horses appeared to influence the students' perception of the horses' behaviour. Qualitative assessments of horse behaviour may be a useful tool for assessing students' knowledge of horse behaviour.

Abstract: Many veterinary and undergraduate equine science students have little previous horse handling experience and a poor understanding of horse behaviour; yet horses are one of the most unsafe animals with which veterinary students must work. It is essential for veterinary and equine students to learn how to interpret horse behaviour in order to understand demeanour and levels of arousal, and to optimise their own safety and the horses' welfare. The study utilised a qualitative research approach to investigate veterinary science and veterinary technology and undergraduate equine science students' interpretation of expressive behaviours shown by horses. The students (N = 127) were shown six short video clips and asked to select the most applicable terms, from a pre-determined list, to describe the behavioural expression of each individual horse. A wide variation of terms were selected by students and in some situations of distress, or situations that may be dangerous or lead to compromised welfare, apparently contradictory terms were also selected (happy or playful) by students with less experience with horses. Future studies should consider the use of Qualitative Behavioural Analysis (QBA) and free-choice profiling to investigate the range of terms used by students to describe the expressive demeanour and arousal levels of horses.

Keywords: horse behaviour; horse welfare; qualitative analysis; expressive behaviour

---

1. Introduction

Horses are one of the most dangerous animals that veterinary students have to learn how to handle correctly [1], in part due the innate flight response of this species [2]. It has been proposed that many of the accidents involving horses can be attributed to breakdowns in human-horse communication [3]. Thompson et al. [4] suggested that people with a poor understanding of horse behaviour may be at an increased risk of injury, as their ability to anticipate unwanted, yet natural, horse behaviours may be lacking. Veterinary students often lack previous horse experience and an understanding of horse behaviour, as many now come from an urban background [5–8]. In New Zealand, it has been reported that 60% of veterinary students were from cities and only 18% were from rural areas [6].

A cross-sectional survey in Australia reported that most mixed animal (59%) and large animal (65%) veterinarians had suffered severe acute injuries or chronic musculoskeletal injuries at work [9–11].

Studies in other countries have reported similar statistics, suggesting that such findings are not unique to Australian veterinarians [12,13]. An improved ability of veterinarians to assess behavioural cues from the horse, including changes in its arousal and affective state, could improve human-horse communication and potentially prevent some of these accidents [3,14].

Accurately assessing horse arousal and affective states would also benefit the horses in regard to their welfare. The affective state of an animal is a reflection of the animal's welfare, according to The Five Domains model [15]. The Five Domains model was initially developed to evaluate welfare compromise in animals used in research, teaching, and testing [16]. This model provides a method for recognising compromise in the four physical domains (nutrition, environment, health, and behaviour) and in one mental domain, which is the animal's affective experiences and reflects the animal's overall welfare state [15]. In order to maximise welfare, it is important for veterinary students to learn how to assess the horse's affective state in addition to the physical states [15,17].

Key features such as previous experience, training, and familiarity with the animal can influence what a person working with horses brings to a situation [18], which can either compromise or enhance the welfare status of the horse [19]. Whilst we might not expect veterinary students to become ethologists, it is important that they have a good understanding of horse behaviour and animal welfare science to ensure the safety of the student, so instances of poor welfare can be recognised [17]. However, there are currently no studies on the baseline level of student awareness or knowledge of animal behaviour on entry to their undergraduate equine science or veterinary courses, and the impact prior experience with horses and other animals may have on this.

Whole animal profiling is a subjective or qualitative technique commonly used in the assessment of animals' demeanour and behaviour (body language) [20,21]. Such methods rely on a human observer's ability to note apparent details of an animal's expressive behaviour, using "whole animal" descriptors such as playful, content, calm, or frustrated [22–24]. As part of this qualitative assessment of behaviour, predefined terms can be provided to the observers who are asked to score the strength of the expressive behaviour on a scale provided [21]. The current study utilised a qualitative analysis to investigate students' interpretation of expressive horse behaviour in various contexts and to determine the influence of previous experience with horses on their interpretation of the behaviours shown.

2. Materials and Methods

The study was conducted using first-year veterinary science or veterinary technology students and students enrolled in undergraduate equine science courses at Massey University. The students (N = 127) were informed that the aim of the project was to investigate the words used by students to describe horse behaviour. Written consent was obtained from each student and the project was evaluated as being a low risk project by the Massey University Human Ethics committee (No. 4000015518).

The study was based on short video recordings, rather than live animals, of everyday situations that were not set-up or specifically created for the purposes of this study. The testing was conducted at the start of the first semester, before the students received lectures or practical training on horse behaviour. Students were briefed verbally and in writing on the methodology of the study, and the testing procedures were explained in detail. The students provided demographic information including: gender ("male", "female", or "other"), age category "<20 years", "21–25 years", "25–30 years", or ">30 years") and previous horse experience ("None—never interacted with a horse prior to the start of this paper", "Little—interacted with or ridden horses a few times under supervision", "Some—interacted with or ridden horses regularly under supervision", "Experienced—interacted with or ridden horses regularly unsupervised" or "Very Experienced—competitive rider or worked in the horse industry").

The protocol consisted of a test video, which was played prior to the start of the testing and 6 short video clips (approximately 10 s long) (Table 1). After each video, the students had 30 s to score the horse behaviour using 15 pre-selected fixed terms [25] adapted from Minero et al. [26] (Table 2).

For this study, the term "responsive" was changed to "alert" to avoid confusion with a "responsive horse", which is used regularly in reference to a horse's response to a rider's aids.

Table 1. Descriptions of the six videos used in a study to assess students' interpretation of expressive horse behaviour.

Video	Description
1	The handler and horse walk towards each other in a paddock. The handler strokes the horse and attaches the lead rope whilst the horse stands still.
2	The handler holds the horse with a halter and lead rope. The handler moves a worming syringe towards the horse's mouth while the horse backs away. When the syringe reaches the mouth, the horse pulls back and canters away.
3	The horse is in an outside yard on its own. The horse initially walks quickly and then trots around the perimeter of the yard. Horse is lifting and lowering its head to the ground as it moves around the yard.
4	The horse stands alone in an arena with a saddle and halter on. The horse is resting with eyes half closed.
5	The horse is loose in an indoor yard on its own. The horse is standing still with its head elevated and ears pointing forward. The horse then moves off at a brisk walk around the yard.
6	The horse is in a yard with several other horses. The horse is partaking in mutual grooming with one other horse.

Table 2. The pre-selected terms and descriptions used in a study to assess students' interpretation of expressive horse behaviour, based on Minero et al. [26].

Term	Description
Aggressive	Behaving in an angry or rude way, fighting or attacking
Agitated	Restless, fidgety, worried or upset, excited, disturbed, troubled
Alert	Receptive, aware of the environment
Anxious	Worried/tense, troubled, apprehensive, distressed
Apathetic	Having or showing little or no emotion, indifferent
At ease	In a relaxed attitude or frame of mind
Curious	Eager to learn, inquisitive, wishing to investigate
Distressed	Much troubled, upset, afflicted, panicking
Fearful	Having fear, afraid, displaying a flight response, looking anxious, back up/away
Friendly	Not hostile, showing positive feelings toward another horse or person
Happy	Feeling, showing or expressing joy, pleased
Playful	Very active, happy, and wanting to have fun, mischievous
Pushy	Offensively assertive or forceful, bossy, dominant
Uncomfortable	Not comfortable, not relaxed
Withdrawn	Secluded or remote, shy, not searching for contact with others

Data collection sheets were provided to the students, and each page contained the 15 pre-selected terms, listed in the same order for each video, and a scale between 1 (Weak) and 5 (Strong). The students were asked to circle the number that represented the strength of the behaviour they perceived to best describe the horses in the individual videos. The students could choose more than one term for each video, and students were not made to select a score for every term for each video.

Statistical Analysis

Data were entered into Microsoft Excel and summarised using Pivot tables to describe the demographic variables. The number and percentage of students selecting each term for each video was calculated. For each video, the median score and interquartile range (IQR) for each term was calculated to describe the strength of the selected term. Multiple Correspondence Analysis (MCA), with the joint option in Stata, was used to visualise the relationship between the students' previous experience and the terms selected (in binary form presence or absence of behaviour) for each video. MCA is a descriptive technique used to visualise binary and categorical variables of interest (student experience) as points on a two-dimensional plot to describe how strongly and in which way the variables are related or cluster together [27]. The variables were represented on the plot by a symbol and one for terms that were selected by students and zero for terms that were not selected. The variables that explain each axis (or dimension) can be determined by their position on the graph. The points that are clustered together are considered similar to each other, whereas those plotted furthest apart are rarely associated with each other. The centre of the plot represents the average profile and is said to be homogeneous, so points in this position are very similar to the average profile [27]. All analyses were conducted in Stata version 14.1

3. Results

Data were collected from 127 students that agreed to take part in the study. Most students were female, aged <20 years old, and enrolled in the veterinary science or technology programme (Table 3). Just under half of the students rated themselves as being either Experienced or Very Experienced, with 10% of students indicating they had no previous experience with horses (Table 3).

Table 3. The number and percentage of first-year veterinary science or veterinary technology and undergraduate equine science students, by gender, age, and level of previous experience with horses, used in a study to assess students' interpretation of expressive horse behaviour.

Demographic Variables	Number	Percentage
Gender		
Male	17	13
Female	109	87
Age		
<20	100	79
21–25	23	18
25–30	2	2
>30	2	2
Level of experience		
None	13	10
Little	45	35
Some	13	10
Experienced	30	24
Very Experienced	26	20
Course studied		
Veterinary science or technology	66	52
Undergraduate equine science	61	48

The number of students selecting each term and the median score for each selected term for each video is shown in Table 4. The students selected 13/15 terms for Video 1, of which most students strongly perceived the horse as being Alert (80%) and Curious (84%), and 81% of students scored the horse as moderately Friendly (Table 4). Just over half of students (54%) gave moderate scores for At Ease, and 30% of students gave moderate scores for Happy. One student with no experience with horses selected strongly Fearful, and two students with little or some experience selected weakly Fearful, whilst five students with little or some experience with horses, and one Experienced student selected Uncomfortable. Withdrawn and Agitated were selected by five and two students, respectively, whose experience with horses was little or none.

Video 2 was characterised by six terms and over 80% of students scored the horse as strongly Agitated, Distressed, and Uncomfortable; just under 80% of students strongly scored the horse as Anxious and Fearful (Table 4). Moderately Playful was selected by one student, and weakly Happy and At Ease were selected by one student each for Video 2. Of these 3 students, one had Little Experience, one was Experienced, and one was Very Experienced with horses. Multiple correspondence analysis for the terms for Videos 1 and 2 did not show any clustering with student experience (data not shown).

Table 4. Number of students that selected each pre-defined term and the median (interquartile range) score for each term, selected by first-year veterinary science or veterinary technology and undergraduate equine science students, for each of the 6 videos used in a study to assess students' interpretation of expressive horse behaviours (- = term not selected by students).

	Video											
Word	1		2		3		4		5		6	
	N	Median	N	Median	N	Median	N	Median	N	Median	N	Median
Aggressive	-	-	14	2 (1–3)	2	3 (1–4)	-	-	1	2 (2–2)	-	-
Agitated	2	2 (1–3)	104	4 (3–5)	75	4 (3–4)	2	2 (1–3)	61	3 (2–4)	-	-
Alert	102	4 (3–4)	67	4 (3–5)	79	4 (3–4)	17	2 (2–3)	115	5 (4–5)	8	3 (2–4)
Anxious	16	2 (2–3)	100	4 (3–5)	69	3 (2–4)	16	2 (1–4)	78	3 (2–4)	-	-
Apathetic	6	3 (3–3)	-	-	8	4 (3–4)	89	4 (4–5)	3	4 (2–5)	3	3 (2–4)
At Ease	68	3 (3–4)	1	1 (1–1)	25	3 (3–4)	88	5 (4–5)	9	3 (2–3)	103	4 (3–5)
Curious	107	4 (3–4)	4	4 (2–5)	47	4 (3–4)	1	4 (4–4)	72	3 (3–4)	20	3 (2–4)
Distressed	-	-	110	4 (4–5)	48	3 (2–4)	7	3 (2–5)	41	2 (1–3)	-	-
Fearful	3	2 (1–4)	100	4 (4–5)	13	2 (2–3)	7	2 (1–3)	17	2 (1–3)	-	-
Friendly	103	3 (3–4)	-	-	11	2 (2–2)	8	3 (1–3)	9	2 (2–3)	119	5 (4–5)
Happy	38	3 (3–4)	1	1 (1–1)	24	3 (2–4)	21	3 (2–3)	10	2 (2–3)	110	4 (3–4)
Playful	8	3 (2–4)	1	3 (3–3)	25	3 (2–3)	-	-	5	2 (1–2)	67	4 (3–4)
Pushy	2	2 (1–3)	8	3 (2–3)	1	4 (4–4)	-	-	1	2 (2–2)	4	3 (2–4)
Uncomfortable	6	3 (2–4)	106	4 (4–5)	53	3 (2–4)	14	3 (1–5)	53	3 (2–3)	-	-
Withdrawn	5	2 (2–2)	13	3 (2–5)	6	3 (2–3)	52	4 (2–4)	7	3 (2–4)	-	-

All terms were selected for Videos 3 and 5 (Table 4). For Video 3 over half of students perceived the horse to be strongly Agitated, Alert, Curious, and moderately Anxious, with 20%, 20%, and 19% of students perceiving the horse to be Moderately At Ease, Playful, and Happy, respectively. Of the students selecting moderately Happy, Playful, or At Ease, 67% (16/24), 68% (17/25) and 72% (18/25) had little or no previous experience with horses, respectively. Multiple Correspondence Analysis for Video 3 showed that Axis 1 (first dimension) was characterised by the terms Distressed, Agitated, Anxious, At Ease, Happy, Playful, and the previous experience of students (No Experience or Very Experienced) (Figure 1). The plot shows that students with no previous experience with horses clustered with the terms Friendly, Playful, Happy, and At Ease (Figure 1) in the same direction away from the average profile (in the centre of the plot). Very Experienced students clustered with the terms Distressed, Anxious, and Agitated, and this cluster was in an extreme opposite position to the other cluster.

Most students selected that the horse in Video 4 was strongly At Ease or Apathetic, with 41% of students selecting moderately Withdrawn and 17% selecting Happy (Table 4). Some students selected Anxious or Uncomfortable, of which 81% (13/16) and 64% (9/14) of the students had little or no experience with horses. Two students with no previous experience with horses perceived the horse in Video 4 to be strongly Anxious, Distressed, and Uncomfortable, whilst four students with Little or Some Experience and one Experienced student rated the horse as Distressed. Five students with little or no experience and one Experienced student rated the horse as fearful.

Most students (91%) perceived the horse in Video 5 to be Alert (highest score), with 61%, 57%, and 48% of students scoring the horse as moderately Anxious, Curious, and Agitated, respectively (Table 4). The horse in Video 5 was given weak scores for Happy, Friendly, At Ease, and Playful by

8%, 7%, 7%, and 4% of students, respectively, of which 6/10, 5/9, and 6/9 students had little or no experience with horses, respectively.

Eight terms were selected for Video 6, with 94%, 87%, and 81% of students perceiving the horse to be Friendly, Happy, and At Ease (Table 4). Pushy was selected by four students, of whom three had Little Experience and one student was Very Experienced with horses, and Apathetic was selected by three students, of whom two had Little Experience and one was Experienced with horses. Multiple correspondence analysis for the terms for Videos 4, 5, and 6 did not show any strong clustering with student experience (data not shown).

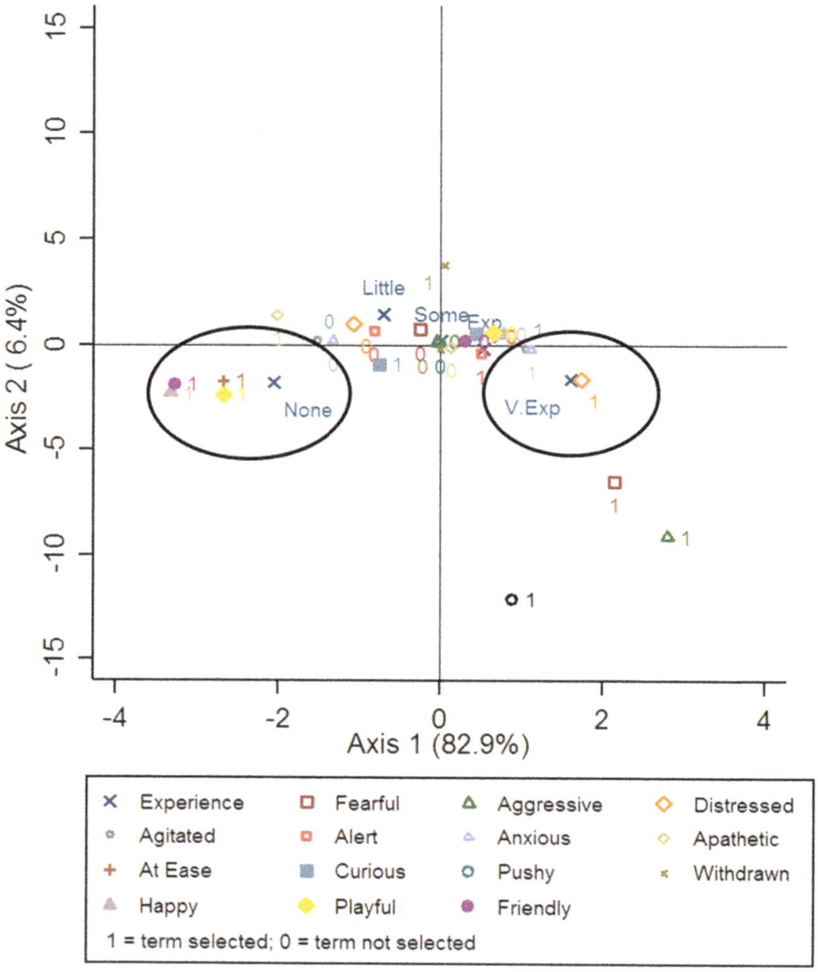

Figure 1. Multiple correspondence analysis showing the terms selected by first-year veterinary science or veterinary technology and undergraduate equine science students to describe the horse's behaviour shown in Video 3, and the level of the students' experience with horses. 1 = term selected at any strength; 0 = term not selected. 'Experience = None—never interacted with a horse prior to the start of this paper', 'Little—interacted with or ridden horses a few times under supervision', 'Some—interacted with or ridden horses regularly under supervision', 'Experienced—interacted with or ridden horses regularly unsupervised' or 'Very Experienced—competitive rider or worked in the horse industry'. Black circles are used to indicate clusters in the data.

4. Discussion

This study aimed to assess first-year veterinary science or veterinary technology and undergraduate equine science students' interpretation of expressive behaviour shown by horses. Across all videos, a range of terms were selected at varying strengths, and in some instances the percentage of students selecting a term was clustered with the students' level of experience with horses. Whilst most students selected similar terms to describe the horse's behaviour in a video, apparently contradictory terms were also selected for some videos. For example, terms describing negative affective states (Agitated, Anxious, Fearful, and Uncomfortable) were selected for the videos showing a horse resting with eyes half closed and the handler catching the horse from the paddock. Of concern was the apparent confusion of the expressive behaviour exhibited by the horse alone in an outside yard with positive affective states (Curious, Playful, Happy, and At Ease).

Misinterpretation and failing to recognise behaviours such as anxiety may be an animal welfare concern. A horse in isolation is unable to carry out interactive behaviours with other horses and experiences constraints on its environment (unable to escape), which results in negative affective states such as loneliness, anxiety, fearfulness, and panic [15,28]. Such negative affective states ultimately result in a negative welfare status for the horse. The importance of veterinary and equine science students learning to identify negative affective states and situations that may be predictive of dangerous or aggressive behaviour, and the potential for welfare compromise, should not be underestimated. Humans have a large influence over the welfare status of horses, and their knowledge, skills, training, and familiarity with the animal can compromise or enhance the horses' welfare status [18,19]. People working with horses need to be able to anticipate and identify problems, as well as ensuring good welfare is maintained [19]. Integrating animal welfare science in the veterinary and equine science curriculum would ensure students can use a more holistic approach to assessing horse behaviour and welfare states [17,29].

An inability of veterinary and equine students to assess horse behaviour could create breakdowns in human-horse communication and subsequently pose a safety risk for the students [3]. The most common injuries during horse handling practicals at an Australian Veterinary School were inflicted by horses stepping on ankles or feet and by bites or hind limb kicks [30]. Results from the Australian study suggest an inability of the students to swiftly and accurately assess the affective state of the horse and respond accordingly. The behaviour expressed by an individual animal is believed to, at least in part, be a reflection of the individual's present arousal level and affective state [29,31]. Furthermore, it is worth noting that there is wide variation in the behaviour expressed and the reactivity of horses. Riley et al. [30] reported that inattention and inexperience were cited by students as the cause of 30% and 39% of horse-related accidents, respectively. Moreover, 30% of students believed that the accident occurred due to the horse being distressed or fearful [30]. An improved ability of veterinary students and professionals to recognise dangerous and threatening, as well as subtle, behavioural cues from the horse, including changes in its arousal and affective state, could help to prevent some of these accidents [3,8].

Previous research using qualitative methods has reported differences between experienced and inexperienced observers [32] and between different stakeholders (farmers vs. animal scientists vs. urban citizens) [33]. In the current study, the students' interpretation of the horses' behaviour in one video was clustered with the students' level of experience with horses. Additionally, across all videos, there were terms selected that appeared to be contradictory to the terms selected by most students and the behaviour expressed in the video, which were often selected by students with less experience. These results suggest that it may be useful to identify 'at risk' students with less experience of horses who may benefit from additional learning activities before practical handling sessions, which have previously been shown to successfully increase student awareness and understanding of animal behaviour [34]. However, it is likely that further studies are required to build on the baseline data provided in this study, in order to investigate the influence of previous experience with horses on students' interpretation of behaviour and safety around horses.

A study examining inter- and intra-observer reliability of experienced and inexperienced observers of behaviour described differences with experience in how the observers perceived the terms and descriptors [32]. Observers with less experience were reported to have less correlation amongst terms compared to observers experienced in assessing dairy cattle [32]. Although future studies are required to specifically investigate the reliability of student observers, the results from the current study support preceding research suggesting that the way in which the observers perceive and interpret the descriptors may be affected by their previous experience [32]. These findings require further exploration to better understand how much and what type of experience with horses is useful (for example, ridden versus husbandry skills) to potential veterinary and undergraduate equine science students.

The expressive behaviours of the horse may have been the driver in the videos where no pattern was demonstrated with the selected terms and student experience. It is possible that in some situations animal behaviours may be easily identified with general knowledge (rather than specific experience with horses) or by applying a human emotional interpretation [32]. Such behaviours may be considered more obvious, rather than subtle, behaviours that people with very little experience with horses (or other animals) may be able to recognise. Mutual grooming between horses appeared to be a behaviour that students with little previous horse experience could recognise. Mutual grooming is a common behaviour seen in a large number of group living species [35]. Perhaps the ease with which students identified mutual grooming was linked to the fact that this behaviour is commonly seen in different social species. Whereas, it may not always be clear to unexperienced students when horses are demonstrating fearful or anxious behaviour; such behaviours may be subtler to students that have no familiarity with identifying the changing arousal levels associated with the natural flight response in horses [29]. The hypothesis that students' recognition of subtle or more obvious behaviours may differ is of importance when teaching students to work safely around horses in order to minimise harm.

Additionally, a number of videos included a handler with the subject horse (Table 1) and it is possible that the handlers in the videos provided the less experienced students with clues as to the arousal level of the horses, resulting in some contextual bias [24]. For example, students may be directed towards certain terms in Video 1 based on the behaviour (posture, facial expressions, behaviour) of the handler when catching the horse. Similarly, knowing that a handler is administering something to the horse in Video 2 may have biased the students towards terms that reflect negative affective states and interpreted the horse's behaviour as Uncomfortable or Anxious. The compounding factors of the handler or the context of the video may be one explanation for the lack of strong clustering with student experience of horses in most videos.

The current study used a fixed set of terms and descriptions adapted from Minero et al. [26] to assess the welfare of donkeys. These terms were selected for use in the current study due a lack of published studies using fixed terms to investigate the welfare, or otherwise, of horses. Furthermore, a thorough consultation process with experts in donkey welfare and behaviour was used by Minero et al. [26] to develop the list of terms and their descriptions. However, during the application to videos of horses in the current study, it became clear that a number of the descriptions overlapped or included other terms in the descriptions, potentially causing confusion amongst students. Minero et al. [26] noted that linguistic barriers may lead to confusion in the interpretation of descriptors and bilingual dictionaries were utilised to reach consensus when characterising the terms. Due to this, it is possible that there was some level of ambiguity in the terms in relation to the behaviour observed, which may have contributed to the wide variation of terms selected by students, and a lack of apparent association with experience, for some videos. Additionally, it is possible that the terms selected were suitable for donkeys, but were not suitable for assessing expressive behaviour in horses. Since the current study was completed a set of pre-defined terms for using Qualitative Behavioural Analysis (QBA) to assess horse welfare have been developed through the AWIN (animal welfare indicators) welfare assessment protocol for horses [36]. More emphasis may need to be put towards choosing the correct terminology in future studies [32], with consideration of the new terms developed specifically for horses [36]. Alternatively, there may be some merit in utilising a QBA approach with free-choice profiling and Generalised Procrustes Analysis [23] to assess students' language used to describe the behaviour of horses.

5. Conclusions

Overall, a wide variation of terms were selected by students to describe the horse's behaviour shown in each of the videos. In some situations, the ability to recognise expressive horse behaviour was associated with students' previous horse experience. Some situations of distress, or those that may lead to compromised welfare or potentially dangerous situations were poorly identified by some students, and the reasons for this requires further investigation. Further improvements can be made to the qualitative approach used in this study to provide a novel mechanism for evaluating the students' interpretation of horse behaviour and arousal level.

Acknowledgments: The authors would like to thank the students that agreed to participate in this study. We would like to thank Mirjam Guesgen and Ngaio Beausoleil for their valuable comments on the manuscript.

Author Contributions: Gabriella Gronqvist, Charlotte Bolwell, Erica Gee, and Chris Rogers designed the study and produced the manuscript. Gabriella Gronqvist collected the data. Audrey Martinez, Chris Rogers, and Charlotte Bolwell conducted analysis and contributed to the interpretation of the results.

Conflicts of Interest: The authors declare no conflict of interest.

References

1. Stafford, K.J.; Erceg, V.H. Teaching animal handling to veterinary students at Massey University, New Zealand. *J. Vet. Med. Educ.* **2007**, *34*, 583–585. [CrossRef] [PubMed]
2. LeGuin, E.; Raber, K.; Tucker, T.J. Man and Horse in Harmony. In *The Culture of the Horse: Status, Discipline, and Identity in the Early Modern World*; Palgrave Macmillan US: New York, NY, USA, 2005.
3. Hausberger, M.; Roche, H.; Henry, S.; Visser, E.K. A review of the human–horse relationship. *Appl. Anim. Behav. Sci.* **2008**, *109*, 1–24. [CrossRef]
4. Thompson, K.; McGreevy, P.; McManus, P. A critical review of horse-related risk: A research agenda for safer mounts, riders and equestrian cultures. *Animals* **2015**, *5*, 561–575. [CrossRef] [PubMed]
5. Heath, T.J. Number, distribution and concentration of Australian veterinarians in 2006, compared with 1981, 1991 and 2001. *Aust. Vet. J.* **2008**, *86*, 283–289. [CrossRef] [PubMed]
6. Wake, A.F.; Stafford, K.J.; Minot, E.O. The experience of dog bites: A survey of veterinary science and veterinary nursing students. *N. Z. Vet. J.* **2006**, *54*, 141–146. [CrossRef] [PubMed]
7. White, P.; Chapman, S. Two students' reflections on their training in animal handling at the university of sydney. *J. Vet. Med. Educ.* **2007**, *34*, 598–599. [CrossRef] [PubMed]
8. Doherty, O.; McGreevy, P.D.; Pearson, G. The importance of learning theory and equitation science to the veterinarian. *Appl. Anim. Behav. Sci.* **2017**, *190*, 111–122. [CrossRef]
9. Lucas, M.; Day, L.; Fritschi, L. Injuries to Australian veterinarians working with horses. *Vet. Rec.* **2009**, *164*, 207–209. [CrossRef] [PubMed]
10. Lucas, M.; Day, L.; Shirangi, A.; Fritschi, L. Significant injuries in Australian veterinarians and use of safety precautions. *Occup. Med.* **2009**, *59*, 327–333. [CrossRef] [PubMed]
11. Fritschi, L.; Day, L.; Shirangi, A.; Robertson, I.; Lucas, M.; Vizard, A. Injury in Australian veterinarians. *Occup. Med. (Oxf. Engl.)* **2006**, *56*, 199–203. [CrossRef] [PubMed]
12. Loomans, J.B.A.; van Weeren-Bitterling, M.S.; van Weeren, P.R.; Barneveld, A. Occupational disability and job satisfaction in the equine veterinary profession: How sustainable is this 'tough job' in a changing world? *Equine Vet. Educ.* **2008**, *20*, 597–607. [CrossRef]
13. BEVA. Survey reveals high risk of injury to equine vets. *Vet. Rec.* **2014**, *175*, 263. [CrossRef]
14. Payne, E.; Boot, M.; Starling, M.; Henshall, C.; McLean, A.; Bennett, P.; McGreevy, P. Evidence of horsemanship and dogmanship and their application in veterinary contexts. *Vet. J.* **2015**, *204*, 247–254. [CrossRef] [PubMed]
15. Mellor, D.J.; Beausoleil, N.J. Extending the 'five domains' model for animal welfare assessment to incorporate positive welfare states. *Anim. Welf.* **2015**, *24*, 241–253. [CrossRef]
16. Mellor, D.J.; Reid, C.S.W. Concepts of animal well-being and predicting the impact of procedures on experimental animals. In *Improving the Well-Being of Animals in the Research Environment*; Baker, R., Jenkin, G., Mellor, D.J., Eds.; Australian and New Zealand Council for the Care of Animals in Research and Teaching: Glen Osmond, Australia, 1994; pp. 3–18.

17. Beausoleil, N.J. Fulfilling public expectation: Training vets as animal welfare experts. In Proceedings of the New Zealand Veterinary Association Conference, Hamilton, New Zealand, 21–24 June 2016.
18. Coleman, G.J.; Hemsworth, P.H. Training to improve stockperson beliefs and behaviour towards livestock enhances welfare and productivity. *Rev. Scientifique Tech. (Int. Off. Epizoot.)* **2014**, *33*, 131–137. [CrossRef]
19. Mellor, D.J. Updating animal welfare thinking: Moving beyond the "five freedoms" towards "a life worth living". *Animals* **2016**, *6*, 21. [CrossRef] [PubMed]
20. Beausoleil, N.J.; Mellor, D. Complementary roles for systematic analytical evaluation and qualitative whole animal profiling in welfare assessment for three RS applications. In Proceedings of the 8th World Congress on Alternatives and Animal Use in the Life Sciences, Montreal, QC, Canada, 21–25 August 2011.
21. Meagher, R.K. Observer ratings: Validity and value as a tool for animal welfare research. *Appl. Anim. Behav. Sci.* **2009**, *119*, 1–14. [CrossRef]
22. Wemelsfelder, F.; Hunter, E.A.; Mendl, M.T.; Lawrence, A.B. The spontaneous qualitative assessment of behavioural expressions in pigs: First explorations of a novel methodology for integrative animal welfare measurement. *Appl. Anim. Behav. Sci.* **2000**, *67*, 193–215. [CrossRef]
23. Wemelsfelder, F.; Hunter, T.E.A.; Mendl, M.T.; Lawrence, A.B. Assessing the 'whole animal': A free choice profiling approach. *Anim. Behav.* **2001**, *62*, 209–220. [CrossRef]
24. Wemelsfelder, F.; Nevison, I.; Lawrence, A.B. The effect of perceived environmental background on qualitative assessments of pig behaviour. *Anim. Behav.* **2009**, *78*, 477–484. [CrossRef]
25. Clarke, T.; Pluske, J.R.; Fleming, P.A. Are observer ratings influenced by prescription? A comparison of free choice profiling and fixed list methods of qualitative behavioural assessment. *Appl. Anim. Behav. Sci.* **2016**, *177*, 77–83.
26. Minero, M.; Dalla Costa, E.; Dai, F.; Murray, L.A.M.; Canali, E.; Wemelsfelder, F. Use of qualitative behaviour assessment as an indicator of welfare in donkeys. *Appl. Anim. Behav. Sci.* **2016**, *174*, 147–153. [CrossRef]
27. Greenacre, M. *Correspondence Analysis in Practice*, 2nd ed.; Taylor and Francis Group: Boca Raton, FL, USA, 2007.
28. Yarnell, K.; Hall, C.; Royle, C.; Walker, S.L. Domesticated horses differ in their behavioural and physiological responses to isolated and group housing. *Physiol. Behav.* **2015**, *143*, 51–57. [CrossRef] [PubMed]
29. Gronqvist, G.; Rogers, C.W.; Bolwell, C.F.; Gee, E.K.; Stuart, G. The challenges of using horses for practical teaching purposes: The capacity of students to interpret equine behaviour and the potential consequences to equine and student welfare. *Animals* **2016**, *6*, 69. [CrossRef] [PubMed]
30. Riley, C.; Liddiard, J.; Thompson, K. A cross-sectional study of horse-related injuries in veterinary and animal science students at an Australian university. *Animals* **2015**, *5*, 951–964. [CrossRef] [PubMed]
31. Barrett, L.F.; Mesquita, B.; Ochsner, K.N.; Gross, J.J. The experience of emotion. *Annu. Rev. Psychol.* **2007**, *58*, 373–403. [CrossRef] [PubMed]
32. Bokkers, E.A.M.; de Vries, M.; Antonissen, I.; de Boer, I.J.M. Inter- and intra-observer reliability of experienced and inexperienced observers for the qualitative behaviour assessment in dairy cattle. *Anim. Welf.* **2012**, *21*, 307–318. [CrossRef]
33. Duijvesteijn, N.; Benard, M.; Reimert, I.; Camerlink, I. Same pig, different conclusions: Stakeholders differ in qualitative behaviour assessment. *J. Agric. Environ. Eth.* **2014**, *27*, 1019–1047. [CrossRef]
34. Old, J.M.; Spencer, R. Development of online learning activities to enhance student knowledge of animal behaviour prior to engaging in live animal handling practical sessions. *Open J. Anim. Sci.* **2011**, *1*, 65–74. [CrossRef]
35. Goodenough, J.; McGuire, B.; Jakob, E. *Perspectives on Animal Behavior*, 3rd ed.; Wiley: Hoboken, NJ, USA, 2009.
36. AWIN. AWIN Welfare Assessment Protocol for Horses. Available online: https://www.google.ch/url?sa=t&rct=j&q=&esrc=s&source=web&cd=1&ved=0ahUKEwjzsZ-N5tXVAhXqAMAKHcmZAK4QFggrMAA&url=http%3A%2F%2Funi-sz.bg%2Ftruni11%2Fwp-content%2Fuploads%2Fbiblioteka%2Ffile%2FTUNI10015665.pdf&usg=AFQjCNFV5KuoTmZYy4c91cA0dRmatmQQFQ (accessed on 7 April 2017).

10

Public Understanding and Attitudes towards Meat Chicken Production and Relations to Consumption

Ihab Erian and Clive J. C. Phillips *

Centre for Animal Welfare and Ethics, School of Veterinary Science, University of Queensland, Gatton, Queensland 4343, Australia; ihab.erian@gmail.com

* Correspondence: c.phillips@uq.edu.au

Academic Editor: Marina von Keyserlingk

Simple Summary: Public knowledge of meat chicken production and how it influences attitudes to birds' welfare and consumer behaviour is poorly understood. We therefore conducted a survey of the public in SE Queensland, Australia, from which we determined that industry knowledge was limited. Where it existed, it related to an empathetic attitude towards chicken welfare and an increase in chicken consumption. This suggests that consumers who eat more chicken believe that they should understand the systems of production of the animals that they are consuming.

Abstract: Little is known about public knowledge of meat chicken production and how it influences attitudes to birds' welfare and consumer behaviour. We interviewed 506 members of the public in SE Queensland; Australia; to determine how knowledge of meat chicken production and slaughter links to attitudes and consumption. Knowledge was assessed from 15 questions and low scores were supported by respondents' self-assessed report of low knowledge levels and agreement that their knowledge was insufficient to form an opinion about which chicken products to purchase. Older respondents and single people without children were most knowledgeable. There was uncertainty about whether chicken welfare was adequate, particularly in those with little knowledge. There was also evidence that a lack of empathy towards chickens related to lack of knowledge, since those that thought it acceptable that some birds are inadequately stunned at slaughter had low knowledge scores. More knowledgeable respondents ate chicken more frequently and were less likely to buy products with accredited labelling. Approximately half of the respondents thought the welfare of the chicken was more important than the cost. It is concluded that the public's knowledge has an important connection to their attitudes and consumption of chicken.

Keywords: animal welfare; attitudes; chicken; knowledge; consumption; poultry

---

1. Introduction

Consumers' selection of food is governed by many factors, including culture, religion, lifestyle, diet, knowledge, health concerns and food trends, often influenced by the media [1]. Because they are no longer intimately involved in the food production process, the public's trust in the product is largely dependent on livestock producers having an empathetic approach to the animals used in the production of food [2]. This involves conforming to ethical standards throughout the breeding, growing and processing of the product. Two important concepts govern the intention to purchase animal-welfare-friendly products: consumer self-identification with ethical issues and Theory of Planned Behaviour, in which the attitudes, subjective norms and perceived level of behavioural control combine to influence the intention to purchase [3]. Self-identification is influenced by socio-demographic factors and the consumer's animal-related experiences [3]. The latter may be closely linked to their understanding of production systems; we chose meat chicken production to investigate this as it is one of the areas in which there is major concern for animal welfare. It is also important to recognize that consumers may identify animal welfare as far from optimal, but continue to buy and eat meat [4]. This is only possible because of distancing themselves from the production process.

Little is known about the public's sources of information on animal welfare, including the role of the media, with the associated problem of the accuracy of reporting. Their understanding of animal production systems may be anthropomorphic, an approach supported by some animal welfare scientists [5]. The media may particularly influence public opinion on contentious issues, such as the phasing out of battery cages [6,7]. Conversely, some possible developments in chicken production, such as breeding blind chickens, are not supported by the public because of their interference with bird integrity, even if they do appear to give welfare advantages in intensive production [8]. Many authors have attempted to alert the public to the welfare impact of intensifying production systems, starting with Ruth Harrison in the 1960s [6]. As production systems have intensified, it has become difficult for the public to assess the animals' welfare [7]. Modern industrial chicken meat production practices are designed to provide low-cost meat to consumers, retailing at less than one-half the price of other meats [9]. Many consumers have a negative perception of intensive farming and say that they are willing to pay more for food produced where animal welfare standards are considered and followed [10–12]. However, many consumers do not purchase the products from animals kept in better welfare because of the high price [13].

Despite a belief that welfare is being sacrificed for industrial-scale production, the demand for chicken meat is increasing worldwide; for example, the Australian market has increased by 160% in the last 20 years and the consumption of chicken has exceeded the consumption of any other kind of meat [9]. As well as low cost, the chicken industry attributes the rapid increase of demand to the versatility and ease of handling and cooking chicken products, and the fact that they are a low-fat protein source [9]. However, some discriminatory buying by consumers is evidenced in their reluctance to buy meat produced from intensive systems if the quality of meat produced is perceived to be adversely affected by the way the animals have been treated [14].

The Australian chicken industry has a "vertically integrated" structure. Individual companies control almost all aspects of production: breeding farms, hatcheries, feed mills, supply of feed to contractors, broiler growing farms/units, medication, transport, and initial and further processing plants [9,15]. The management from day-old stock until the day of processing, including staffing, housing and equipment, is mostly contracted out to growers (for example, 800 growers in Australia produce 80% of meat chickens, nearly all under contract to just two integrated national companies [9]. Growers are paid a negotiated monetary return per 100 birds or per weight at the end of the growing cycle. The typical modern unit has 3–10 poultry sheds that are tunnel ventilated, each about 150 × 15 m, with a capacity of 40–60,000 birds [9,16]. Aspects of production that relate to welfare, such as stocking densities, lighting regime and general husbandry practices, are usually determined by the companies' regulatory quality control systems or the industry code of practice [17]. Apart from contract rearing, other meat chickens are produced by large company farms, or on farms owned and managed by intermediary companies, where each is controlled by a manager who is contracted to a processing company. Breeding farms owned by the major chicken companies are strategically located across Australia, with a trend towards siting of great-grandparent and grandparent breeder farms in areas isolated from traditional poultry rearing places to reduce the risk of exposure to disease agents [9,16].

The aim of this study was to assess the public's knowledge of chicken production systems and its influence on attitudes towards animal welfare and chicken consumption. We hypothesized that, in line with the Theory of Planned Behaviour, the knowledge of the public about chicken production systems would have an impact on attitudes and the way consumers choose chicken products, and that these would also be influenced by key demographic factors. We anticipated that low levels of knowledge would cause uncertainty in choosing chicken products and in attitudes towards chicken welfare.

2. Materials and Methods

We used a quantitative questionnaire that addressed (1) public knowledge of intensive chicken meat production, including transport and slaughtering systems; (2) the attitude of the public towards meat chicken welfare and (3) their choice of chicken products. Socio-demographic questions were also included, which were used to further explain consumer behaviour and identify potential market segments. The questionnaire was designed taking into account the literature on public knowledge and attitudes towards chicken production systems [18–24]. We defined knowledge as "facts, information, and skills acquired through experience or education; the theoretical or practical understanding of a subject" and attitude as a "relatively stable favourable or unfavourable feeling or belief about a concept, person, or object" [25].

A total of 2663 consumers were approached in a face-to-face survey conducted with respondents who were randomly selected from the public in shopping centres, social clubs, cultural events and professional gatherings in Brisbane CBD, Surfers Paradise on the Gold Coast and a suburb of north Brisbane (Strathpine) during April and May 2013. Locations were selected to obtain a broad spectrum of views, in order to most accurately determine the relationships between knowledge and attitudes/consumption. Only respondents aged 18 years and over were eligible to take part. Clearances were obtained from the Brisbane City Council and the University of Queensland Human Ethics Committee (reference number 2013000458).

2.1. Questionnaire Design

A pilot survey was conducted by randomly selecting 15 individuals in Brisbane on each of three successive days. Following this, adjustment was made to the order and language of some of the questions to avoid any possible bias or leading responses.

The final questionnaire focused on the public knowledge and attitude to meat chicken production systems and the consumption of chicken products. Demographic questions were included to determine the respondent's gender, age, education level, place of residence, income, marital status and religion. Two initial questions addressed subjective knowledge, asking respondents about their level of knowledge about chicken production systems (options: expert, good, some, little or no knowledge) and how they gained it (formal qualifications, farm employment, personal interest, friends and acquaintances and all of these). Then the questions covered three topics specifically related to meat chicken production: (a) objective knowledge of common practices during rearing, transport and slaughter of broilers, how much knowledge they thought they had and whether it was sufficient for choosing which products to purchase, (b) attitudes towards the welfare of birds on farm and during transport and slaughter, and (c) frequency of consumption of different chicken products, their attitude towards labelling systems and willingness to pay more for accredited chicken products (defined as accredited by *FREPA* (Free Range Egg and Poultry Association), *RSPCA* approved farming, *ACO* (Australian Certified Organic) farming, *NASAA* (National Association for Sustainable Agriculture, Australia) or *OGA* (Organic Growers of Australia)). In total there were 15 knowledge questions (Appendix A), 13 attitude questions, four consumption questions and 10 demographic questions. In the knowledge section, there were eight initial questions, all of which were marked correct (score 1) or incorrect (score 0), except one which asked the normal distance that chickens travel from their place of rearing to the abattoir. Respondents were given a score of 1 for a distance of 5–100 km, this being normal in Australia, and 0.5 for a distance of 100–200 km. Respondents were also asked how long it takes intensively-reared meat chickens to reach a slaughter weight of 2 kg (1 point for 35–45 days; 0.5 point for 30–35 days, otherwise no score). We were also able to explore the relationship between knowledge score and respondents' self-reported understanding of chicken production systems as a validity check. Participants were asked to express their three biggest welfare concerns during transport and in production barns, with each valid welfare concern given a score of one, up to a maximum of 3. The maximum score in the knowledge section was 15.

2.2. Statistical Analysis

The questionnaire data was analysed in Minitab Version 16. The demographic background of participants was matched to the categories of the most recent Queensland Census 2011. Knowledge scores (K score) were determined for each respondent by the total number of correct answers out of 15 questions. This assumes that each question was the most relevant to test information on this aspect of the production system and contributes equally to a respondent's total knowledge. Numerical distribution of the K scores was examined and, to determine its influence on attitude and consumption, the total score was regressed against 21 predictors describing attitude and consumer behaviour using forwards backwards stepwise regression with alpha levels of 0.015 and fitted intercepts. Effects of the predictors found to be significantly ($p \leq 0.05$) correlated with knowledge scores were entered into a General Linear Model to examine the differences between levels. For this purpose, Knowledge scores were transformed to square root to approximate a normal distribution of residuals. Both back-transformed and untransformed means are provided. Pairwise comparisons were carried out using Tukey's test.

Logistic regression analyses (either binary, nominal or ordinal, as appropriate to the response structure) were used to analyse the effects of demographic variables and knowledge scores on the attitude and behaviour questions. For example, to evaluate whether place of residence influenced responses to attitude questions, urban residents were used as the referent base group and compared to the other three groups, acreage/large blocks, rural (country town) and rural (farming property) using nominal logistic regression. Referent base groups were selected as those with the most responses, except that males were used rather than females since either group can be chosen without affecting the analysis if there are only two. A principal component analysis was used to cluster responses to attitude (13 questions) and knowledge (15 questions) questions.

3. Results

Of the 2663 eligible participants approached, 506 answered the survey, a response rate of 19%. The average response time was estimated at 18 min. There were 205 males and 286 females, a higher female proportion compared with Queensland census data (Table 1). Fifteen respondents chose not to disclose their gender. The most common age bracket was 30–49 (50.7%), and the most common education level was to college or university degree level. Our respondents were more numerous in the 40–49 year old category and less numerous in the 60 and over category compared with the Queensland census data of 2011, and represented a more educated sample of the population. Most were urban dwellers (87%), with few rural town dwellers (6%) or acreage (large block) dwellers (5%). Respondents' income status was similar to the Australian average of AUD $78,000/year [26]. Most ($n$ = 305, 64%) were partnered with children, more than in the state of Queensland, and almost half (n = 234, 46%) were of the Christian religion, fewer than in the state.

Table 1. Demographics of respondents compared with data from Queensland, Australia (n = 506).

		Number of Respondents	% of Survey Sample	Queensland Data, % *
Gender	Male	205	41.7	49.6
	Female	286	58.2	50.4
Age	18–19	36	7.4	27.0 **
	20–29	66	13.6	13.7
	30–39	111	22.9	13.7
	40–49	135	27.8	14.2
	50–59	98	20.2	12.7
	60 & over	39	8.0	18.7

Table 1. *Cont.*

		Number of Respondents	% of Survey Sample	Queensland Data, % *
Education	No formal schooling	0	0	0
	Primary	10	2.06	29.7
	Secondary	74	15.2	20.2
	Technical College	61	12.5	6.2
	University	184	37.9	13.5
	Higher University Degree	141	29.0	30.4
	Other	16	3.3	
Dwelling	Urban	421	86.6	
	Acreage	26	5.3	
	Rural–town	27	5.6	
	Rural–farm	9	1.8	
	Other	3	0.62	
Annual Income	Less than $20,000	95	22.3	Mean $78,000
	$20,000–$39,000	47	11.0	
	$40,000–$59,000	75	17.6	
	$60,000–$80,000	81	19.0	
	>$80,000	128	30	
Marital Status	Single, no children	100	20.9	39.2
	Single, with children	20	4.2	7.8
	Married/De Facto no children	43	9.0	6.0
	Married/De Facto with children	305	63.7	42.0
	Widowed	11	2.3	5.0
Religion	Christian	234	46.1	64.8
	Jewish	12	2.4	0.1
	Hindu	1	0.2	0.7
	Buddhist	7	1.4	1.5
	Muslim	22	4.3	0.8
	Atheist	53	10.5	22.1
	Other	36	7.1	10.0 ***
	No response	141	27.9	

* [26,27]; ** [26] lists only 15–19 years of age; *** Includes other religions and/or not stated.

3.1. Respondents' Knowledge

Respondents' level of understanding of chicken production systems was most commonly reported as no, little or some knowledge, with fewer than 10% responding that their knowledge was good or expert (Table 2). Most gained their knowledge from the Internet and media, with a significant number gaining it from friends. Most had never visited a chicken farm, and, of the approximately one-third that had visited one before, it was not recent for most.

The distribution of K scores was not normal, but $\sqrt{K}$ score approximated a normal distribution, except that there was a higher than expected number of zero values ($n = 28$) (Figure 1). The mean value for $\sqrt{K}$ score was 1.99 (K score 3.96/15), median 2.0 (K score 4.0/15), with a Standard Deviation of 1.24. Given that the mean and medians were very similar, $\sqrt{K}$ score values were used for analysis.

Table 2. Number and % of respondents with answers to knowledge questions that were not significantly ($p < 0.05$) related to respondents' knowledge (K score).

Questions and Response Options	Number of Respondents	% of Respondents
Knowledge of chicken production systems		
Self-rated understanding of chicken production system		
Expert	7	1.4
Good knowledge	38	7.5
Some knowledge	134	26.5
Little knowledge	191	37.7
No knowledge	136	26.9

Table 2. *Cont.*

Questions and Response Options	Number of Respondents	% of Respondents
Source of knowledge		
Formal qualifications—relevant degree, training course	15	3.7
Farm employment—hands-on experience, relevant training course	23	5.7
Personal interest, e.g., internet, journals, newspaper articles, television programmes	223	55.1
Friends and acquaintances	136	33.6
All of the above	8	2.0
Visits to a chicken production farm?		
Yes, in the last two years	25	4.9
Yes, more than two years ago or on a school trip	153	30.2
I live on a chicken production farm	4	0.8
Never	324	64.1

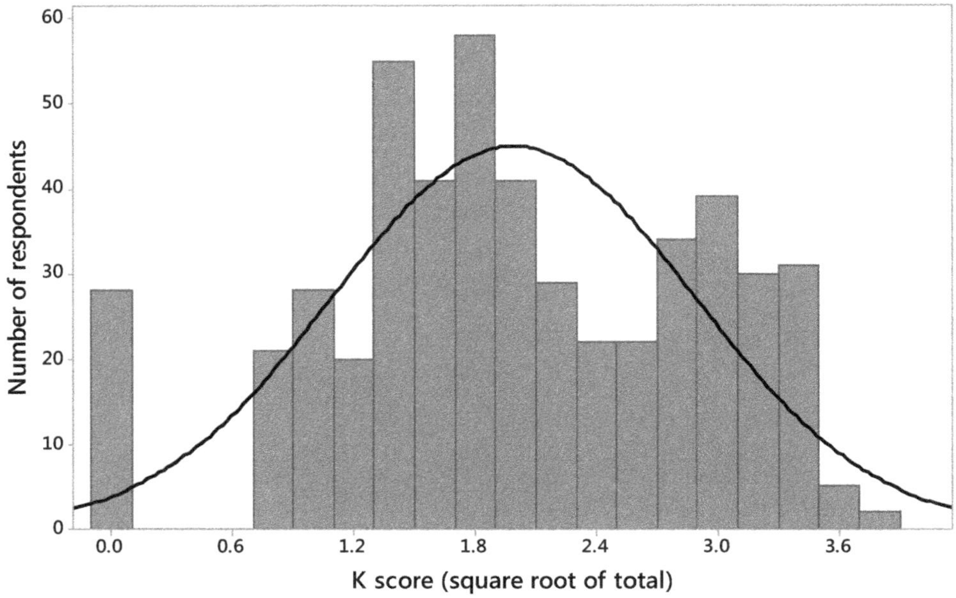

Figure 1. Distribution of K scores (out of 15) approximated a normal distribution curve with a higher than expected number of zero values (n = 28). The mean value was 1.99 (K score 3.96/15), Standard Deviation 1.24, and Median Value 2.00 (K score 4/15).

After eight knowledge questions (Appendix A), respondents were asked how long it takes for intensively-reared meat chickens to reach a slaughter weight of 2 kg. Only 47 respondents (9%) gave the correct answer of 35–45 days, 32 (6%) said 30–35 or 45–50 days and 427 (84%) gave answers outside of these choices. Respondents were asked, what are three of the biggest welfare problems for meat chickens in barns? A total of 156 respondents (31%) gave three valid responses, a further 65 respondents (13%) gave two valid responses and a further 48 respondents (9%) gave one valid response. The remaining 47% (n = 237) did not respond. The most common responses were poor lighting systems, too little space per bird, unable to reach feeders, unable to spread wings and too rapid growth. When the same question was asked for chickens in transport, 143 respondents (28%) gave three valid responses, a further 63 respondents (12%) gave two responses and a further 50 respondents (10%) gave one response. The most common responses were overcrowding, hot temperatures, odour, absence of food and water, and long distances.

3.2. Attitudes towards Welfare in Chicken Rearing System

A cluster analysis of the attitude questions produced 4 components with eigen values > 1, explaining 61% of the variation in total. A biplot of the first two components demonstrated that there were similar responses to questions about how good or bad animal welfare was on the farm, during transport and in the abattoir (Figure 2). Similar responses were observed at the opposite end of the scale for the first component for questions relating to attitudes to consumption, and to the two questions about chickens being conscious (religious slaughter and stunning acceptability). The first component appears to relate to purchasing issues, with Cost versus Animal Welfare (the cost of chicken meat is more important to me than the chicken's welfare) at one end and willingness to pay more, including of accredited products, at the other end. The second component appears to relate to providing for animal welfare (most positive) versus pragmatic issues of cost and religious concerns (least positive).

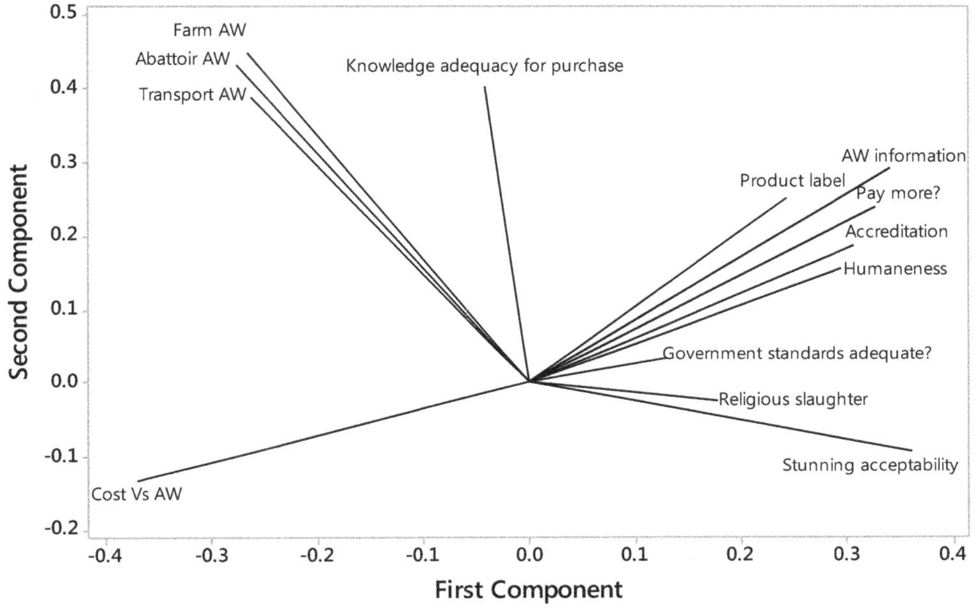

Figure 2. Biplot of Principal Component Analysis of attitude questions, showing the first two components. The first component appears to relate to purchasing issues and the second to pragmatic issues of providing for animal welfare. AW = animal welfare.

Most respondents were unsure whether chickens reared in meat production systems are protected by government standards which ensure that the welfare of birds is adequate (Table 3). They were also either unsure what they thought about meat chicken welfare on farms and during transport, or they thought it was good, bad or neither good nor bad in approximately equal numbers. Very few thought it was very good or very bad. In the abattoir (Table 4) most were unsure, but many thought it was neither good nor bad and a significant proportion (17%) thought that it was good.

Most respondents (54%) felt that it was unacceptable or very unacceptable that 1% of birds do not get adequately stunned by normal abattoir practices (Table 4). A similar proportion (58%) felt that it was unacceptable that some Australian abattoirs are allowed to kill chickens that are conscious, for religious reasons (Table 4). Most respondents (83%) agreed or strongly agreed with the statement "Food must be produced and processed from chickens that are treated humanely" (Table 3).

There was no consensus among respondents about whether their knowledge of the welfare of meat chickens was sufficient to allow them to form an opinion about which chicken products they should purchase (Table 4), but 49% disagreed or strongly disagreed that the cost of chicken meat was more important to them than the chicken's welfare, compared with only 27% agreeing or strongly agreeing (Table 3).

Table 3. Number and % of respondents in each category for attitudinal and consumption questions, for questions that were not significantly ($p < 0.05$) related to respondents' knowledge (K score).

Questions and Response Options	Number of Respondents	% of Respondents
Attitudes regarding chicken rearing systems		
Australian meat chickens are not protected by government welfare standards		
Strongly agree	32	6.5
Agree	84	16.9
Neither agree nor disagree	233	47.0
Disagree	137	27.6
Strongly disagree	10	2.0
Welfare of Australian meat chickens on the farm		
Very good	23	4.6
Good	105	20.8
Neither good nor bad	120	23.8
Bad	99	19.6
Very bad	17	3.4
Unsure	141	27.9
Welfare of Australian meat chickens during transport		
Very good	15	3.0
Good	108	21.3
Neither good nor bad	101	20.0
Bad	89	17.6
Very bad	44	8.7
Unsure	149	29.5
Killing chickens that are conscious for religious reasons in Australian abattoirs		
Very unacceptable	173	36.1
Unacceptable	106	22.1
No strong feelings	89	18.5
Acceptable but with some reservations	80	16.7
Perfectly acceptable	32	6.7
Food must be produced and processed from chickens that are treated humanely		
Strongly agree	250	51.4
Agree	155	31.9
Neither agree nor disagree	64	13.2
Disagree	15	3.1
Strongly disagree	2	0.41
Cost of chicken meat is more important to me than the chicken's welfare		
Strongly agree	32	6.6
Agree	101	20.7
Neither agree nor disagree	114	23.4
Disagree	166	34.1
Strongly disagree	74	15.2
Consumption of chicken products		
What brands of chicken meat are you most likely to buy?		
Free range	213	42.1
Corn or whole grain fed	46	9.1
Cheapest/home brand/on special	95	18.8
Products from a known producer	71	14.0
Products with heart foundation tick	41	8.10
Whole chicken	187	37.0
Chicken portions	177	35.0
Processed chicken products	72	14.2

Table 3. *Cont.*

Questions and Response Options	Number of Respondents	% of Respondents
Consumption of chicken products		
What type of chicken products do you usually buy?		
Whole chicken	275	54.4
Chicken pieces	343	67.8
Flavoured chicken meals	72	14.2
Processed chicken meat	84	16.6
Importance of rearing system on the product label when purchasing chicken products		
Very important	144	29.3
Quite important	164	33.3
Neither important nor unimportant	97	19.7
Not very important	64	13.01
Not important at all	23	4.7
Need for chicken welfare information wherever they are sold?		
Yes	308	63.1
No	82	16.8
Not interested	98	20.1
Amount you would be willing to pay to set up animal welfare ratings on animal products		
50 c/product if cost is ≤ $20	118	45.5
$1.00/product if cost is ≤ $20	42	16.2
$2.00/product if cost is ≤ $20	19	7.3
Whatever it costs to include	37	14.3
Should be done but I shouldn't pay	43	16.6

Table 4. Number and % of respondents in each category for attitudinal and consumption questions, for those questions with significant relationship to knowledge (K) score, together with the K score for responders to each option and probability of these being different (Standard Error of the Difference between any two $\sqrt{K}$ score means = 0.042).

Questions and Response Options	Number of Respondents	% of Respondents	$\sqrt{K}$ Score	K Score/15
Attitudes				
Welfare of Australian meat chickens at the abattoir				
Very good	8	1.6	2.34 [a]	5.47
Good	87	17.2	1.92 [ab]	3.69
Neither good nor bad	141	27.9	1.87 [b]	3.50
Bad	62	12.3	2.13 [ab]	4.54
Very bad	34	6.7	2.35 [a]	5.52
Unsure	174	34.4	1.57 [c]	2.46
p value			0.001	
1% of birds do not get adequately stunned in abattoir practices				
Very unacceptable	92	19.3	2.99 [a]	6.15
Unacceptable	164	34.5	2.20 [a]	4.84
No strong feelings	130	27.3	1.74 [b]	3.03
Acceptable with reservation	70	14.7	2.21 [a]	4.88
Very acceptable	20	4.2	1.52 [b]	2.31
p value			0.001	

Table 4. *Cont.*

Questions and Response Options	Number of Respondents	% of Respondents	√K Score	K Score/15
		Attitudes		
Killing chickens that are conscious for religious reasons in Australian abattoirs				
Very unacceptable	173	36.0	1.73 [b]	2.99
Unacceptable	106	22.1	2.14 [a]	4.58
No strong feelings	89	18.5	2.22 [a]	4.93
Acceptable with reservation	80	16.7	1.93 [ab]	3.72
Very acceptable	32	6.7	2.13 [ab]	4.53
p value			0.007	
Self-rated knowledge of chicken welfare is enough to form opinion about buying chicken products				
Strongly agree	35	7.3	2.14 [ab]	4.58
Agree	138	28.6	2.39 [a]	5.71
Disagree	91	18.9	2.04 [b]	4.16
Strongly disagree	37	7.7	1.73 [b]	2.99
p value			0.001	
		Consumption/labelling		
Number of times per week you eat chicken				
Never/I'm vegetarian	7	1.5	2.32 [bc]	5.38
Never/Don't like chicken	21	4.4	1.79 [bcd]	3.2
<1/Week	133	28.0	0.66 [d]	0.43
Once/Week	299	63.0	1.74 [c]	3.03
2 or 3/Week	11	2.3	2.14 [b]	4.58
Daily	4	0.8	3.53 [a]	12.46
p value			0.001	
Type of chicken meat consumers buy				
Fresh	288	60.8	1.93 [ab]	3.72
Frozen	37	8.8	2.26 [a]	5.11
Mix of Both	149	31.4	1.90 [b]	3.61
p value			0.05	
Labelling—would you purchase a product with accredited labelling?				
Yes	307	*63.2*	1.91	3.65
No	179	*36.8*	2.15	4.62
p value			0.002	

Means with different superscripts differ significantly ($p < 0.05$) by the Tukey's test. √K Score = square root of the K score.

3.3. Consumption and Attitudes towards Labelling

The most common chicken products purchased were chicken pieces or whole chicken, not flavoured or processed products (Table 3). Free range chicken was the most common branded product purchased, followed by whole chicken and chicken portions (Table 3). Chicken was most commonly eaten weekly, and if not it was most likely to be eaten less than once a week (Table 4). Most bought it fresh (61.0%), not frozen (Table 4). Most respondents considered labelling of production systems important (Table 4), and an overwhelming majority (63%) wanted to see information regarding welfare wherever chicken products are sold (Table 3) and were specifically looking to buy accredited chicken products (Table 4). Just over half (56%) said that they were prepared to pay to set up animal welfare ratings but most commonly at the lowest option, $0.50 (AUD) per product item (Table 3), although some (14%) were willing to pay whatever it costs.

3.4. *Relationships between Respondents' Knowledge and:*

3.4.1. Demographics

Respondents' K scores increased with age from 1.9/15 for respondents ≤19 years to 5.5/15 for respondents aged 50–59 (Table 5). College certificate or diploma graduates had higher levels of knowledge than either respondents with high school certificates or university graduates. Acreage dwellers had higher knowledge scores than urban, rural town and other dwellers. K score was greatest for single people with no children.

Table 5. Number and % of respondents to questions with significant relationship to knowledge (K) score, together with the K score for responders to each option (Standard Error of the Difference between two means = 0.042) and probability of these being different.

Questions and Response Options	√K Score	K Score/15
Demographics		
Age		
≤19	1.39 [c]	1.93
20–29	1.97 [ab]	3.88
30–39	1.99 [b]	3.96
40–49	2.30 [a]	5.29
50–59	2.35 [a]	5.52
≥60	2.18 [ab]	4.75
p value	<0.001	
Highest level of education		
Primary	1.45 [abc]	2.10
High school	2.20 [b]	4.84
Technical college certificate/diploma	2.73 [a]	7.45
College/university degree	2.17 [b]	5.88
Higher university degree	2.31 [b]	5.34
Other	1.32 [c]	1.74
p value	0.001	
Place of residence		
Urban—city/town	2.06 [b]	2.24
Acreage/large block	2.61 [a]	6.81
Rural—country town	1.88 [b]	3.5
Rural—farming property	2.06 [ab]	4.24
Other	1.45 [b]	2.10
p value	0.002	
Marital status		
Single, no children	2.30 [a]	5.29
Single, children	1.82 [ab]	3.31
Partnered/de facto, no children	2.13 [ab]	4.53
Partnered/de facto, children	1.83 [b]	3.35
Widowed	2.08 [ab]	4.32
p value	0.001	

Means with different superscripts differ significantly ($p < 0.05$) by the Tukey's test.

3.4.2. Attitudes

When asked about the welfare of meat chickens at Australian abattoirs, respondents who rated it very bad or very good had higher K scores than those with intermediate ratings (Table 4). Those who were unsure had the lowest K score. When told that 1% of birds do not get adequately stunned by normal abattoir practices prior to slaughter, respondents who regarded the practice as unacceptable had high K scores of 4.58 compared with respondents with no strong feelings regarding the issue with a K score of 4.93. When asked if their knowledge about the welfare of meat chickens was sufficient to allow them to form an opinion about which chicken products to purchase, respondents who agreed had higher K scores than those who disagreed.

3.4.3. Consumption

Respondents' K scores increased with the frequency of eating chicken, from <1/week to daily. However, those who did not eat chicken because they were vegetarian or they did not like chicken had intermediate K scores, lower than those with the highest consumption rate. Frozen chicken purchasers tended to have higher K scores than consumers who bought fresh products or a mixture of fresh and frozen (Table 4). Consumers who were willing to buy products with accredited labelling had lower K scores than those that were not. When told that some Australian abattoirs are allowed to kill chickens without them being unconscious for religious reasons, respondents who rated the practice as very unacceptable had a low K score of 2.99 (Table 5), compared with other acceptability ratings.

3.5. Relationships between K Score and Attitude/Consumption

In the stepwise regression, K scores were regressed against 21 predictors about attitudes to chicken meat production system, consumption of chicken and demographics (Appendix B). The final model included 13 significant predictors and had an R^2 of 43%. The most important predictor was that as people said they had a greater understanding of chicken production systems, their K score increased. The second most important predictor was that high K scores were closely correlated with a self-reported low level of education, and the third most important predictor was that it was acceptable to kill chickens without stunning for religious purposes. Those with high K scores were more likely to be older, single with no children and agreeing that their knowledge is sufficient to form an opinion when purchasing chicken products. Of next importance was that they ate chicken frequently, they purchased frozen products and that they did not purchase chicken products with accredited labelling. They also regarded the welfare of meat chickens at the abattoir as good, and they were more likely to live in rural areas.

3.6. Gender Effects

3.6.1. Knowledge

For most questions females had the same level of understanding as males, however females were more likely to incorrectly identify food fed to chickens as not being of vegetable origin and more males than females thought that chickens' diets would include grass and hay (Appendix C).

3.6.2. Attitudes

More females than males thought that the welfare of meat chickens at Australian abattoirs was bad (Table 6). Females considered it less acceptable than males that approximately 1% of chickens do not get adequately stunned at the abattoir and that some Australian abattoirs are allowed to kill chickens without them being unconscious for religious reasons. More females than males agreed with the statement "Food must be produced and processed from chickens that are treated humanely."

Table 6. Significant differences in attitudes and consumer behaviour between the gender groups. Mean values are shown for the referent group for gender, male respondents, and the comparative group, female respondents, as well as Odds Ratio and *p* value for the difference.

Questions and Response Options	Males	Females	Coefficient	Odds Ratio	*p* Value
	Attitudes				
Chicken welfare at the abattoir, 1 vg–5 vb	2.80	3.17	−0.79	0.45	0.001
1% of birds do not get adequately stunned in abattoir practices, 1 vu–5 va	2.66	2.38	0.57	1.77	0.007
Abattoirs slaughter birds without stunning, 1 vu–5 va	2.49	2.26	0.91	2.49	0.001
Chicken must be treated humanely, 1 sa–5 sd	1.78	1.63	0.63	1.87	0.007
Cost of chicken is more important than chicken's welfare, 1 sa–5 sd	3.09	3.46	−0.74	0.48	0.001
	Consumption/labelling				
What chicken products do you buy?					
Free Range (no. respondents)	70	143	0.90	2.46	0.001
Processed (no. respondents)	15	57	−1.11	0.33	0.003
Whole (no. respondents)	79	108	−0.53	0.59	0.004
Chicken consumption (1 never, 6 daily).	3.67	3.62	0.699	2.01	0.006
Need information on chicken welfare (1 yes, 2 dk, 3 no).	1.60	1.49	0.53	1.69	0.04
Willing to pay more for animal welfare (1 yes, 2 no).	1.49	1.41	−0.43	0.65	0.04

vg = very good, vb = very bad, vu = very unacceptable, va = very acceptable, sa = strongly agree, sd = strongly disagree, dk = don't know.

3.6.3. Consumption

Females were less likely than males to regard the cost as more important than welfare in chicken production. Females were more likely than males to buy free range chicken products (143 compared to 70), whole chicken (108 compared to 79) and processed chicken products, e.g., chicken schnitzel (57 compared to 15). Females said that they ate chicken less frequently than males. Males were more interested than females in seeing information regarding the welfare of chickens at the point of sale and were more prepared to contribute to the cost of setting up animal welfare ratings on animal products, by paying extra for the product.

3.7. *Place of Residence Effects*

3.7.1. Knowledge

Compared to acreage dwellers, urban dwellers thought that birds in barns had greater space availability (Table 7). Participants living on acreage were more likely to be incorrect in questions about housing, gender determination and stunning, compared to urban dwellers. Rural dwellers were more likely to be correct in relation to gender determination, but incorrect in relation to housing systems, they were also more likely to believe that chickens travelled further to the abattoir, compared to urban dwellers.

Table 7. The difference in attitudes towards meat chicken welfare according to dwelling place: urban (city/town) (referent group, 1), acreage/large block (AC) (group 2), rural (country town) (group 3), rural (farming property) (group 4) and other dwellers (group 5).

Questions and Response Options	Base Line Group	Comparative Group	Coefficient	Odds Ratio	*p* Value
		Knowledge			
Space for each bird in barn, (1, 0.25 m^2–4, 5 m^2)	1: 1.61	2: 1.34	2.08	8.00	0.003
Housing the same for egg and meat production, 1 T, 2 DK, 3 F	1: 2.14	2: 2.64 3: 2.32	−2.06 −2.52	0.13 0.08	0.000 0.003
Feather sexing of chicken, 1 T, 2 DK, 3 F	1: 2.17	2: 2.41 3: 1.69	−0.94 2.24	0.39 9.43	0.47 0.001
Chicken travelling distance to abattoir	1: 2.23	3: 2.39	−1.46	0.23	0.02
Normal practice for meat chickens to be stunned before slaughter? (1 yes, 2 Dk, 3 No)	1: 1.92	2: 3.19	−2.37	0.09	0.001
		Attitudes			
Chicken welfare not adequately protected by government standards, 1 sa–5 sda	1: 2.96	2: 3.38	−0.89	0.41	0.038
Chicken welfare on farm, 1 vg–5 vb	1: 2.99	2: 2.69	1.48	4.34	0.001
Unstunned birds at abattoir, 1 vu–5 spa	1: 2.44	2: 3.19	−1.51	0.22	0.001
Abattoirs slaughter birds without stunning, 1 vu, 5 pa	1: 2.34	2: 2.92	−1.11	0.33	0.007
Chicken must be treated humanely, 1 sa, 5 sd–21	1: 1.69	2: 1.90	−1.55	0.21	0.001
My chicken welfare knowledge is adequate, 1 sa, 5 sd	1: 2.92	2: 3.27	−1.35	0.26	0.001
Cost of chicken is more important than chicken's welfare, 1 sa, 5 sd	1: 3.33	2: 2.63	1.73	5.65	0.001
		Consumption			
Chicken consumption, 1 never, 6 daily	1: 3.61	2: 3.93 3: 4.00	−1.06 −1.87	0.35 0.15	0.049 0.017
The importance of labelling chicken kept, 1 VI, 5 NI	1: 2.26	4: 3.55	−2.22	0.11	0.001
Need information on chicken welfare, 1 yes, 2 Dk, 3 no	1: 1.49	2: 2.21 4: 1.88	−1.91 −1.64	0.15 0.19	0.00 0.013
Buy chicken with accredited labelling, 1 yes, 2 no	1: 1.36	4: 1.66	1.67	5.28	0.025

Vg = very good, vb = very bad, vu = very unacceptable, va = very acceptable, sa = strongly agree, sd = strongly disagree, T = true (1), 2 = unsure/do not know (DK), F = False (3); VI = very important; NI = not important (5).

3.7.2. Attitudes

Acreage dwellers more strongly agreed than urban dwellers that the welfare of chickens reared for meat production systems is inadequately protected by government standards to ensure the welfare of the birds. Urban respondents were more likely to believe that the welfare of meat chickens on the farm is bad. Acreage dwellers found the facts that 1% of birds are not adequately stunned and abattoirs slaughter birds without stunning for religious reasons more acceptable than did urban dwellers. They more strongly disagreed that chickens must be treated humanely, and agreed that their meat chicken welfare knowledge is adequate to form an opinion about purchases, compared with urban dwellers. They agreed more than urban dwellers that the cost of chicken was more important than chicken welfare.

3.7.3. Consumption/Labelling

Acreage and rural dwellers said that they were likely to eat chicken more often than urban dwellers. Rural dwellers considered product labelling less important than urban dwellers, and both acreage and rural dwellers were less interested in seeing information regarding the welfare of chickens at the point of sale or to seek to purchase chicken products with accredited labelling systems.

3.8. Marital Status Effects

3.8.1. Knowledge

Single respondents with no children thought that they had more limited understanding of chicken production systems than those who were partnered with no children, and were most likely to be correct for three questions (Table 8).

Table 8. Significant differences in attitudes of respondents towards meat chicken welfare and consumption of respondents according to marital status. Means are shown for single, no children (referent group, 1) and the comparative groups, single with children (group 2), married/de facto, no children (group 3), married/de facto with children (group 4) and widowed (group 5), as well as coefficients of the regression, odds ratios and *p* values.

Questions and Response Options	Single, no Children (Referent)	Comparative Group	Coefficient	Odds Ratio	*p* Value
Knowledge					
Understanding chicken production system (1 little K to 4 expert)	1: 0.98	3: 1.60	−1.39	0.25	0.001
Attitude					
Chicken welfare not protected by government standards, 1 sa–5 sda	1: 2.89	3: 2.23	1.17	3.22	0.009
Chicken welfare on farm, 1 vg–5 vb	1: 2.95	5: 2.56	4.22	68.26	0.001
Chicken Welfare during transport 1 vg–5 vb	1: 4.36	2: 4.98 4: 3.55 5: 3.73	−1.32 0.71 2.12	0.27 2.03 8.33	0.009 0.023 0.009
Abattoir welfare rating, 1 vg–5 vb	1: 4.58	3: 4.21 5: 2.82	−1.17 4.26	0.31 70.89	0.01 0.001
Unstunned birds at abattoir, 1 vun–5 va	1: 2.31	5:3.37	−2.17	0.11	0.008
Abattoirs slaughter birds without stunning, 1 vun–5 va	1: 2.41	3: 2.88 5: 1.50	−1.14 2.02	0.32 7.55	0.008 0.03
Chickens must treated humanely, 1 sa–5 sd	1: 1.48	2: 2.15 4: 1.72	−2.05 −1.12	0.13 0.33	0.001 0.003
Consumption					
My chicken welfare knowledge sufficient 1 sa–5 sd	1: 3.01	5: 2.00	2.19	8.90	0.009
Chicken consumption rate, 1 never, 6 daily	1: 3.70	5: 3.27	3.65	38.55	0.000
The importance of chicken rearing system on 1 vi–5 ni	1: 2.06	4: 2.47	0.75	0.47	0.02
Information on chicken welfare 1 yes, 3 no	1: 1.27	2: 2.00 3: 1.48 4: 1.63 5: 2.00	−3.49 −1.53 −1.64 −3.72	0.03 0.22 0.19 0.02	0.001 0.009 0.001 0.001
Willing to pay how much more for animal welfare rating, 1 no money–5 whatever it takes	1: 2.33	2: 3.67 3: 3.32 4: 2.35 5: 5.00	−2.76 −1.69 0.15 −22.3	0.06 0.18 1.16 0.00	0.02 0.07 0.56 0.00

Vg = very good, vb = very bad, vun = very unacceptable, va = very acceptable, sa = strongly agree, sd = strongly disagree, T = true (1), Unsure/Do Not Know (2), F = False (3); K = knowledge, vi = very important; ni = not important (5).

3.8.2. Attitudes

Respondents who were single with no children were more likely than partnered respondents with no children to agree that the welfare of meat chickens is adequately protected by government standards. Widowers rated meat chicken welfare on the farm and in the abattoir to be worse and during transport to be better than single respondents without children; they also were more accepting of inadequate stunning procedures than single respondents without children. Partnered respondents with children rated welfare during transport to be better as well but single respondents with children rated it worse. Partnered respondents without children rated welfare worse in the abattoir. Partnered respondents with no children found killing without stunning for religious reasons more acceptable than widowers and single respondents without children. Single respondents without children agreed more with the statement that food must be produced and processed from chickens that are treated humanely than single or partnered respondents with children.

3.8.3. Consumption

Single respondents with children agreed more than widowers that their welfare knowledge is sufficient for chicken product purchase, and they considered labelling information about chicken farming systems more important than did partnered respondents with children. They wanted information on welfare of chickens at point of sale more than any other group, and they, and widowers, were more likely than those without children to say that they would pay for the cost of setting up animal welfare ratings on products.

3.9. Religion Effects

3.9.1. Attitudes

Muslims thought that the fact that 1% of birds are not adequately stunned was more acceptable than Christians (Table 9). Jews and atheists found it less acceptable than did Christians. Muslims also found it much more acceptable to kill chickens without stunning for religious reasons than did Christians. Christians more than Muslims, Jews and atheists agreed with the statements that food must be from chickens that are treated humanely and that cost was more important than the chicken's welfare more than Muslims, Jews and atheists. Compared to Muslims, atheists and Buddhists, Christians more strongly believed that their welfare knowledge about meat chickens was not sufficient for food purchasing, compared to Muslims, atheists and Buddhists.

Table 9. Differences between religion groups (Christian, Group 1), compared with other groups, Jewish (Group 2), Hindu (Group 3), Buddhist (Group 4), Muslim (Group 5), Atheist (Group 6) and others (Group 7).

Questions and Response Options	Christian Group	Comparative Groups	Coefficient	Odds Ratio	*p* Value
		Attitude			
Chicken welfare on farm, 1 vg–5 vb	1:5.22	5:3.54 6:3.71	5:−1.34 6:−0.83	0.26 6:0.43	0.003 6:0.019
Unstunned birds at abattoir, 1 vu–5 pa	1:2.88	2:2.58 5:3.08 6:2.78	2.08 −1.009 0.95	7.96 0.36 2.59	0.011 0.020 0.005
Chicken welfare not protected by government standards, 1 sa–5 sd	1:2.89	6:3.00	1.03	2.79	0.021
Chicken welfare during transport, 1 vg–5 vb, 6 us	1:3.63	6:4.11	−3.71	0.49	0.03
Abattoir welfare rating, 1 vg–5 vb, 6 us	1:3.91	5:4.14 6:4.04	−1.28 −1.00	0.28 0.37	0.006 0.007
Abattoirs slaughter birds without stunning, 1 vu–5 pa	1:2.88	5:3.78	−1.80	0.16	0.001

Table 9. *Cont.*

Questions and Response Options	Christian Group	Comparative Groups	Coefficient	Odds Ratio	p Value
		Attitude			
Chicken must be treated humanely, 1 sa–5 sd	1:1.43	2:1.04	2.39	10.94	0.036
		5:0.89	2.41	11.12	0.002
		6:0.98	1.79	5.97	0.001
My chicken welfare knowledge is sufficient for food choice, 1 sa–5 sd, 6 us	1:2.45	4:2.45	−2.27	0.10	0.045
		5:2.09	1.52	4.62	0.001
		6:2.18	1.42	4.16	0.001
Cost of chicken is more important than chicken's welfare, 1 sa–5 sd, 6 us	1:4.14	2:4.92	−2.93	0.05	0.000
		5:4.77	1.12	0.33	0.013
		6:4.58	−0.79	0.45	0.022
		Consumption/labelling			
The importance of labelling chicken kept, 1 vi–5 ni, 6 us	1:1.97	5:1.29	2.60	13.49	0.001
		6:1.66	1.10	3.00	0.002

Ni, not important, vg = very good, vb = very bad, vi = very important, vu = very unacceptable, va = very acceptable, sa = strongly agree, sd = strongly disagree, us = unsure

3.9.2. Consumption/Labelling

Christians thought that product labels giving details of chicken rearing systems were more important when making purchases than did Muslims and atheists.

3.10. Age Effects

As age increased, respondents were willing to pay less for an animal welfare rating (Table 10); they were more likely to select products with a Heart Foundation approval and more likely to choose chicken portions or corn/whole grain-fed chickens.

Table 10. Significant effects of age on responses.

	Coefficient	Odds Ratio	p Value
Willing to pay more for animal welfare rating, 1 yes, 2 no	−0.39	0.68	0.0001
Which kind of chicken products are you most likely to buy: products with heart foundation tick	−0.38	0.69	0.03
chicken portions	−0.35	0.71	0.003
Corn- or whole grain-fed	−0.36	0.70	0.0001

3.11. Income Effects

As income increased, respondents were more likely to believe that the welfare of meat chickens on the farm (Regression Coefficient 0.16, OR 1.18, $p = 0.04$) and during transport (Regression Coefficient 0.16, OR 1.18, $p = 0.04$) was bad. They were also more likely to believe that chickens can be killed for religious reasons without stunning (Regression Coefficient 0.24, OR 1.27, $p = 0.002$) and to know that gender could be determined from feathers (Regression Coefficient −0.28, OR 0.75, $p = 0.02$) and to know the distance that chickens travelled to the abattoir (Regression Coefficient 0.35, OR 1.42, $p < 0.0001$). However, they were less likely to know that birds are usually stunned before slaughter (Regression Coefficient −0.32, OR 0.73, $p < 0.0001$).

4. Discussion

The response rate of 19% was similar to other farm animal welfare surveys [21]. Randomly approaching members of the public who were not aware of the nature of the survey helped to minimize any potential bias [28,29]. However, some selection bias is evident and in particular the higher education level of the respondents compared to the Australian population could potentially influence people's understanding of chicken production systems. The preponderance of middle-aged respondents, compared with the Australian population, may have influenced our results on consumption (Table 9) and knowledge scores (Table 4). Most respondents indicated that they were urban dwellers, which is representative of the Australian population. Gaining contemporary knowledge about the industry was through the internet, journals, newspaper articles, television programmes and more noticeably, through friends and acquaintances. Further work on knowledge sources is warranted as the Australian public spends about $5.6 billion per year on poultry products [9].

4.1. Knowledge

K scores generally increased with self-rated knowledge of chicken welfare, adding validity that objective knowledge matched subjective knowledge assessment. The disproportionately high number of zero values in the knowledge score suggests that some respondents deliberately avoided answering all knowledge questions, but this may also have been because they genuinely did not know the answers. The majority of urban respondents (87%), compared with the Australian average of 63% [30], would be less likely than rural dwellers to be familiar with farming systems, which could contribute to low knowledge scores.

The knowledge questions demonstrated that public knowledge of chicken production systems was limited, with many participants possessing little or no knowledge of the industry and a median knowledge score of 4 out of 15, indicating that they answered four questions correctly out of 15, i.e., 27% (and mean of 3.96/15, Section 3.1). Bergman and Maller [20] studied the factors leading Australians to support or reject factory farming, especially poultry and pig productions, and concluded that Australian consumers knew little about these systems, there was significant confusion and scepticism about 'organic' & 'free range' labelling and limited trust in the RSPCA labelling systems. Napolitano et al. [18] examined the effect of information about animal welfare, expressed in terms of rearing conditions, on acceptability of lamb for consumption. Prior knowledge of rearing conditions influenced their perceived acceptability, with worse scores given to meat if they knew it had been reared artificially, rather than by their mother. According to Costell et al. [21], the hedonic acceptability of food items is related to whether our perception of food differs from the expected, which in turn may be influenced by understanding of production and processing systems involved in producing the food item.

Some responses indicated that chicken welfare problems become a banality as K score increases, for instance believing that it was acceptable to kill chickens without stunning for religious purposes. Although increasing K score was associated with increased self-reported knowledge of chicken meat production systems, it was associated with low levels of general education. The latter may indicate that high K score respondents lacked the broad education necessary to empathize with chickens in poor welfare conditions.

4.2. Attitudes

The attitude of 52% of respondents was that meat chicken welfare on the farm was neither good nor bad, or they were unsure, suggesting that there is a great deal of uncertainty about this issue. Similarly, 49 and 62% of respondents had no definite attitude regarding the welfare of meat chickens during transport or at the abattoir, respectively. By contrast, in Europe most (77%) of the public believe that improvement in animal welfare is needed, with meat chickens being one of the systems of production most in need of reform [31]. Similarly, there was little agreement about whether the existing Australian standards ensured that the welfare of reared meat chickens is adequate. Mench [32] and Sumner et al. [33] suggested that standards should not only minimise animal suffering during transport and slaughter but maintain quality of life for animals throughout their production life. Such uncertainty appeared to link to respondents' lack of knowledge, with those that were unsure about the chickens' welfare in abattoirs having lowest knowledge scores. Similarly those that had no strong feelings in relation to inadequate stunning, which may indicate uncertainty, also had low scores. Lowest scores in this question were given by those finding it very acceptable, giving some credence to a relationship between empathy towards the chicken and knowledge.

4.3. Consumption

With the exception of those that avoided chicken because they were vegetarian or they did not like it, K score increased considerably with the frequency of consumption, and more knowledgeable respondents were less likely to buy products with accredited labelling. The latter may be explained by those with knowledge believing accreditation to be unnecessary for their choice of chicken product. Regarding the frequency of chicken consumption, one possibility is that people consuming more chicken are interested to learn about the industry. Another is that people of higher socioeconomic status were more knowledgeable about farming systems and ate more chicken because they are more aware of its health benefits. However, the more knowledgeable respondents were less willing to pay for accredited labelling for chicken welfare, which would not be expected of high socioeconomic respondents. A third possibility is that the more knowledgeable, frequent chicken consumers were connected with the industry, however, we considered this unlikely as only 1% lived on a chicken farm, 5% had visited one in the last two years and 6% indicated that they had gained their knowledge as farm employees.

The type of chicken meat that respondents said they were most likely to buy was free range chicken products, whereas Australian free range chicken meat production accounts for only 10 to 15% of the total production [34]. This properly reflected an intention or desire, rather than actuality. Furthermore, a total of 63% of respondents sought to purchase a chicken product with accreditation, particularly if they had little knowledge about chicken production systems. This suggests that consumers are using accreditation as a means of ensuring products are of high welfare, replacing their limited knowledge, even though accredited labels exhibit no information regarding the conditions where birds were raised or processed and no reference to animal welfare [35]. Consumers have put pressure on retailers to properly label products and on producers, manufacturers and supermarkets to have an animal welfare labelling system [23] as well as the country of origin, production techniques [36] and conditions of rearing [19]. Fifty-six percent of respondents were prepared to contribute to the cost of setting up animal welfare ratings by paying extra for the products, particularly females, and the most common increase in cost that would be accepted was 2.5%. A study in Chile indicated a willingness to pay up to 15% more for meat produced to improved animal welfare standards [19]. European consumers have indicated their willingness to change their usual place of shopping to be able to purchase more animal-welfare-friendly products [37]. Consumers are also willing to pay more for natural or organic chicken [23], with the latter being perceived as safer, healthier and having fewer pesticides, hormones and antibiotics than other meat [38]. Labelling systems are based on transparency, informing consumers that the products have satisfied the welfare conditions where animals were reared, transported and processed [39].

4.4. Demographic Effects

On ethical issues, socio-demographics as explanatory variables of behaviour may be less influential than values, attitudes, motives and lifestyles. In our study we had major effects of gender and dwelling place on attitudes and consumption, whereas religion had the most influence on attitudes but little on consumption.

4.4.1. Gender

There was no evidence that females had a better understanding of the chicken production system than males. However, it is recognised elsewhere that females have greater knowledge of animal welfare concerns, with males being more traditional in their purchasing habits for animal products [40]. Females displayed greater sensitivity to chicken welfare than males, confirming much previous research [41,42]. Females were more ethical about their chicken consumption intentions, and reported being twice as likely to buy free range but only slightly more likely to buy whole chicken. They were also much more likely to buy processed products, which may reflect their role in managing the nutrition of children. They reported buying less chicken than males, confirming a Eurasian survey which found that female students reported that they ate poultry less commonly than male students [43]. Males' showed greater interest than females in seeing information regarding the welfare of chicken at the point of sale than females and even being more prepared to pay for this conflicts with other studies [44,45] which found that females were willing to pay extra for certified food products. Females reporting less frequent consumption of chicken than did males probably reflects the fact that women show more health-related behaviours and considered attitudes towards food than men [46,47].

4.4.2. Place of Residence

Dwellers on acreage/large blocks were more knowledgeable than most other groups, but they were generally less sympathetic to chicken welfare than urban dwellers, in relation to stunning practices and treating birds humanely. The acreage or large block dwellers are more likely to keep chickens and gain their attitudes towards chicken welfare from this practice, rather than through the media, which would be the case for urban dwellers. Acreage/large block dwellers also ate more chicken than urban dwellers and were less interested in labelling about chicken welfare, even though they thought government standards were less than adequate to protect welfare. The latter suggests a better knowledge but less concern in acreage/large block dwellers, compared with urban dwellers.

4.4.3. Marital Status

Single respondents with children were most likely to want information on the welfare of chicken at point of sale, probably reflecting their limited time for shopping, and they considered this information more than some other groups. Other research has identified that single parents with children spend more of their food budget eating away from the home, compared to partnered respondents with children [48]. The study suggested that the most sympathetic consumers were single respondents without children, as they rated welfare worse on the farm and in the abattoir and agreed most that chickens must be treated humanely. They were also least accepting of inadequate stunning or avoiding stunning for religious reasons. Results for widowers should be treated with caution as they are confounded with age.

4.4.4. Religion

Muslims knew more about stunning than Christians and they were less likely to find it acceptable, reflecting their belief that animals must be alive when their throats are cut and must die from loss of blood [49]. Overall 54% of our respondents' believed that the practice of slaughtering birds without adequate stunning was unacceptable, probably because of the welfare impact [50]. There was an apparent contradiction between Christians having greater regard for cost than an animal's welfare but also requiring chicken to be from animals that are treated humanely, compared to Muslims, Buddhists and atheists.

4.4.5. Age, Income and Education

The reduced willingness to pay for animal welfare ratings as respondents aged may reflect reduced disposable cash for this purpose, or it may reflect changing attitudes, this not being a longitudinal study. Greater tolerance to not stunning the chicken for religious reasons was evident in higher income respondents, confirming previous findings in Chinese studies [43]. A greater willingness to recognise poor welfare on farm and during transport in high income respondents may reflect a greater ability to pay for high welfare products. Respondents with a low level of education had high K scores. This suggests that there was a cohort of poorly-educated respondents who had knowledge of the poultry industry.

5. Conclusions

Public knowledge of the Australian poultry production systems was limited. Most was indirectly gained from the media, and few respondents had direct experience with chicken farming. Our finding that knowledge related to an improved attitude towards chicken welfare is valuable, since it suggests that informing the public about chicken welfare could increase levels of concern. However, this was not associated with increased consumption of high-welfare products; in fact, high-level consumers had a natural suspicion of accreditation programmes that would make it difficult to improve animal welfare through this method. The observed positive relationship between chicken consumption and knowledge may derive from a belief in respondents who ate relatively more chicken that they should understand the systems of production of the animals that they are consuming. The connection between knowledge and attitudes suggests that educating consumers might help to improve their empathy towards meat chickens, but the lack of relationship between empathy and consumption and the suspicion of accreditation systems suggests that any increased empathy will not necessarily have an impact on the sales of high-welfare products.

More scientific studies are needed to support public demand for improving the welfare conditions of chickens, as they were at least willing to contribute a small amount (median about 5%) to establish labelling systems that take into account the welfare of birds. The study also identified those consumers who were most concerned about the welfare of chickens in this context: females, urban dwellers and relatively high-income respondents.

Acknowledgments: The authors are grateful to the volunteers who assisted in data collection.

Author Contributions: Both authors designed the survey. Ihab Erian organized the collection of data, which was analysed by Clive Phillips.

Conflicts of Interest: The authors declare no conflict of interest.

Appendix A

Table A1. Number and % of respondents responding to eight knowledge questions (*correct answer in italics*).

Questions and Response Options	Number of Respondents	% of Respondents
What type of housing is most commonly used to rear meat chickens in Australia?		
Multi-tier battery cages in barns	207	49.3
No housing, free range on pasture is normal	30	7.1
Single tier battery cages on the floor of barns	81	19.3
Loose in barns	*102*	*24.3*

Table A1. *Cont.*

Questions and Response Options	Number of Respondents	% of Respondents
How much space is it usual to give each bird in barns?		
About 1 m^2	168	38.9
About the size of a piece of A4 paper [[34]]	*224*	*51.9*
About 5 m^2	25	5.8
About 2 m^2	15	3.5
Housing for egg production chickens is the same as for meat production chickens		
True	84	16.7
False	*179*	*35.5*
Don't know	241	47.8
The sex of a chicken is usually determined from the feathers on their wings		
True	*61*	*12.4*
False	133	27.00
Don't know	299	60.6
Chickens are usually fed food of vegetable origin		
True	*197*	*39.2*
False	108	21.5
Don't know	197	39.2
The usual feed for meat chickens in barns is:		
Hay	24	5.4
Pelleted cereal feed	*266*	*59.9*
Cut grass	35	7.9
Household waste food	15	3.4
All of these	104	23.4
What is the normal distance that chickens travel from their place of rearing to the abattoir?		
Up to 5 km	85	20.8
5 to 100 km	*187*	*45.7*
100 to 200 km	92	22.5
200 to 500 km	29	7.1
500 km or more	16	3.9
Is it normal practice for meat chickens to be rendered unconscious (stunned) before slaughter?		
Yes	*116*	*23.9*
No	279	57.4
Don't know	91	18.7

Appendix B

Table A2. Stepwise regression of 21 attitude, consumption and demographic predictors on $\sqrt{K}$ score values for 378 respondents.

	Step 1	Step 2	Step 3	Step 4	Step 5	Step 6	Step 7	Step 8	Step 9	Step 10	Step 11	Step 12	Step 13
Constant	3.35	6.38	5.45	6.64	5.59	5.92	7.48	5.20	6.10	4.95	6.00	5.54	6.52
Self-rated understanding of chicken production systems	1.61	1.74	1.65	1.60	1.52	1.50	1.23	1.20	1.19	1.27	1.24	1.23	1.17
t-Value	8.40	9.18	8.96	8.80	8.42	8.61	6.34	6.25	6.26	6.68	6.57	6.55	6.18
p-Value	0.000	0.000	0.000	0.000	0.000	0.000	0.000	0.000	0.000	0.000	0.000	0.000	0.000
Highest level of education		−0.65	−0.75	−0.76	−0.80	−0.65	−0.56	−0.48	−0.49	−049	−0.49	−0.45	−0.46
t-Value		−4.43	−5.20	−5.40	−5.74	−4.68	−3.99	−3.43	−3.47	−3.57	−3.58	−3.21	−3.30
p-Value		0.000	0.000	0.000	0.000	0.000	0.000	0.001	0.001	0.000	0.000	0.001	0.001
Killing chickens that are conscious for religious reasons in Australian abattoirs			0.64	0.95	0.98	0.87	0.88	0.78	0.82	0.81	0.83	0.82	0.81
t-Value			5.04	6.43	6.69	6.06	6.22	5.38	5.69	5.69	5.82	5.80	5.77
p-Value			0.000	0.000	0.000	0.000	0.000	0.000	0.000	0.000	0.000	0.000	0.000
1% of birds do not get adequately stunned in abattoir practices				−0.70	−0.80	−0.70	−0.77	−0.76	−0.71	−0.81	−0.90	−0.91	−0.81
t-Value				−3.91	−4.46	−4.01	−4.42	−4.41	−4.13	−4.72	−5.11	−5.21	−4.42
p-Value				0.000	0.000	0.000	0.000	0.000	0.000	0.000	0.000	0.000	0.000
Age					0.44	0.87	0.90	0.92	0.90	0.92	0.90	0.88	0.91
t-Value					3.46	5.81	6.07	6.24	6.13	6.33	6.23	6.08	6.26
p-Value					0.000	0.000	0.000	0.000	0.000	0.000	0.000	0.000	0.000
Marital status						−0.80	−0.83	−0.83	−0.87	−0.89	−0.94	−0.97	−0.98
t-Value						−5.00	−5.22	−5.32	−5.59	−5.77	−6.04	−6.22	−6.32
p-Value						0.000	0.000	0.000	0.000	0.000	0.000	0.000	0.000
Self-rated knowledge of chicken welfare is enough to form opinion about buying chicken products							−0.54	−0.62	−0.61	−0.59	−0.54	−0.52	−0.50
t-Value							−3.17	−3.62	−3.59	−3.52	−3.21	−3.09	−2.96
p-Value							0.002	0.000	0.000	0.000	0.001	0.002	0.003
Number of times per week you eat chicken								0.64	0.60	0.63	0.65	0.63	0.62
t-Value								2.88	2.70	2.88	2.98	2.90	2.83
p-Value								0.004	0.007	0.004	0.003	0.004	0.005

Table A2. *Cont.*

	Step 1	Step 2	Step 3	Step 4	Step 5	Step 6	Step 7	Step 8	Step 9	Step 10	Step 11	Step 12	Step 13
Type of chicken meat consumers buy									−0.45	−0.56	−0.52	−.55	−0.59
t-Value									−2.66	−3.27	−3.05	−3.23	−3.45
p-Value									0.008	0.001	0.002	0.001	0.001
Would you purchase a product with accredited labelling?										1.04	1.01	1.00	0.89
t-Value										3.13	3.04	3.02	2.64
p-Value										0.002	0.003	0.003	0.009
Chicken welfare at the abattoir											−0.22	−0.22	−0.22
t-Value											−2.22	−2.13	−2.14
p-Value											0.027	0.034	0.033
Place of residence												0.43	0.48
t-Value												1.81	1.98
p-Value												0.071	0.048
Australian meat chickens not protected by government welfare standards													−0.32
t-Value													−1.92
p-Value													0.056
S	3.36	3.28	3.18	3.12	3.07	2.98	2.94	2.91	2.89	2.86	2.84	2.83	2.82
R-Sq	15.81	19.99	25.08	28.03	30.27	34.67	36.40	37.80	38.97	40.56	41.34	41.87	42.45

Appendix C

Significant differences ($p < 0.05$) in knowledge according to demographics that are not presented in the paper. Mean values are shown for the referent group and the comparative group, as well as Odds Ratio and p value for the differences.

Table A3. Significant differences between males and females.

	Males	Females	Coefficient	Odds Ratio	p Value
Space for each bird in barn, (1, 0.25 m^2–4, 5 m^2)	1.61	1.88	−0.71	0.49	0.006
Chickens are usually fed food of vegetable origin (1 T, 2 DK, 3 F)	1.82	2.04	−0.80	0.45	0.001
The usual feed for meat chickens in barns, those answering:					
Cut grass	0.52	0.25	−1.89	0.15	0.001
Hay	0.52	0.10	−2.54	0.08	0.001
Pelleted Cereal food	3.8	2.0	−0.78	0.46	0.02

T = true, DK = Do Not Know, F = False.

Table A4. Significant differences according to marital status, between Single, no children (referent group, 1) and the comparative groups, single with children (group 2), married/de facto, no children (group 3), married/de facto with children (group 4) and widowed (group 5).

Single, no Children (Referent)		Comparative Group	Coefficient	Odds Ratio	p Value
		Knowledge			
Space for each barn bird (1, 0.25 m^2–4, 5 m^2)	1:1.79	4:1.57	0.72	2.00	0.049
Feather sexing of chicken (1 T, 2 DK, 3 F)	1:2.03	2:2.45	−1.95	0.14	0.001
Chicken food is of vegetable origin (1 T, 2 DK, 3 F)	1:1.86	2:1.15 5:2.1	3.55 −2.69	34.82 0.02	0.001 0.003
Chicken travelling distance to abattoir 1 < 5 km–5500 km +	1:2.62	2:2.16 3:1.94 4:2.20	1.59 1.02	4.87 2.77	0.004 0.03
Stunned meat chicken (*1 yes, 2 DK, 3 no*)	1:2.27	5:2.70	1.10 −2.33	3.01 0.10	0.002 0.012

Table A5. Significant differences according to religion, between the Referent group: Christian, Group 1, compared with comparative groups, Jewish (Group 2), Hindu (Group 3), Buddhist (Group 4), Muslim (Group 5), Atheist (Group 6) and others (Group 7).

	Christian (Referent)	Comparative Group	Coefficient	Odds Ratio	p Value
		Knowledge			
Understanding chicken production systems	1:1.13	6:1.00	−0.72	0.49	0.037
Space for each barn bird (1, 0.25 m^2–4, 5 m^2)	1:1.69	5:1.47	1.17	0.03	0.03
Housing the same for egg & meat production	1:2.18	2:1.47 5:1.82	1.86 1.13	6.44 3.08	0.02 0.02
Chicken travelling distance to abattoir	1:0.72	5:0.54	1.22	0.30	0.01
Meat chickens stunned	1:0.21	5:0.50	1.19	3.3	0.01

References

1. Popa, A.; Draghici, M.; Popa, M. Consumer choice and food policy: A literature review. *J. Environ. Prot. Ecol.* **2011**, *12*, 708–717.
2. Fraser, D. Could animal production become a profession? *Livest. Sci.* **2014**, *169*, 155–162. [CrossRef]
3. Gracia, A. The determinants of the intention to purchase animal welfare-friendly meat products in Spain. *Anim. Welf.* **2013**, *22*, 255–265. [CrossRef]
4. Te Velde, H.; Aarts, N.; Van Woerkum, C. Dealing with ambivalence: Farmers' and consumers' perceptions of animal welfare in livestock breeding. *J. Agric. Environ. Ethics* **2002**, *15*, 203–219. [CrossRef]
5. Korte, S.M.; Oliver, B.; Koolhaas, J.M. A new animal welfare concept based on allostasis. *Physiol. Behav.* **2007**, *92*, 422–428. [CrossRef] [PubMed]
6. Sayer, K. Animal machines: The public response to intensification in Great Britain, c 1960—c 1973. *Agric. Hist.* **2013**, *87*, 473–501. [CrossRef]
7. Daigle, C.L. Incorporating the philosophy of technology into animal welfare assessment. *J. Agric. Environ. Ethics* **2014**, *27*, 633–647. [CrossRef]
8. Verrinder, J.; Phillips, C.J.C. Author's Response: Response to Letter to the Editor, "The VetDIT and Veterinary Ethics Education". *J. Vet. Med. Educ.* **2015**, *42*, 174–175. [CrossRef] [PubMed]
9. Australian Chicken Meat Federation (ACMF) Inc. Industry Facts and Figures 2013. Available online: http://www.chicken.org.au/page.php?id=4 (accessed on 8 March 2016).
10. Köbrich, K.; Maino, M.; Diaz, C. El bienestar animal como atributo de diferenciación en la compra de alimentos de origen animal. *Econ. Agraria* **2001**, *6*, 251–260.
11. Maria, G.A. Public perception of farm animal welfare in Spain. *Livest. Sci.* **2006**, *103*, 250–256. [CrossRef]
12. Bonamigo, A.; Bonamigo, C.B.S.S.; Molento, C.F.M. Broiler meat characteristics relevant to the consumer: Focus on animal welfare. *Revista Brasileira Zootecnia-Braz. J. Anim. Sci.* **2012**, *41*, 1044–1050. [CrossRef]
13. Akaichi, F.; Revoredo-Giha, C. Consumers demand for products with animal welfare attributes—Evidence from homescan data for Scotland. *Brit. Food J.* **2016**, *118*, 1682–1711. [CrossRef]
14. Schröder, M.J.A.; McEachern, M.G. Consumer value conflicts surrounding ethical food purchase decisions: A focus on animal welfare. *Int. J. Consum. Stud.* **2004**, *28*, 168–177. [CrossRef]
15. Kvaløy, O.; Tvetera, R. Cost structure and vertical integration between farming and processing. *J. Agric. Econ.* **2008**, *59*, 296–311. [CrossRef]
16. Australian Chicken Meat Federation (ACMF) Inc. An Industry in Profile: ACMF 2011. Available online: http://www.chicken.org.au/industryprofile/downloads/The_Australian_Chicken_Meat_Industry_An_Industry_in_Profile.pdf (accessed on 7 March 2016).
17. Robins, A.; Phillips, C.J.C. International approaches to the welfare of meat chickens. *World Poultry Sci. J.* **2011**, *67*, 351–369. [CrossRef]
18. Napolitano, F.; Caporale, G.; Carlucci, A.; Monteleone, E. Effect of information about animal welfare and product nutritional properties on acceptability of meat from Podolian cattle. *Food. Qual. Preference* **2007**, *18*, 305–312. [CrossRef]
19. Schnettler, B.M.; Vidal, R.M.; Silva, R.F.; Vallejos, L.C.; Sepúlveda, N.B. Consumer perception of animal welfare and livestock production in the Araucania region, Chile. *Chile J. Agric. Res.* **2008**, *68*, 80–90. [CrossRef]
20. Bergmann, I.; Maller, C.J. What factors lead Australians to support or actively reject factory farming? Presentation at the International Inaugural Minding Animals Conference, Newcastle University, Newcastle Civic Precinct, Newcastle, NSW, Australia, 13–18 July 2009.
21. Costell, E.; Tárrega, A.; Bayarri, S. Food acceptance: The role of consumer perception and attitudes. *Chemosens. Percept.* **2009**, *3*, 42–50. [CrossRef]
22. Prickett, R.W.; Norwood, F.B.; Lusk, J.L. Consumer preferences for farm animal welfare: Results from a telephone survey of US households. *Anim. Welf.* **2010**, *19*, 335–347.
23. Sismanoglou, A.; Tzimitra-Kalogianni, I. Consumer perception of poultry meat in Greece. *World. Poultry Sci. J.* **2011**, *67*, 269–276. [CrossRef]
24. Gifford, K.; Bernard, J.C. The effect of information on consumers' willingness to pay for natural and organic chicken. *Int. J. Consum. Stud.* **2011**, *35*, 282–289. [CrossRef]
25. Last, J.M. *A Dictionary of Public Health*; Oxford University Press: Oxford, UK, 2007.

26. Zoethout, C.M. Ritual slaughter and the freedom of religion: Some reflections on a stunning matter. *Hum. Rights Q.* **2013**, *35*, 651–672. [CrossRef]
27. Australian Bureau of Statistics 2011 Census QuickStats. Available online: http://www.censusdata.abs.gov.au/census_services/getproduct/census/2011/quickstat/SOS30?opendocument&navpos=220 (accessed on 8 March 2016).
28. Manfreda, K.L.; Bosnjak, M.; Berzelak, J.; Haas, I.; Vehovar, V. Web Surveys versus Other Survey Modes: A Meta-Analysis Comparing Response Rates. *Int. J. Market. Res.* **2008**, *50*, 79–104.
29. Fan, W.; Yan, Z. Factors affecting response rates of the web survey: A systematic review. *Comput. Hum. Behav.* **2010**, *26*, 132–139. [CrossRef]
30. National Church Life Survey, 2016 Urban and Rural Dwellers. Available online: http://www.ncls.org.au/default.aspx?sitemapid=2294 (accessed on 6 March 2016).
31. Martelli, G. Consumers' perception of farm animal welfare: An Italian and European perspective. *Ital. J. Anim. Sci.* **2009**, *8*, 31–41. [CrossRef]
32. Mench, J.A. Farm animal welfare in the U.S.A: Farming practice, research, education, regulation, and assurance programs. *Appl. Anim. Behav. Sci.* **2008**, *13*, 298–312. [CrossRef]
33. Sumner, D.A.; Matthews, W.A.; Mench, J.A.; Rosen-Molina, T. The economics of regulations on hen housing in California. *J. Agric. Appl. Econ.* **2010**, *42*, 429–438. [CrossRef]
34. Australian Chicken Meat Federation (ACMF) Inc. Media Release, 6 March 2015. Available online: http://www.chicken.org.au/files/ACMF%20Media%20Release%20-%20Chicken%20Meat%20Outlook%20-%206%20March%202015.pdf (accessed on 8 March 2016).
35. Caswell, J.A.; Mojduszka, E.M. Using informational labelling to influence the market for quality in food products. *Am. J. Agric. Econ.* **1996**, *78*, 1248–1253. [CrossRef]
36. Marian, L.; Thøgersen, J. Direct and mediated impacts of product and process characteristics on consumers' choice of organic vs. conventional chicken. *Food Qual. Preference* **2013**, *29*, 106–112. [CrossRef]
37. Velarde, A.; Dalmau, A. Animal welfare assessment at slaughter in Europe: Moving from inputs to outputs. *Meat Sci.* **2012**, *92*, 244–251. [CrossRef] [PubMed]
38. Van Loo, E.; Caputo, V.; Nayga, M.; Rodolfo, J.; Meullenet, J.F.; Crandall, P.G.; Ricke, S.C. Effect of organic poultry purchase frequency on consumer attitudes towards organic poultry meat. *J. Food Sci.* **2010**, *75*, 384–397. [CrossRef] [PubMed]
39. Tonser, G.T.; Olynk, N.; Wolf, C. Consumer preferences for animal welfare attributes: The case of gestation crates. *J. Agric. Resour. Econ.* **2009**, *41*, 713–730. [CrossRef]
40. Beardworth, A.; Brynan, A.; Leil, T.; Goode, J.; Haslam, C.; Haslam, E. Women, men and food: The significant of gender for nutritional attitudes and choices. *Br. Food J.* **2002**, *104*, 470–491. [CrossRef]
41. Vanhonacker, F.; Verbeke, W.; Van Poucke, E.; Tuyttens, A.M. Do citizens and farmers interpret the concept of farm animal welfare differently? *Livest. Sci.* **2008**, *116*, 126–136. [CrossRef]
42. Kendall, H.; Lobao, L.; Sharp, J. Public concern with animal well-being: Place, social structural location and individual experience. *Rural Soc.* **2006**, *7*, 399–428. [CrossRef]
43. Phillips, C.J.C.; Izmirli, S.; Aldavood, S.J.; Alonso, M.; Choe, B.I.; Hanlon, A.; Handziska, A.; Illmann, G.; Keeling, L.; Kennedy, M.; et al. Students' attitudes to animal welfare and rights in Europe and Asia. *Anim. Welf.* **2012**, *21*, 87–100. [CrossRef]
44. Henson, S. Consumer willingness to pay for reductions in the risk of food poisoning in the UK. *J. Agric. Econ.* **1996**, *47*, 403–420. [CrossRef]
45. Tsakiridou, E.; Tsakiridou, H.; Mattas, K.; Arvaniti, E. Effects of animal welfare standards on consumers' food choices. *Food Econ. Acta Agric. Scand. Sect. C* **2010**, *7*, 234–244. [CrossRef]
46. Harvey, J.; Erdos, G.; Challinor, S.; Drew, S.; Taylor, S.; Ash, R.; Ward, S.; Gibson, G.; Scarr, C.; Dixon, F.; et al. The relationship between attitudes, demographic factors and perceived consumption of meats and other proteins in relation to the BSE crisis: A regional study in the United Kingdom. *Health Risk Soc.* **2001**, *3*, 181–197. [CrossRef]
47. Yen, S.T.; Huang, C.L. Cross-sectional estimation of US demand for beef products: A censored system approach. *J. Agric. Resour. Econ.* **2002**, *27*, 320–334.
48. Ziol-Guest, K.M.; Deleire, T.; Kalil, A. The allocation of food expenditure in married and single-parent families. *J. Consum. Aff.* **2006**, *40*, 347–371. [CrossRef]

49. Ozari, R. Rituelles Schlachten bei Juden (Schechita), Muslimen (Dhabḥ) und Sikhs (Jhatkā). Ph.D. Dissertation, Ludwig-Maximilians-Universität, München, Germany, 1984.
50. Gregory, N.G.; von Wenzlawowicz, M.; Von Hollenben, K.; Fielding, H.R.; Gibson, T.J.; Mirabito, L.; Kolesar, L. Complications during Shechita and halal slaughter without stunning in cattle. *Anim. Welf.* **2012**, *21*, 81–86. [CrossRef]

11

Statistical Evaluations of Variations in Dairy Cows' Milk Yields as a Precursor of Earthquakes

Hiroyuki Yamauchi [1,*], Masashi Hayakawa [2], Tomokazu Asano [2], Nobuyo Ohtani [1] and Mitsuaki Ohta [3]

1. Department of Animal Science and Biotechnology, Azabu University Graduate School of Veterinary Science, 1-17-71 Fuchinobe, Chuo-ku, Sagamihara, Kanagawa 252-5201, Japan; ohtani@azabu-u.ac.jp
2. Hayakawa Institute of Seismo Electromagnetics Co. Ltd., UEC (University of Electro-Communications) Incubation Center, 1-5-1 Chofugaoka, Chofu, Tokyo 182-8585, Japan; hayakawa@hi-seismo-em.jp (M.H.); tomokei0929@gmail.com (T.A.)
3. Department of Human and Animal-Plant Relationships, Tokyo University of Agriculture, 1737 Funako, Atsugi, Kanagawa 243-0034, Japan; mo205684@nodai.ac.jp

* Correspondence: h-yamauchi@azabu-u.ac.jp

Academic Editor: Clive J. C. Phillips

Simple Summary: There are many reports of abnormal changes occurring in various natural systems prior to earthquakes. Unusual animal behavior is one of these abnormalities; however, there are few objective indicators and to date, reliability has remained uncertain. We found that milk yields of dairy cows decreased prior to an earthquake in our previous case study. In this study, we examined the reliability of decreases in milk yields as a precursor for earthquakes using long-term observation data. In the results, milk yields decreased approximately three weeks before earthquakes. We have come to the conclusion that dairy cow milk yields have applicability as an objectively observable unusual animal behavior prior to earthquakes, and dairy cows respond to some physical or chemical precursors of earthquakes.

Abstract: Previous studies have provided quantitative data regarding unusual animal behavior prior to earthquakes; however, few studies include long-term, observational data. Our previous study revealed that the milk yields of dairy cows decreased prior to an extremely large earthquake. To clarify whether the milk yields decrease prior to earthquakes, we examined the relationship between earthquakes of various magnitudes and daily milk yields. The observation period was one year. In the results, cross-correlation analyses revealed a significant negative correlation between earthquake occurrence and milk yields approximately three weeks beforehand. Approximately a week and a half beforehand, a positive correlation was revealed, and the correlation gradually receded to zero as the day of the earthquake approached. Future studies that use data from a longer observation period are needed because this study only considered ten earthquakes and therefore does not have strong statistical power. Additionally, we compared the milk yields with the subionospheric very low frequency/low frequency (VLF/LF) propagation data indicating ionospheric perturbations. The results showed that anomalies of VLF/LF propagation data emerged prior to all of the earthquakes following decreases in milk yields; the milk yields decreased earlier than propagation anomalies. We mention how ultralow frequency magnetic fields are a stimulus that could reduce milk yields. This study suggests that dairy cow milk yields decrease prior to earthquakes, and that they might respond to stimuli emerging earlier than ionospheric perturbations.

Keywords: dairy cows; earthquake precursors; unusual animal behavior; milk yields; subionospheric very low frequency/low frequency (VLF/LF) propagation

---

1. Introduction

There have been numerous studies on precursors of earthquakes [1–11]. These studies have mainly focused on pre-seismic unusual physical and/or chemical variations near the epicenters, such as electromagnetic emissions, ionospheric perturbations, radiation belt electron precipitation and radon gasses, that emerge prior to earthquakes [1–11]. In particular, it has recently been reported that some precursors such as electromagnetic and radon anomalies showed statistical correlations with earthquakes [12–16]. Additionally, there have been many reports of unusual animal behavior (UAB) prior to earthquakes. UABs accounted for approximately half of all reports of macroscopic anomalies identified in posteriori surveys, which also included abnormal sounds, earthquake lights, earthquake clouds, ground deformation, and abnormalities in the ground water [17–19]. However, most reports on UAB are based on qualitative rather than quantitative observations. As an example of quantitative UAB, changes in the locomotive activities of mice before large earthquakes were reported by Yokoi et al. [20] and Li et al. [21]. Grant et al. [22] recently revealed, by the use of motion-triggered cameras, that wild animal activity in various species declined prior to the Contamana earthquake, with a magnitude (M) of 7.0. However, these reports were case studies for single large earthquakes. Few studies have found statistical correlations between earthquakes and UAB using longitudinal quantitative observations.

UAB prior to earthquakes includes stress or emotional responses to physical or chemical anomalies, although the mechanism by which these anomalies are sensed remains unknown. We hypothesize that the milk yields of dairy cows, while not a behavior, might be useful as an earthquake precursor because milk yields are decreased by various stressors [23–25]. Specifically, Regalma et al. [25] reported that cows exhibited responses including decreases in milk yields in response to small stray voltage in the ground, and Rushen et al. [23,24] reported decreases following handling by unfamiliar people, moves to novel places and social isolation. In addition, milk yields are measured every day by farmers as a normal part of their work. There are also reports that cows exhibited UABs prior to earthquakes [26,27]. Prior evidence in support of this hypothesis comes from a study that reported milk yields from cows located 340 km from the epicenter decreased from three to six days before the 2011 earthquake off the Pacific coast of Tohoku in Japan (Tohoku earthquake; Mw 9.0) [28]. However, this was a case study of one extremely large earthquake. To assess whether milk yields decrease prior to earthquakes, a time-series analysis of the correlation between milk yields and earthquakes with various magnitudes (Ms) using long term data is necessary. To confirm the value of a certain phenomenon as an earthquake precursor, it is not only necessary to evaluate by the time-series analysis but also to estimate the performance level using retrospective earthquake prediction. Mathematical methodologies and parameters to evaluate the performance of predictors have been described in previous studies [29,30].

An electromagnetic anomaly in the ionosphere prior to earthquakes is one well-known precursor. Very low frequency/low frequency (VLF/LF) subionospheric propagation data have recently been used to monitor lower ionospheric perturbations associated with earthquakes. Statistical correlations with earthquakes have been confirmed in previous studies [13,14], and there are some theoretical mechanisms regarding this anomaly [31,32]. Interestingly, on the same day the milk yields decreased [28], anomalies in the VLF/LF propagation signal emerged [33] in studies on the Tohoku earthquake. A comparison of the milk yield anomalies and other well-known precursors could help elucidate mechanisms associated with UAB prior to earthquakes.

The aims of this study are to elucidate whether the milk yields of dairy cows decreased before various earthquakes and to confirm whether precursory decreases in milk yields are an explainable phenomenon from a scientific point of view by comparison with VLF/LF propagation data. To achieve our aims, we estimated the relationships between milk yields and earthquakes by time-series analyses and preliminarily estimated the performance level using milk yields, with some definitions of anomalies, and earthquakes with varying thresholds regarding distance from epicenters and M.

2. Materials and Methods

2.1. Milk Yields

With the aid of a farm in Ibaraki prefecture of Japan, we analyzed the daily milk yields of 48 Holstein cows from 1 January 2014 to 31 December 2014. The location of the farm is shown in Figures 1 and 2. The milking process was as follows; the cows were brought into the milking parlor and then milked by machines. They were individually identified by tags, and milk yields were assessed using electronic milk meters. The milking frequency was twice a day. The data measured were transferred to computers, and we used the total milk yield per cow per day. Variability in milk yield due to some physiological and environmental effects were corrected for before statistical analyses, based on our previous work [28]. Milk yields from dairy cows increase for approximately four to eight weeks postpartum and gradually decrease thereafter, with lactation being complete by approximately forty weeks. First, we removed the effect of the number of days after calving using Wood's lactation curve model [34]. Individual milk yields during one lactation period were estimated by the use of Wood's model, as follows:

$$y_t = a \times n_t^b \times e^{-c \times n_t}, \tag{1}$$

where y_t is the expected milk yield in time t, n_t is the number of days after calving, e is natural logarithm, and a, b and c are parameters. These parameters were estimated using a general linear model (glm function in R; version 3.3.0, The R Foundation for Statistical Computing, Vienna, Austria) after a logarithmic transformation [35,36]. Table 1 shows estimated parameters for individual lactation curves. Then, residual values I of milk yields (dM-I) were calculated by subtracting these predicted values from the actual measurement values of the current day:

$$dM\text{-}I_t = M_t - pM\text{-}I_t, \tag{2}$$

where M_t is the milk yield for a current day and $pM\text{-}I_t$ is the predicted value for the current day estimated by Wood's model.

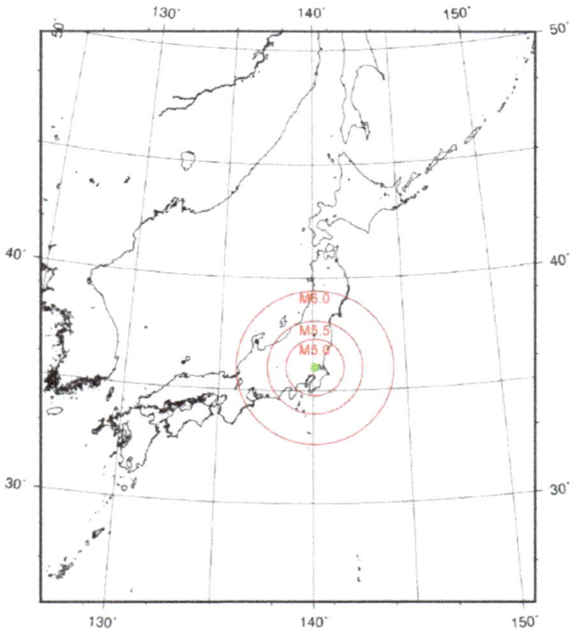

Figure 1. The radius (km) calculated according to the Dobrovolsky radius condition with magnitudes (Ms) = 5.0, 5.5 and 6.0. The red circles represent the maximum range for earthquakes of each M. The solid green circle represents the location of the farm used to observe milk yields.

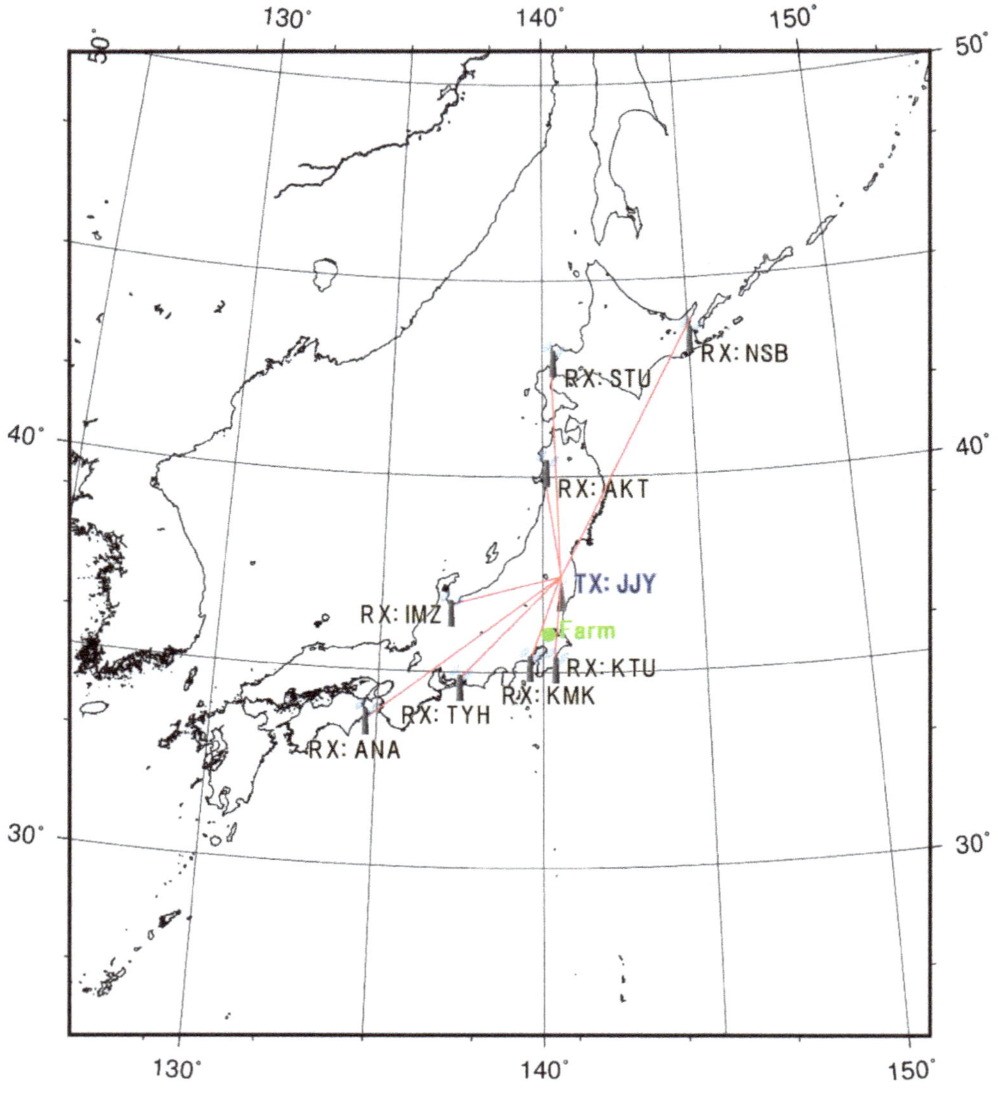

Figure 2. The location of the farm used to observe milk yields, eight observatcries and the Japanese low frequency transmitting station (JJY) for very low frequency/low frequency (VLF/LF) propagation data. Red lines represent the propagation path from JJY to the eight observatories. NSB = Nakashibetsu; AKT = Akita; IMZ = Imizu; KTU = Katsuura; KMK = Kamakura; TYH = Toyohashi ANA = Anan; STU = Suttu.

Table 1. Estimated parameters in Equation (1) for individual lactation curves (Mean ± Standard deviation).

a	b	c
27.1 ± 15.9	0.183 ± 0.520	0.00644 ± 0.01981

Factors affecting milk yields in dairy cows include ambient temperature and humidity as well as the number of days after calving. The temperature–humidity index (THI) is often used as an index of heat stress for cows. West et al. [37] reported that milk yields begin to decrease when the THI exceeds 72. These effects from environmental conditions stem from decreases in food consumption due to heat stress. Therefore, increased temperature might affect milk yields after several days rather than on the current day. West et al. reported that THI during hot periods had the greatest impact on milk yields two days later [37]. This effect on heat stress should be removed to elucidate the relevance to earthquakes. Second, we calculated the daily THI values during the observation period using mean temperature and mean relative humidity from Japan Meteorological Agency's meteorological observatory closest to the farm. The formula to calculate the THI is

$$THI_t = (1.8\ T_t + 32) - (5.5 + 0.055\ H_t) \times (1.8\ T_t - 26), \quad (3)$$

where T is the dry bulb temperature in °C and H is the relative humidity in % [38]. The critical point in THI at which milk yields start to decrease and the linear relationship between THI (above the critical point) and milk yields were estimated using a two-phased regression model [39]. The model formula is:

$$\begin{aligned} y_t &= a + bx_{t-2}, \quad (x \geq x0) \\ y_t &= c, \quad (x < x0) \end{aligned} \quad (4)$$

where y_t is the mean dM-I at time t, x_{t-2} is the THI at the relevant lag, x0 is the critical point in THI, and a, b and c are parameters. The estimated model was used to calculate the expected daily milk yields from the THI values. Table 2 shows estimated parameters and the critical point. The residual values of the milk yields (dM-II) were calculated by subtracting these predicted values from dM-I:

$$\text{dM-II}_t = \text{dM-I}_t - \text{pM-II}_t, \quad (5)$$

where dM-I_t is the milk yield after removing the effect of calving time and pM-II_t is the predicted value based on the THI values. The dM-II data might have trend variation, which is inappropriate for time-series analyses. Therefore, final variations in milk yields were calculated by the partly-changed equation described by Maekawa et al. [14] and Hayakawa et al. [33]. That is, we calculated residual values in the milk yields as:

$$\text{dM-III}_t = \text{dM-II}_t - \langle \text{dM-II}_t \rangle, \quad (6)$$

where dM-II_t is the milk yield after removal of two effects at time t and $\langle \text{dM-II}_t \rangle$ is the 7-day backward moving average at the same time t. These final variations (dM-III) were used in analyses regarding the relevance of the data to earthquakes.

Table 2. Estimated parameters in Equation (4) and the critical point in the temperature–humidity index (x0).

a	b	c	x0
−0.24	17.86	0.27	74.60

2.2. Cross-Correlation Analyses

To examine statistical correlations between changes in milk yields and earthquakes, a cross-correlation analysis was applied. The information on occurrence date, location, M, and depth of earthquakes were obtained from the Japan Meteorological Agency. To examine the relationships between milk yields and earthquakes, it was necessary to select thresholds regarding the distance from the epicenter. We used earthquakes that satisfied the Dobrovolsky radius condition (DRC). According to the DRC, the effective precursory manifestation zone depends on the M of earthquakes and can be calculated as:

$$r\ (\text{km}) = 10^{0.43M}, \quad (7)$$

where r is the radius from the epicenter and M is the magnitude [40]. Figure 1 shows the Dobrovolsky radius for some Ms.

The presence or absence of the occurrence of earthquakes on a given day was treated as a binary outcome (0/1). It was necessary to count multiple earthquakes on a given day as a single occurrence, because the milk yield data were collected only once per day. In a previous study, Maekawa et al. [14] used effective magnitude (Meff), which was calculated by integrating the released energy of earthquakes that occurred within one day and converting it back into M. The released energy was calculated using the Richter scale [41], as follows:

$$E = 10^{(4.8+1.5M)}, \quad (8)$$

where E is the released energy and M is the magnitude.

We also calculated this Meff from earthquakes with M > 2.0 as the active level of earthquakes. Then, we used dates with Meff > 5.0, 5.5 and 6.0 in the analyses. The analyses were performed using the ccf function in R (version 3.3.0). The 95% confidence intervals were set to $1.96/\sqrt{(n - |k|)}$.

2.3. Performance Evaluations of Binary Earthquake Forecasts

We also used a binary earthquake forecasting approach to evaluate the reliability of decreases in milk yields as a precursor for earthquakes. We obtained cross-tabulation tables consisting of the presence or absence of earthquakes and alarm or no alarm days from anomalies in milk yields (Table 3). To make these tables, it was necessary to determine the critical point that defines milk yields as being anomalous. In previous studies, values farther from the mean than twice the standard deviation (σ) were defined as anomalies [42]. We defined milk yields more than 1.5σ below the mean as anomalies to examine the relationship between relatively small decreases in milk yields and earthquakes. If milk yields were at abnormally low levels for more than two consecutive days, the dates were summarized as one anomalous occurrence because decreased milk yields were followed for a few days in a previous study [28]. To evaluate the continuity of anomalies in milk yields, we set two criteria regarding the definition of anomalies. In one criterion, we defined an anomaly as a decrease of more than 1.5σ for one day; in the other criterion, an episode was defined as an anomaly when the decreases continued over two days. 'Alarm days' were defined by the lag with the lowest cross-correlation coefficient. Assuming the temporal relationship between the presence of precursors and the occurrence of earthquakes can vary by a few days, the 'alarm period' included a margin of $\pm$ 4 days. To make the cross-tabulation tables, it was also necessary to determine the thresholds for M and the distance from the epicenters of earthquakes. We set three criteria on M, (i.e., >5.0, 5.5 or 6.0) because whether earthquakes with lower values of M led to decreases in milk yields was unclear. In addition, to confirm the appropriateness of the DRC as the sensitive area for milk yields, we considered two criteria regarding the distance from epicenters for targeted earthquakes. In one criterion, we defined an earthquake as relevant when it occurred within the DRC; in the other criterion, we defined an earthquake as relevant when it occurred within the DRC + 250 km. Finally, we made twelve cross-tabulation tables based on the varying criteria for the duration of anomalous milk yields and the targeted earthquakes. The details for each criterion are shown in Table 4.

Table 3. The format of cross-tabulation tables consisting of the presence or absence of earthquakes and anomalies in milk yields; a represents the number of earthquakes occurring in alarm days, b represents the number of earthquakes occurring in no alarm days; c represents the number of alarm days without targeted earthquakes; d represents the number of no alarm days without targeted earthquakes.

Earthquake	Alarm		Total
	Yes	No	
Yes	a	b	a + b
No	c	d	c + d
Total	a + c	b + d	a + b + c + d

Table 4. The definitions in anomalies of milk yields and targeted earthquakes based on each criterion; M = magnitude; DRC = Dobrovolsky radius condition; σ = standard deviation.

	Anomalies in Milk Yields		Targeted Earthquakes	
	σ <	Duration (Days) ≥	M >	Distance from Epicenters <
Criterion 1	−1.5	1	5.0	DRC
Criterion 2	−1.5	1	5.5	DRC
Criterion 3	−1.5	1	6.0	DRC
Criterion 4	−1.5	2	5.0	DRC
Criterion 5	−1.5	2	5.5	DRC
Criterion 6	−1.5	2	6.0	DRC
Criterion 7	−1.5	1	5.0	DRC + 250 km
Criterion 8	−1.5	1	5.5	DRC + 250 km
Criterion 9	−1.5	1	6.0	DRC + 250 km
Criterion 10	−1.5	2	5.0	DRC + 250 km
Criterion 11	−1.5	2	5.5	DRC + 250 km
Criterion 12	−1.5	2	6.0	DRC + 250 km

These tables were used to calculate three scores indicating their performance level for binary earthquake forecasting.

Following Holliday et al. [30], we used hit rate (H), defined as

$$H = a/(a + b). \tag{9}$$

Probability gain (PG), the ratio of the probability of an earthquake occurring in an alarm period divided by the probability of an earthquake occurring on any given day, is also an important parameter [29]. The formula for PG is:

$$PG = [a/(a + c)]/[(a + b)/(a + b + c + d)], \tag{10}$$

Additionally, we calculated the successful rate of earthquake prediction (SEP) using the following equation:

$$SEP = a/(a + c), \tag{11}$$

2.4. Comparison with Anomalies of VLF/LF Propagation Data

We collected subionospheric VLF/LF propagation data with the aid of the Earthquake Analysis Laboratory in Japan. This network for VLF/LF propagation has been in place since 2001 and revealed that anomalous propagations emerge prior to earthquakes [13,43]. In this study, we used eight observatories to confirm there were physical anomalies before relevant earthquakes: (1) Nakashibetsu (NSB); (2) Suttu (STU); (3) Akita (AKT); (4) Imizu (IMZ); (5) Katsuura (KTU); (6) Kamakura (KMK); (7) Toyohashi (TYH); and (8) Anan (ANA). We used the received signal from each observatory in the Japanese low frequency transmitting station (JJY; in Fukushima, 40 kHz). The locations of each observatory are shown in Figure 2, which also illustrates the propagation path from JJY to each observatory. To detect anomalies in VLF/LF propagation data, we analyzed the obtained propagation data in accordance with previous studies [44]. That is, we used daily average amplitude at nighttime and calculated residual values as $dA(t) = A(t) - \langle A(t) \rangle$, where $A(t)$ is the amplitude at time t on the current day, and $\langle A(t) \rangle$ is the running average at the same time t over ±15 days (i.e., 15 days before and after the relevant day). In parallel with milk yields, we defined the anomalies in VLF/LF propagation data by the occurrence of values that were decreased by more than 1.5σ.

3. Results

3.1. Cross-Correlation Analyses

The number of lactating cows per day ranged from 18 to 36 (mean ± Standard deviation = 25.3 ± 4.2). The data on milk yields used in the analyses (mean ± S.D. = 0.016 ± 0.812) followed a normal distribution (one-sample Kolmogorov–Smirnov test, $p = 0.47$, ks.test function in R version 3.3.0) and exhibited stationarity (Phillips-Perron Unit Root Test, $p < 0.01$, p.p test function of tseries package in R version 3.3.0). Figure 3 shows the distribution of the milk yields. Figure 4 shows the mean variations of milk yields in the total observation period (361 days). Meff exceeded 5.0 on ten days, 5.5 on seven days and 6.0 on four days. The detail of earthquakes which occurred during the days exceeding Meff 5.0 is shown in Figure 5. Figure 6 shows the results of cross-correlation analyses between milk yields and each earthquake activity level. Significant negative coefficients were revealed approximately three weeks before the days exceeding each earthquake activity level, and the significances became clear for Meff > 5.5. Following this period, we found positive correlation coefficients; the coefficients 11 days before Meff 5.5 and 10 days before Meff 6.0 were statistically significantly positive. The positive coefficients continued until the days on which earthquakes occurred, although no other days' coefficients were significant.

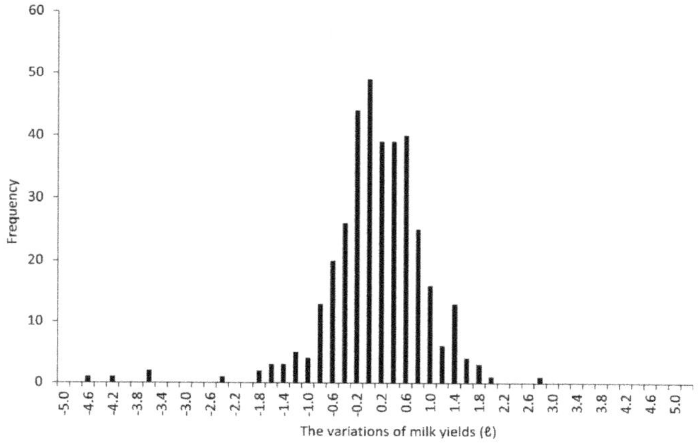

Figure 3. The distribution of the variations of milk yields.

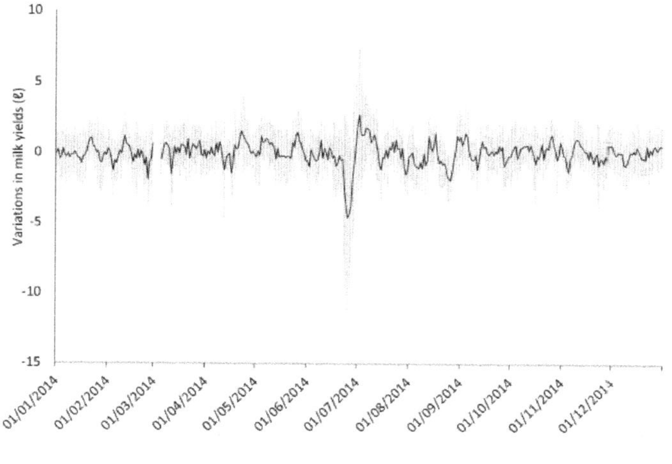

Figure 4. The variations of milk yields in the total observation period (Mean ± Standard deviation). The blank area indicates no observation.

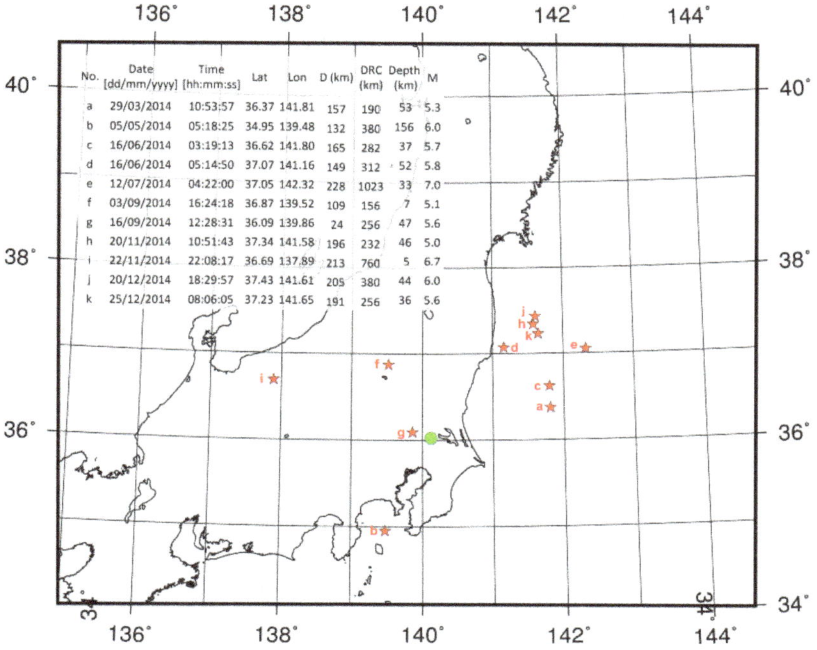

No.	Date [dd/mm/yyyy]	Time [hh:mm:ss]	Lat	Lon	D (km)	DRC (km)	Depth (km)	M
a	29/03/2014	10:53:57	36.37	141.81	157	190	53	5.3
b	05/05/2014	05:18:25	34.95	139.48	132	380	156	6.0
c	16/06/2014	03:19:13	36.62	141.80	165	282	37	5.7
d	16/06/2014	05:14:50	37.07	141.16	149	312	52	5.8
e	12/07/2014	04:22:00	37.05	142.32	228	1023	33	7.0
f	03/09/2014	16:24:18	36.87	139.52	109	156	7	5.1
g	16/09/2014	12:28:31	36.09	139.86	24	256	47	5.6
h	20/11/2014	10:51:43	37.34	141.58	196	232	46	5.0
i	22/11/2014	22:08:17	36.69	137.89	213	760	5	6.7
j	20/12/2014	18:29:57	37.43	141.61	205	380	44	6.0
k	25/12/2014	08:06:05	37.23	141.65	191	256	36	5.6

Figure 5. The detail of earthquakes which occurred during the days exceeding effective magnitude (Meff) 5.0. The red stars represent the locations of earthquakes. The green circle represents the location of the farm used to observe milk yields. The upper table shows the detail of earthquakes. The alphabets in the map correspond with those in the table. Lat = Latitude; Lon = Longitude; D = Distance from epicenters; DRC = Dobrovolsky radius condition.

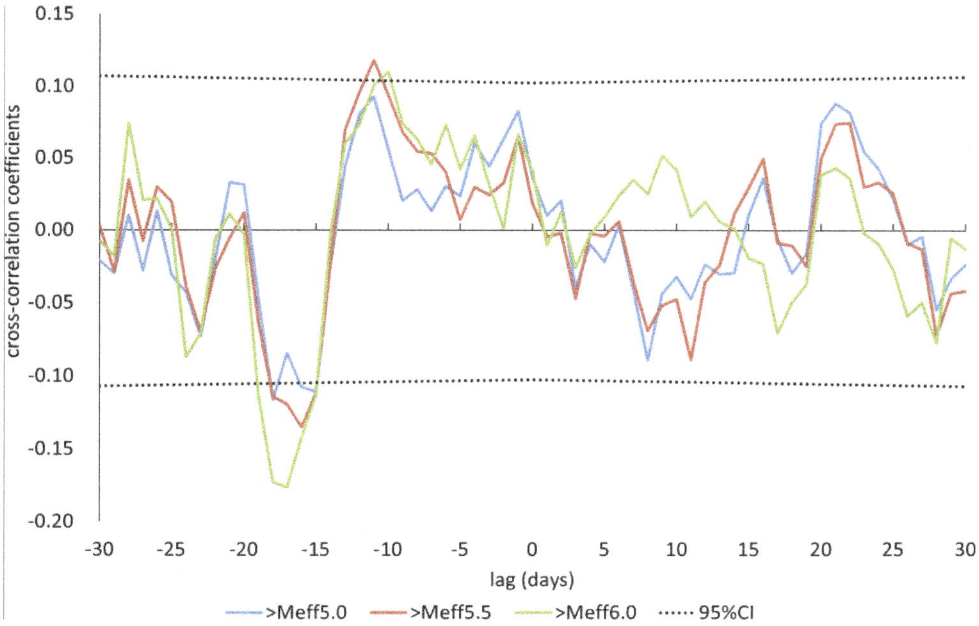

Figure 6. The results of cross-correlation analyses between milk yields and the dates exceeding each Meff; CI = confidence interval.

3.2. Binary Earthquake Forecasts

On 16 June 2014, two earthquakes of similar M occurred in nearby regions (see Figure 5). We treated these earthquakes as one event with M greater than 5.5 that satisfied the DRC to prevent over- or under-estimation. On 31 days, milk yields decreased below the mean by more than 1.5σ. In 17 instances, the decrease persisted for more than two consecutive days. Table 5 shows cross-tabulation tables using the twelve aforementioned criteria. The scores of H, PG and SEP from the cross-tabulation tables are shown in Figure 7. Criteria 2 and 5 had highest H (85.7%). With respect to PG, criterion 5 had the highest score (6.8). The SEP for criteria 5 was 14.0%, which was not the highest value.

Table 5. The cross-tabulation tables based on twelve criteria.

Criterion 1				**Criterion 7**			
Earthquake	**Alarm**		**Total**	**Earthquake**	**Alarm**		**Total**
	Yes	**No**			**Yes**	**No**	
Yes	7	3	10	Yes	11	21	32
No	103	231	334	No	66	246	312
Total	110	234	344	Total	77	267	344
Criterion 2				**Criterion 8**			
Earthquake	**Alarm**		**Total**	**Earthquake**	**Alarm**		**Total**
	Yes	**No**			**Yes**	**No**	
Yes	6	1	7	Yes	7	8	15
No	109	228	337	No	90	239	329
Total	115	229	344	Total	97	247	344
Criterion 3				**Criterion 9**			
Earthquake	**Alarm**		**Total**	**Earthquake**	**Alarm**		**Total**
	Yes	**No**			**Yes**	**No**	
Yes	3	1	4	Yes	4	2	6
No	124	216	340	No	116	222	338
Total	127	217	344	Total	120	224	344
Criterion 4				**Criterion 10**			
Earthquake	**Alarm**		**Total**	**Earthquake**	**Alarm**		**Total**
	Yes	**No**			**Yes**	**No**	
Yes	6	4	10	Yes	7	25	32
No	35	299	334	No	30	282	312
Total	41	303	344	Total	37	307	344
Criterion 5				**Criterion 11**			
Earthquake	**Alarm**		**Total**	**Earthquake**	**Alarm**		**Total**
	Yes	**No**			**Yes**	**No**	
Yes	6	1	7	Yes	6	9	15
No	37	300	337	No	33	296	329
Total	43	301	344	Total	39	305	344
Criterion 6				**Criterion 12**			
Earthquake	**Alarm**		**Total**	**Earthquake**	**Alarm**		**Total**
	Yes	**No**			**Yes**	**No**	
Yes	3	1	4	Yes	3	3	6
No	52	288	340	No	52	286	338
Total	55	289	344	Total	55	289	344

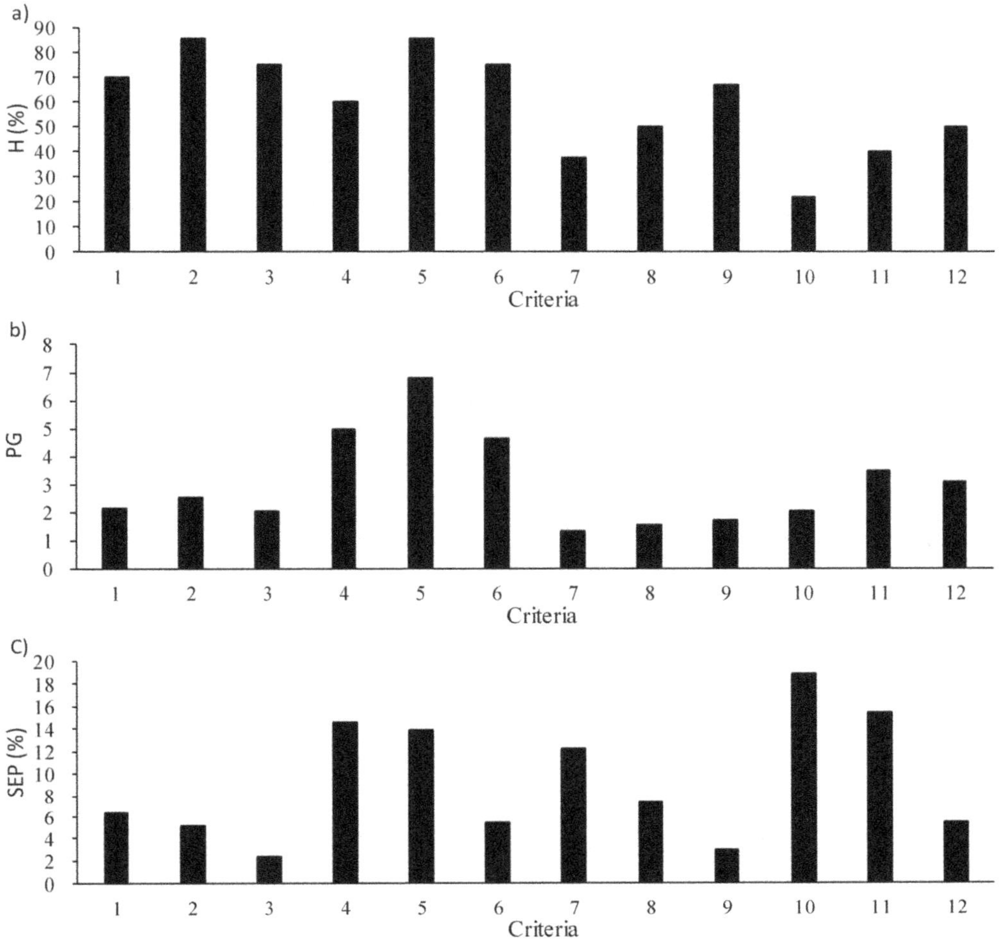

Criteria	1	2	3	4	5	6	7	8	9	10	11	12
Duration (days) ≥	1	1	1	2	2	2	1	1	1	2	2	2
M >	5	5.5	6	5	5.5	6	5	5.5	6	5	5.5	6
Distance from epicenters (km) <	DRC	DRC	DRC	DRC	DRC	DRC	DRC + 250	DRC + 250	DRC + 250	DRC + 250	DRC + 250	DRC + 250

Figure 7. The scores used to estimate the cross-tabulation tables using the twelve criteria. (**a**) shows hit rate (H, %); (**b**) shows probability gain (PG) and (**c**) shows the successful rate of earthquake prediction (SEP, %). The lower table shows the details of the criteria.

3.3. Comparison with Anomalies in VLF/LF Propagation Data

Based on the results of the binary earthquake forecasts, we compared milk yields with VLF/LF propagation data prior to earthquakes with M > 5.5 that satisfied the DRC and had high scores in both H and PG. Milk yields were defined as anomalous when values more than 1.5σ below the mean continued for more than two days. Table 6 shows the correspondence between the anomalies in milk yields, targeted earthquakes, and anomalies in VLF/LF propagation data. The anomalies in VLF/LF propagation data were presented in this table only if targeted earthquakes followed, because the focus of this study is milk yield in dairy cows. The results suggest that the anomalies in VLF/LF propagation data emerged prior to all of the earthquakes that accompanied anomalies in milk yields. However, in all cases, the lag times between anomalies in milk yields and earthquake occurrence were longer than those associated with anomalies in VLF/LF propagation. The lag times for milk yields and VLF/LF propagation were 17.7 and 10.5 days on average, respectively.

Table 6. The correspondence table of the anomalous milk yields, the observed targeted earthquakes and VLF data. Lat = Latitude, Lon = Longitude, D = Distance from epicenters, DRC = Dobrovolsky radius condition.

Anomalies of Milk Yields					Earthquake Data							VLF Data		
Start	End	Duration (Days)	Lead Time (Days)	σ (min)	Date (dd/mm/yyyy)	Lat	Lon	D (km)	DRC (km)	Depth (km)	M	Anomalies (Yes or No)	Lead Time (Days)	Path (Anomalous Day)
11/4/2014	12/4/2014	2	-	−3.57	-	-	-	-	-	-	-	-	-	-
16/04/2014	17/04/2014	2	18–19	−3.30	5/5/2014	34.95	139.48	132	380	156	6.0	Yes	13	JJY-KTU (22/04/2014)
31/05/2014	2/6/2014	3	14–16	−2.13	16/06/2014	36.62	141.80	165	282	37	5.7	Yes	10	JJY-KTU (06/06/2014)
					16/06/2014	37.07	141.16	149	312	52	5.8	Yes	10	JJY-NSB (06/06/2014)
23/06/2014	27/06/2014	5	15–19	−4.78	12/7/2014	37.05	142.32	228	1023	33	7.0	Yes	11	JJY-IMZ (01/07/2014)
29/07/2014	31/07/2014	3	-	−2.31	-	-	-	-	-	-	-	-	-	-
24/08/2014	26/08/2014	3	21–23	−1.90	16/09/2014	36.09	139.86	24	256	47	5.6	Yes	11	JJY-TYH (05/09/2014)
4/11/2014	5/11/2014	2	17–18	−2.93	22/11/2014	36.69	137.89	213	760	5	6.7	Yes	10	JJY-IMZ (12/11/2014)
-	-	-	-	-	20/12/2014	37.43	141.61	205	380	44	6.0	Yes	11	JJY-STU (09/12/2014)
8/12/2014	9/12/2014	2	16–17	−1.74	25/12/2014	37.23	141.65	191	256	36	5.6	Yes	8	JJY-NSB (17/12/2014)

4. Discussion

4.1. Cross-Correlation Analyses

These analyses showed that milk yields decreased approximately three weeks before earthquakes. This result differed from the result reported regarding the leading time of decreases prior to the Tohoku earthquake of three to six days [28]. In our observation period, there were no earthquakes occurring near the hypocenter of the Tohoku earthquake; the earthquake nearest the epicenter of the Tohoku earthquake was 150 km away. This difference could be due to physical or chemical differences such as geostructural features in the epicentral zones. In addition to epicentral zones, two other differences stand out between these two studies. The M of the Tohoku earthquake was 9.0 while the M of targeted earthquakes in this study was a maximum of 7.0. The Tohoku earthquake also had many foreshocks starting two days before the main shock. Therefore, there are limitations regarding the comparability of the results in this study with those from the case study of the Tohoku earthquake. The significant cross-correlation coefficients became larger as Meff became larger. This indicates that the probability of decreasing milk yields increased, or the degree of decrease increased as the earthquake activity level increased. Reports have also suggested that anomalies in LF propagation signals before earthquakes increase with Meff or M [14,45]. However, several statistical studies suggest that anomalies in VLF/LF propagation signal emerge about one week before earthquakes [33]. This study suggests that the stimuli causing decreases in milk yields might occur before ionospheric anomalies. The increase in the cross-correlation coefficients appeared approximately 11 days before earthquakes. We subtracted the 7-day backward moving average from residuals in milk yields to remove trend variations in Equation (6). Therefore, the increase of the coefficients around -11 days likely indicates that actual milk yields did not increase but recovered. However, the positive coefficients lasted for approximately two weeks (i.e., until the day of the earthquake's occurrence), although they did not exceed the significance threshold after day -10. The increased variability in milk yields seemed to gradually decline as the day of the earthquake approached. In addition to the pronounced decreases in milk yields, our study suggests there might also be an increase prior to earthquakes. Adverse effects prior to earthquakes have also been found in a study on the correlation between mental health and earthquakes [46].

4.2. Binary Earthquake Forecasts

Criterion 5 (i.e., in which anomalies included dates when milk yields were reduced by more than 1.5σ for more than two consecutive days and earthquakes with $M > 5.5$ that satisfied the DRC were included) had the highest H and PG scores. On the other hand, the highest SEP occurred in criterion 10, in which anomalies were defined as the dates when milk yields were reduced by at least 1.5σ for more than two consecutive days and targeted earthquakes were those with $M > 5.0$ and that satisfied the DRC + 250 km criterion. However, PG is obtained by dividing the SEP by the probability of an earthquake occurring, and earthquakes included in criterion 5, which included only earthquakes with $M > 5.5$ and those that satisfied the DRC, were stronger than those included in criterion 10 (i.e., $M > 5.0$ and DRC + 250 km). Therefore, it is likely that increasing the number of earthquakes by lowering the threshold of M and expanding the threshold distance led to an increase in accidental successful earthquake predictions. PG is approximately one when earthquakes and anomalies are uncorrelated. In this study, the PG of criterion 5 was 6.8. This value suggests that decreases in milk yields are related to subsequent earthquakes, in agreement with the result of the cross-correlation analyses. The aim of this study is to examine the reliability of decreases in daily cows' milk yields as a precursor of earthquakes. Binary earthquake forecasts were used to evaluate the performance of this putative precursor. To identify the optimal precursor, we shifted the thresholds of several parameters regarding earthquakes and milk yields; however, the parameters considered were discrete and there were wide gaps between them. Therefore, to accurately understand the relationship between decreases in milk yields and subsequent earthquakes, it will be necessary to consider threshold values at smaller intervals. The SEP of criterion 5 was 14.0%, which is not high. The alarm period defined in these analyses was nine days per anomaly. If targeted earthquakes occurred during the alarm period, the remainder of the nine days were no longer considered alarm days (i.e., they were included in "d" in Table 3). Alarm days before the earthquakes were included in "c" in Table 3 (i.e., they were considered as days on which there was an alarm but no earthquake). Therefore, the unsuccessful rate of 86.0% consisted of not only alarm periods that were unrelated to earthquakes but also alarm days just prior to earthquakes.

4.3. Comparison with Anomalies of VLF/LF Propagation Data

In our observation period, anomalies in VLF/LF propagation data from regions that satisfied the DRC were observed prior to all earthquakes with M > 5.5. In previous studies, we found that the lag time of between decreased milk yield and the Tohoku earthquake [28] was similar to those of VLF/LF propagation anomalies [33]. However, in this study, the milk yields decreased approximately one week before VLF/LF propagation anomalies in all earthquakes that followed decreases in milk yields. These differences in the lag times were supported by the results of the cross-correlation analyses. The results suggest that the stimuli that cause decreases in milk precede ionospheric perturbations. Therefore, the similarity between the lag times of the different precursors before the Tohoku earthquake might not indicate a direct causal relationship between milk yields and ionospheric perturbations but rather that triggers of decreased milk yields and ionospheric perturbations can occur simultaneously. Identifying stimuli that cause decreases in milk yield is important to help clarify the mechanism by which earthquake precursors lead to UAB. Anomalies in ultralow frequency (ULF, <10Hz) magnetic fields from the lithosphere are a candidate signal because the lag time between these anomalies and earthquakes is similar to that of decreased milk yields. Many studies have examined ULF radiation prior to earthquakes [7,8,47], and some studies reported the leading times were between a few weeks and a month [4,48]. Some reports suggest that the ULF magnetic field affects behavior and hormones [49–53]. The electromagnetic field with a frequency of 10 Hz is known to affect circadian activity rhythms [52,53]. Mahdavi et al. [49,51] have reported that exposures to electromagnetic fields with 5 Hz or 12 Hz elevated activity levels and adrenocorticotropic hormone concentrations in rats. Additionally, there have been reports that cattle aligned their body axes along geomagnetic field lines, which indicates they have the sense of magnetoreception [54–56]. To verify this hypothesis, observations of daily cow milk yields and ULF radiation from the same period in the same region need to be performed. There is an interesting report that discussed possible mechanisms for UAB prior to earthquakes based on observational data [22]. This study shows that the amount of wildlife (i.e., mammals and birds) captured by motion-triggered cameras in a national park decreased prior to the Contamana earthquake (M = 7.0) in the Peruvian Andes, and the lag time between these behavioral changes and the earthquake was coincident with the VLF propagation anomalies. They have suggested that air ionization due to positive hole carriers is a possible trigger for UAB and VLF propagation anomalies. Our results seem to conflict with their report. However, the observed behaviors and animal species differed between the two studies. We have shown the leading time of UAB differed depending on species [28]. Thus, it should be noted that "milk yield in dairy cows" could decrease due to ULF radiation.

Finally, we discuss the applicability of daily cow milk yields as an objectively observable UAB prior to earthquakes. Milk yields in each individual have been measured every day by many institutes in the animal husbandry industry to manage and improve productivity. As cows are reared in various places, at least in Japan, it is possible to elucidate relationships between milk yields and M, distance from epicenters or depth from hypocenters by more long term observations of milk yields in many regions. In this study, we were only able to evaluate the relevance to earthquakes in a limited region because we only had data on milk yields from one location. Further studies targeted at earthquakes in various regions should be conducted to confirm whether the results of this study are more generally applicable.

5. Conclusions

Our key finding is that daily cow milk yields decreased approximately three weeks before earthquakes. The probability that earthquakes with M > 5.5 that satisfy the Dobrovolsky radius condition occurred was highest 14 to 21 days after decreases in milk yields of greater than 1.5σ for more than two consecutive days (PG = 6.8; H = 85.7%). However, future studies that include more earthquakes in various regions and more detailed analyses need to be performed to confirm the reliability of these statistical estimates; this study only included a maximum of 32 earthquakes. All earthquakes that followed decreases in milk yields also followed VLF/LF propagation anomalies, one of the major precursory phenomena; however, the milk yields decreased approximately one week earlier than anomalies in VLF/LF propagation.

Acknowledgments: The authors are grateful to the staff of the farm for their cooperation in this study. Thanks are also due to Earthquake Analysis Laboratory for VLF/LF data. We thank ScienceDocs Inc. (https://www.sciencedocs.com) for language editing.

Author Contributions: Hiroyuki Yamauchi conceived and designed the study, analyzed the data and drafted the manuscript. Masashi Hayakawa and Tomokazu Asano helped with analyses of the data. Mitsuaki Ohta and Nobuyo Ohtani helped with the design and were involved in interpreting results.

Conflicts of Interest: The authors declare no conflict of interest.

References

1. Wattananikorn, K.; Kanaree, M.; Wiboolsake, S. Soil gas radon as an earthquake precursor: Some considerations on data improvement. *Radiat. Meas.* **1998**, *29*, 593–598. [CrossRef]
2. Hayakawa, M.; Fujinawa, Y. *Electromagnetic Phenomena Related to Earthquake Prediction*; Terra Scientific Publishing Company: Tokyo, Japan, 1994.
3. Hayakawa, M.; Kawate, R.; Molchanov, O.A.; Yumoto, K. Results of ultra-low-frequency magnetic field measurements during the Guam earthquake of 8 August 1993. *Geophys. Res. Lett.* **1996**, *23*, 241–244. [CrossRef]
4. Athanasiou, M.; Anagnostopoulos, G.; Iliopoulos, A.; Pavlos, G.; David, C. Enhanced ULF radiation observed by DEMETER two months around the strong 2010 Haiti earthquake. *Nat. Hazards Earth Syst. Sci.* **2011**, *11*, 1091–1098. [CrossRef]
5. Hartmann, J.; Levy, J.K. Hydrogeological and gasgeochemical earthquake precursors—A review for application. *Nat. Hazards* **2005**, *34*, 279–304. [CrossRef]
6. Park, S.K.; Johnston, M.J.S.; Madden, T.R.; Morgan, F.D.; Morrison, H.F. Electromagnetic precursors to earthquakes in the ULF band: A review of observations and mechanisms. *Rev. Geophys.* **1993**, *31*, 117–132. [CrossRef]
7. Ohta, K.; Izutsu, J.; Schekotov, A.; Hayakawa, M. The ULF/ELF electromagnetic radiation before the 11 March 2011 Japanese earthquake. *Radio Sci.* **2013**, *48*, 589–596. [CrossRef]
8. Schekotov, A.; Zhou, H.; Qiao, X.; Hayakawa, M. ULF/ELF Atmospheric Radiation in Possible Association to the 2011 Tohoku Earthquake as Observed in China. *Earth Sci. Res.* **2016**, *5*, 47–58. [CrossRef]
9. Athanasiou, M.; Anagnostopoulos, G.; David, C.; Machairides, G. The ultra low frequency electromagnetic radiation observed in the topside ionosphere above boundaries of tectonic plates. *Res. Geophys.* **2014**, *4*. [CrossRef]
10. Anagnostopoulos, G.C.; Vassiliadis, E.; Pulinets, S. Characteristics of flux-time profiles, temporal evolution, and spatial distribution of radiation-belt electron precipitation bursts in the upper ionosphere before great and giant earthquakes. *Ann. Geophys.* **2012**, *55*, 21–36.
11. Sidiropoulos, N.; Anagnostopoulos, G.; Rigas, V. Comparative study on earthquake and ground based transmitter induced radiation belt electron precipitation at middle latitudes. *Nat. Hazards Earth Syst. Sci.* **2011**, *11*, 1901–1913. [CrossRef]
12. Silva, H.G.; Bezzeghoud, M.; Oliveira, M.M.; Reis, A.H.; Rosa, R.N. A simple statistical procedure for the analysis of radon anomalies associated with seismic activity. *Ann. Geophys.* **2013**, *56*. [CrossRef]

13. Hayakawa, M.; Kasahara, Y.; Nakamura, T.; Muto, F.; Horie, T.; Maekawa, S.; Hobara, Y.; Rozhnoi, A.; Solovieva, M.; Molchanov, O. A statistical study on the correlation between lower ionospheric perturbations as seen by subionospheric VLF/LF propagation and earthquakes. *J. Geophys. Res.* **2010**, *115*. [CrossRef]
14. Maekawa, S.; Horie, T.; Yamauchi, T.; Sawaya, T.; Ishikawa, M.; Hayakawa, M.; Sasaki, H. A statistical study on the effect of earthquakes on the ionosphere, based on the subionospheric LF propagation data in Japan. *Ann. Geophys.* **2006**, *24*, 2219–2225. [CrossRef]
15. Němec, F.; Santolík, O.; Parrot, M. Decrease of intensity of ELF/VLF waves observed in the upper ionosphere close to earthquakes: A statistical study. *J. Geophys. Res.* **2009**, *114*. [CrossRef]
16. Zhang, X.; Fidani, C.; Huang, J.; Shen, X.; Zeren, Z.; Qian, J. Burst increases of precipitating electrons recorded by the DEMETER satellite before strong earthquakes. *Nat. Hazards Earth Syst. Sci.* **2013**, *13*, 197–209. [CrossRef]
17. Wadatsumi, K. *1591 Witnesses Phenomena Prior to Earthquakes*; Tokyo Publisher: Tokyo, Japan, 1995. (In Japanese)
18. Wadatsumi, K.; Haraguchi, R.; Okamoto, K.; Koga, H. Macroscopic anomaly on the Taiwan Earthquake, the Western Tottori Prefecture Earthquake and Geiyo Earthquake. *Geoinformatics* **2001**, *12*, 130–133. [CrossRef]
19. Ulsoy, Ü.; Ikeya, M. Retrospective Statements of Earthquake Precursors by Eye-Witnesses. In *Future Systems for Earthquake Early Warning*; Ulsoy, Ü., Kundu, K.H., Eds.; Nova Science Publishers, Inc.: New York, NY, USA, 2008; pp. 3–53.
20. Yokoi, S.; Ikeya, M.; Yagi, T.; Nagai, K. Mouse circadian rhythm before the Kobe earthquake in 1995. *Bioelectromagnetics* **2003**, *24*, 289–291. [CrossRef] [PubMed]
21. Li, Y.; Liu, Y.; Jiang, Z.; Guan, J.; Yi, G.; Cheng, S.; Yang, B.; Fu, T.; Wang, Z. Behavioral change related to Wenchuan devastating earthquake in mice. *Bioelectromagnetics* **2009**, *30*, 613–620. [CrossRef] [PubMed]
22. Grant, R.A.; Raulin, J.P.; Freund, F.T. Changes in animal activity prior to a major (M = 7) earthquake in the Peruvian Andes. *Phys. Chem. Earth* **2015**, *85–86*, 69–77. [CrossRef]
23. Rushen, J.; de Passille, A.M.B.; Munksgaard, L. Fear of people by cows and effects on milk yield, behavior, and heart rate at milking. *J. Dairy Sci.* **1999**, *82*, 720–727. [CrossRef]
24. Rushen, J.; Munksgaard, L.; Marnet, P.; DePassillé, A. Human contact and the effects of acute stress on cows at milking. *Appl. Anim. Behav. Sci.* **2001**, *73*, 1–14. [CrossRef]
25. Rigalma, K.; Duvaux-Ponter, C.; Barrier, A.; Charles, C.; Ponter, A.; Deschamps, F.; Roussel, S. Medium-term effects of repeated exposure to stray voltage on activity, stress physiology, and milk production and composition in dairy cows. *J. Dairy Sci.* **2010**, *93*, 3542–3552. [CrossRef] [PubMed]
26. Fidani, C. Biological Anomalies around the 2009 L'Aquila Earthquake. *Animals* **2013**, *3*, 693–721. [CrossRef] [PubMed]
27. Nikonov, A.A. Abnormal animal behaviour as a precursor of the 7 December 1988 Spitak, Armenia, earthquake. *Nat. Hazards* **1992**, *6*, 1–10. [CrossRef]
28. Yamauchi, H.; Uchiyama, H.; Ohtani, N.; Ohta, M. Unusual Animal Behavior Preceding the 2011 Earthquake off the Pacific Coast of Tohoku, Japan: A Way to Predict the Approach of Large Earthquakes. *Animals* **2014**, *4*, 131–145. [CrossRef] [PubMed]
29. Aki, K. A probabilistic synthesis of precursory phenomena. In *Earthquake Prediction*; American Geophysical Union: Washington, DC, USA, 1981; pp. 566–574.
30. Holliday, J.R.; Rundle, J.B.; Tiampo, K.F.; Klein, W.; Donnellan, A. Systematic procedural and sensitivity analysis of the pattern informatics method for forecasting large (M > 5) earthquake events in southern California. *Pure Appl. Geophys.* **2006**, *163*, 2433–2454. [CrossRef]
31. Molchanov, O.A.; Hayakawa, M. *Seismo-Electromagnetics and Related Phenomena: History and Latest Results*; Terra Scientific Publishing Company: Tokyo, Japan, 2008.
32. Pulinets, S.; Boyarchuk, K. *Ionospheric Precursors of Earthquakes*; Springer Science & Business Media: Berlin, Germany, 2004.
33. Hayakawa, M.; Hobara, Y.; Yasuda, Y.; Yamaguchi, H.; Ohta, K.; Izutsu, J.; Nakamura, T. Possible precursor to the March 11, 2011, Japan earthquake: Ionospheric perturbations as seen by subionospheric very low frequency/low frequency propagation. *Ann. Geophys.* **2012**, *55*, 95–99.
34. Wood, P.D.P. Algebraic model of the lactation curve in cattle. *Nature* **1967**, *216*, 164–165. [CrossRef]
35. Catillo, G.; Macciotta, N.P.P.; Carretta, A.; Cappio-Borlino, A. Effects of age and calving season on lactation curves of milk production traits in Italian water buffaloes. *J. Dairy Sci.* **2002**, *85*, 1298–1306. [CrossRef]

36. Olori, V.E.; Brotherstone, S.; Hill, W.G.; McGuirk, B.J. Fit of standard models of the lactation curve to weekly records of milk production of cows in a single herd. *Livest. Prod. Sci.* **1999**, *58*, 55–63. [CrossRef]
37. West, J.; Mullinix, B.; Bernard, J. Effects of hot, humid weather on milk temperature, dry matter intake, and milk yield of lactating dairy cows. *J. Dairy Sci.* **2003**, *86*, 232–242. [CrossRef]
38. National Oceanic and Atmospheric Administration. *Livestock Hot Weather Stress. Operations Manual Letter C-31-76*; Department of Commerce, NOAA, National Weather Service Central Region: Kansas City, MO, USA, 1976.
39. Vitali, A.; Segnalini, M.; Bertocchi, L.; Bernabucci, U.; Nardone, A.; Lacetera, N. Seasonal pattern of mortality and relationships between mortality and temperature-humidity index in dairy cows. *J. Dairy Sci.* **2009**, *92*, 3781–3790. [CrossRef] [PubMed]
40. Dobrovolsky, I.P.; Zubkov, S.I.; Miachkin, V.I. Estimation of the size of earthquake preparation zones. *Pure Appl. Geophys.* **1979**, *117*, 1025–1044. [CrossRef]
41. Gutenberg, B.; Richter, C.F. Magnitude and energy of earthquakes. *Ann. Geophys.* **2010**, *53*, 7–12.
42. Molchanov, O.; Hayakawa, M. Subionospheric VLF signal perturbations possibly related to earthquakes. *J. Geophys. Res.* **1998**, *103*, 17489–17504. [CrossRef]
43. Hayakawa, M.; Molchanov, O.A. Summary report of NASDA's earthquake remote sensing frontier project. *Phys. Chem. Earth* **2004**, *29*, 617–625. [CrossRef]
44. Hayakawa, M.; Hobara, Y.; Rozhnoi, A.; Solovieva, M.; Ohta, K.; Izutsu, J.; Nakamura, T.; Kasahara, Y. The ionospheric precursor to the 2011 march 11 earthquake based upon observations obtained from the Japan-Pacific subionospheric VLF/LF network. *Terr. Atmos. Ocean Sci.* **2013**, *24*, 393–408. [CrossRef]
45. Němec, F.; Santolík, O.; Parrot, M.; Berthelier, J. Spacecraft observations of electromagnetic perturbations connected with seismic activity. *Geophys. Res. Lett.* **2008**, *35*. [CrossRef]
46. Anagnostopoulos, G.C.; Basta, M.; Stefanakis, Z.; Vassiliadis, V.G.; Vgontzas, A.N.; Rigas, A.G.; Koutsomitros, S.T.; Baloyannis, S.J.; Papadopoulos, G. A study of correlation between seismicity and mental health: Crete, 2008–2010. *Geomat. Nat. Hazards Risk* **2015**, *6*, 45–75. [CrossRef]
47. Fraser-Smith, A.C.; Bernardi, A.; McGill, P.R.; Ladd, M.E.; Helliwell, R.A.; Villard, O.G. Low-frequency magnetic field measurements near the epicenter of the Ms 7.1 Loma Prieta earthquake. *Geophys. Res. Lett.* **1990**, *17*, 1465–1468. [CrossRef]
48. Hayakawa, M.; Hattori, K.; Ohta, K. Monitoring of ULF (ultra-low-frequency) geomagnetic variations associated with earthquakes. *Sensors* **2007**, *7*, 1108–1122. [CrossRef]
49. Mahdavi, S.M.; Sahraei, H.; Yaghmaei, P.; Tavakoli, H. Effects of Electromagnetic Radiation Exposure on Stress-Related Behaviors and Stress Hormones in Male Wistar Rats. *Biomol. Ther.* **2014**, *22*, 570–576. [CrossRef] [PubMed]
50. Mahdavi, S.M.; Sahraei, H.; Tavakoli, H.; Yaghmaei, P. Effect of 5Hz electromagnetic waves on movement behavior in male wistar rats (in vitro). *J. Paramedical Sci.* **2013**, *5*, 1.
51. Mahdavi, S.M.; Rezaei-Tavirani, M.; Nikzamir, A.; Ardeshirylajimi, A. 12 Hz electromagnetic field changes stress-related hormones of rat. *J. Paramedical Sci.* **2014**, *5*, 4.
52. Engelmann, W.; Hellrung, W.; Johnsson, A. Circadian locomotor activity of Musca flies: Recording method and effects of 10 Hz square-wave electric fields. *Bioelectromagnetics* **1996**, *17*, 100–110. [CrossRef]
53. Dowse, H. The effects of phase shifts in a 10 Hz electric field cycle on locomotor activity rhythm of Drosophila melanogaster. *Biol. Rhythm Res.* **1982**, *13*, 257–264.
54. Begall, S.; Cerveny, J.; Neef, J.; Vojtech, O.; Burda, H. Magnetic alignment in grazing and resting cattle and deer. *Proc. Natl. Acad. Sci. USA* **2008**, *105*, 13451–13455. [CrossRef] [PubMed]
55. Begall, S.; Burda, H.; Červený, J.; Gerter, O.; Neef-Weisse, J.; Němec, P. Further support for the alignment of cattle along magnetic field lines: Reply to Hert et al. *J. Comp. Psychol.* **2011**, *197*, 1127–1133. [CrossRef] [PubMed]
56. Slaby, P.; Tomanova, K.; Vacha, M. Cattle on pastures do align along the North-South axis, but the alignment depends on herd density. *J. Comp. Physiol. A Neuroethol. Sens. Neural Behav. Physiol.* **2013**, *199*, 695–701. [CrossRef] [PubMed]

12

Corporate Reporting on Farm Animal Welfare: An Evaluation of Global Food Companies' Discourse and Disclosures on Farm Animal Welfare

Rory Sullivan [1,*], Nicky Amos [2] and Heleen A. van de Weerd [3]

[1] Centre for Climate Change Economics and Policy, School of Earth and Environment, University of Leeds, Leeds LS2 9JT, UK

[2] Nicky Amos CSR Services Ltd., Old Broyle Road, Chichester, West Sussex PO19 3PR, UK; nicky@nicky-amos.co.uk

[3] Cerebrus Associates Ltd., The White House, 2 Meadrow, Godalming, Surrey GU7 3HN, UK; heleen@cerebrus.org

* Correspondence: rory@rorysullivan.org

Academic Editor: Paul Koene

Simple Summary: Companies that produce or sell food products from farm animals can have a major influence on the lives and welfare of these animals. The Business Benchmark on Farm Animal Welfare (BBFAW) conducts an annual evaluation of the farm animal welfare-related disclosures of some of the world's largest food companies. The programme looks at companies' published policies and commitments and examines whether these might lead to actions that can improve animal welfare on farms. It also assesses whether companies show leadership in this field. The BBFAW found that, in 2012 and 2013, around 70% of companies acknowledged animal welfare as a business issue, and that, between 2012 and 2013, there was clear evidence of an increased level of disclosure on farm animal welfare awareness in the companies that were assessed. However, only 34% (2012) and 44% (2013) of companies had published comprehensive farm animal welfare policies, suggesting that many companies have yet to report on farm animal welfare as a business issue or disclose their approach to farm animal welfare to stakeholders and society.

Abstract: The views that food companies hold about their responsibilities for animal welfare can strongly influence the lives and welfare of farm animals. If a company's commitment is translated into action, it can be a major driver of animal welfare. The Business Benchmark on Farm Animal Welfare (BBFAW) is an annual evaluation of farm animal welfare-related practices, reporting and performance of food companies. The framework evaluates how close, based on their disclosures, companies are to best practice in three areas: Management Commitment, Governance & Performance and Leadership & Innovation. The BBFAW analysed information published by 68 (2012) and 70 (2013) of the world's largest food companies. Around 70% of companies acknowledged animal welfare as a business issue. Between 2012 and 2013, the mean BBFAW score increased significantly by 5% ($p < 0.001$, Wilcoxon Signed-Rank test). However, only 34% (2012) and 44% (2013) of companies published comprehensive animal welfare policies. This increase suggests that global food companies are increasingly aware that farm animal welfare is of interest to their stakeholders, but also that many companies have yet to acknowledge farm animal welfare as a business issue or to demonstrate their approach to farm animal welfare to stakeholders and society.

Keywords: animal welfare; farm animals; global food companies; CSR; risk management

1. Introduction

Nearly 70 billion animals are farmed globally each year for meat, milk and eggs; the share of intensive landless systems (mainly pork and chicken) is about 45% of total meat output [1]. Animal production has intensified for a variety of reasons, in particular increased demand for food and ongoing pressure to reduce the costs of production for producers and food companies [2]. However, modern commercial practices have also raised concerns about farm animal welfare [3]. The underlying causes are a mix of the systems and processes used to manage farm animals (e.g., the type of housing) and of the competence and diligence of the individuals charged with managing these animals. The most important welfare issues are associated with housing conditions, genetic selection and breeding, management methods (e.g., mutilations), transport and slaughter [4–17]. Issues with housing includes the close confinement of pigs (sow stalls, farrowing crates, single penning, tethering, high stocking densities), cattle (feedlots or concentrated animal feeding operations (CAFOs), tethering, veal crates), poultry (conventional non-enriched cages, high stocking densities) [5–8] and finfish (high stocking densities, solitary close confinement) [9]. Genetic selection and intensive breeding has produced fast-growing farm animals but puts enormous strain on these animals' skeletal structure and physiology [7], and severe effects on health and welfare and high rates of mortality can be caused by genetic engineering (cloning) techniques [10,11]. Management procedures associated with adverse welfare (pain, distress) that are routinely used in farming systems are applied to pigs (castration, teeth clipping, tail docking), poultry (toe clipping, beak trimming, desnooding, de-winging), cattle (disbudding, dehorning, castration), sheep (mulesing, castration) and finfish (fin clipping) [9,12,13]. Transport and the associated handling during loading and unloading exposes farm animals to multiple stressors (such as hunger, thirst, discomfort, pain, frustration, fear, distress, injury, disease and death), which can negatively affect their welfare [9,14]. As the journey length increases, animals become increasingly hungry, fatigued and dehydrated and the risk of morbidity and mortality increases. Killing farmed animals in a humane way involves pre-slaughter stunning, rendering the animal unconsciousness and insensible to pain, discomfort and stress, until death occurs; the induction of unconsciousness should be non-aversive and should not cause anxiety, pain, distress or suffering [15]. Finally, the over-use of antibiotics has been directly linked to the global increase in antibiotic resistance [16]. This issue is also associated with animal production, as antibiotics are used to improve growth and production (e.g., promote abnormal muscle growth or milk production), often putting excessive strain on the animal's physiological capabilities. Furthermore, antibiotics are used to prevent infection before it occurs. The need to use antibiotics in this way is exacerbated by the large numbers of animals living in close proximity in intensive farming environments, often in non-hygienic conditions. This can act as a reservoir of resistance with many opportunities for the transfer of drug-resistant bacteria, thereby accelerating spread of resistance [17].

The views that food companies hold about their responsibilities for animal welfare and their management practices and processes have a critical influence on the lives and welfare of farm animals. An individual company's commitment can be a major driving force to influence the welfare of animals, especially if the commitment expressed is translated into actual behaviour. This is not just a matter of ethical concern. There are compelling business reasons (or 'business case arguments') why companies should be concerned about farm animal welfare. These include regulation and legislation, pressure from animal welfare organisations, and brand and market opportunities for companies with higher farm animal welfare standards (see, for example, [2,18,19]). Consumer pressures are important too, with animal welfare exerting an increasingly strong influence on food purchasing decisions [20] (for a challenge to this view, see [21]). However, there is variation in the pace at which businesses involved in animal production have acted to address welfare issues. For example, within the U.S. food industry, retailers have been quicker to react than producers, in large part because of consumer pressure, which has a direct (or potential) impact on their business [22].

Weaknesses in corporate practices and performance on ethical issues such as animal welfare can be a threat to good business performance (e.g., through impacting on reputation, brand, costs);

therefore, managing these issues effectively should be an integral part of companies' risk and cost management processes [18]. Similarly, the potential for these issues to affect costs, revenues, asset values and brand can have knock-on effects on the cost of debt and of equity and so are a subject of interest for investors [2,23,24]. When we look at the reported actions taken by companies—the subject of this paper—we need to recognise that the actions taken will be critically influenced by the ethical views that food companies hold about the welfare of animals and by the pressures (or lack of pressure) on them to adopt high standards of farm animal welfare.

Despite the importance of corporate practices to animal welfare and the business case for corporate action on animal welfare, relatively little is known about how food companies (either individually or as a sector), manage farm animal welfare, for example as part of their corporate social responsibility (CSR) [2]. There may be various reasons for this. The most important one is the limited information provided by companies on their overall farm animal welfare management practices and processes [19]. Another reason is the lack of tools or frameworks that enable a meaningful assessment of individual company performance (in either absolute terms or relative to its industry peers), despite Maloni and Brown providing an expansive framework for the management of social, environmental and ethical (including animal welfare) issues in the food supply chain [2].

This paper addresses some of the gaps in knowledge by presenting the results of the first structured evaluation of the farm animal welfare-related policies, practices, processes, systems, reporting and performance of 70 of the world's largest food companies. The paper considers whether these companies report that they have established—or are establishing—a management infrastructure (e.g., policies, management accountabilities, objectives and targets) necessary to manage farm animal welfare, and whether they report on the performance outcomes that they are achieving.

The data presented are derived from research conducted by the Business Benchmark on Farm Animal Welfare (BBFAW). The Benchmark is a tool for investors seeking to evaluate the relative performance of food companies on farm animal welfare management. To that end, it assesses company reporting on farm animal welfare using a framework that broadly aligns with the manner in which companies report to investors on other corporate responsibility issues [25] (pp. 14–21); a specific example is the climate change reporting framework developed by the CDP—previously the Carbon Disclosure Project (https://www.cdp.net/en, accessed on 28 October 2016).

BBFAW's main objective (see Table 1) is to improve farm animal welfare standards in the world's leading food businesses by providing investors and, albeit to a lesser extent, other stakeholders with an independent, impartial and reliable assessment of food companies' reported practices and performance. The overarching objective and aims are listed in Table 1, with the underlying assumption that the process of disclosure and the dialogue between investors and companies will stimulate companies' efforts to adopt higher farm animal welfare standards and practices.

Table 1. The overall objective and three aims of the Business Benchmark on Farm Animal Welfare.

Objective	To Drive Higher Farm Animal Welfare Standards in the World's Leading Food Businesses
Aim 1	To provide investors with the information they need to understand the business implications of farm animal welfare for the companies in which they are invested
Aim 2	To provide investors, governments, academics, NGOs, consumers and other stakeholders with an independent, impartial and reliable assessment of individual company efforts to adopt higher farm animal welfare standards and practices
Aim 3	To provide guidance * to companies interested in improving their management and reporting on farm animal welfare issues

* BBFAW produces a range of materials on issues such as the business case for farm animal welfare, best practices in management and reporting, and new/forthcoming farm animal welfare-related regulations and policies. Furthermore, BBFAW conducts structured and extensive engagement programmes, encouraging investors to pay more attention to farm animal welfare in their investment processes and companies to improve their practices, performance and reporting on farm animal welfare.

The central deliverable of BBFAW's work is an annual public benchmark of how global food companies report on how they are managing farm animal welfare. This paper reports on the first two assessments of company performance, the 2012 Benchmark [23] and the 2013 Benchmark [26], and discusses what the results tell us about corporate practices on the reporting on farm animal welfare management. Subsequent Benchmark reports have been produced, but with somewhat different questions and across a significantly extended universe; their data are therefore not included in the current analysis.

2. Materials and Methods

The annual BBFAW public benchmark uses information published by global food companies on farm animal welfare to assess how these companies report on how they manage farm animal welfare. This approach is consistent with the idea that companies need to provide sufficient information to enable their stakeholders to hold them to account for their practices and performance. However, there may also be a disconnect between the information reported and actual performance, manifesting itself in two ways. First, companies with poor disclosure but relatively good performance may find themselves penalized; second, companies with good disclosure may receive scores or evaluations that are somewhat better than their underlying performance. We discuss this potential disconnect in Section 4.

2.1. Selection of Companies

The overarching objective of the company selection process was to provide a broadly representative sample of the larger (in terms of their turnover and their farm animal footprint) companies active in the European food sector. The primary criterion for selecting these companies was the size of their animal footprint within Europe, with economic significance or turnover used as a crude proxy for this footprint. Specifically, data from Euromonitor's analysis of the top 50 EU food businesses by sector (www.euromonitor.com/consumer-foodservice) and Deloitte's annual Global Powers of Retailing report [27] were used to develop a list of the companies to be covered by the Benchmark.

The selected companies were broadly spread across the three food industry subsectors: Food Retailers & Wholesalers, Restaurants & Bars and Food Producers. The coverage included major food companies in most European markets, as well as some North American and Brazilian companies. The companies included in the 2013 Benchmark were broadly the same as those included in 2012, although some adjustments had to be made as a consequence of a large company splitting into smaller ones ('demerger'). The effect of these changes was an increase in the total number of companies covered by the Benchmark from 68 in 2012, to 70 in 2013.

2.2. The Benchmarking Process

The core principle of the benchmarking process was that companies were only assessed on the basis of their published information on farm animal welfare, to encourage better disclosure of information.

The first step in the assessment process was a desktop review of each company's published information. This involved a detailed review of the material on the company's corporate (i.e., parent company) websites, the material contained in formal publications such as annual reports, corporate responsibility reports and other publications, the material on subsidiary company websites, and the information provided in press releases, consumer brochures and similar publications.

For each individual company, a summary report that recorded its scores, an explanation for the score awarded and details of the sources of information used for the assessment was prepared. Each company was sent its (confidential) summary spreadsheet with supporting information and given 3–4 weeks to review the information and provide feedback and/or additional information. Company scores were only revised if (a) the company could demonstrate that the assessment had not taken account of information that was in the public domain at the time of the assessment (i.e., credit was not given for information published after the time of the assessment); and/or (b) where the assessor had made an error of interpretation or fact in the assessment. The final individual company reports, showing individual scores and comments for each question, as well as overall company scores and comparable sector scores, were sent (in confidence) to the companies at the time of issuing the final Benchmark report.

The final Benchmark report presented the overall outcome of the benchmarking process with a summary graph indicating in which tier (level of performance in terms of animal welfare) a company was placed. There were six tiers in total, with Tier 1 as the highest level possible, showing leadership in animal welfare.

The same benchmarking process was used for the 2012 and 2013 Benchmarks, with a few exceptions (see below). The assessments were conducted in August and September of each year. A restricted assessment period was used in order to ensure that companies were assessed at broadly the same point in time. The choice of timing was deliberate as the vast majority of firms publish their annual reports and accounts, and their sustainability reports in the first half of the year. That is, by conducting the assessments in August and September, we were able to use the most up-to-date information for most companies. In both years, the assessments were done by the same trained assessors who regularly calibrated their scoring with the framework criteria and also reviewed each other's assessment reports. This safeguarded consistency in the assessment methodology.

2.3. The Assessment Framework

A framework (assessment questions, guidance, scoring) against which all companies were assessed was used. The framework focused on the management systems and processes related to farm animal welfare. The framework was developed by a Technical Working Group comprising experts on food businesses and animal welfare, supported by expert advisors on investment and corporate social responsibility. The draft framework was subsequently subjected to extensive consultation with investors, companies and other stakeholders invited to comment and offer suggestions for improvement (for a detailed description, see [28,29]). The lists with proposed companies for the 2012 and 2013 Benchmarks were also reviewed as part of this consultation process. The consultation was done prior to the framework being used for the 2012 and 2013 benchmarking process and led to some minor changes, primarily the addition of some companies to ensure that relevant comparator companies within particular sub-sectors were included. For more details of the framework development and review process, including details of the organisations that commented on the draft framework, (see [23,26,28,29]).

The assessment framework focused on the management systems and processes that companies should have in place to identify, assess, understand and manage the risks and opportunities associated with farm animal welfare. The primary focus was on the corporate entity (or parent company) as a whole rather than subsidiary companies (or brands), to prevent the 'filtration effect' whereby part of the company communication is left behind in the transition from parent company to subsidiary [30]. The Benchmark did consider how companies reported on their management of farm animal welfare issues in specific markets or geographic regions and did give credit for reported innovative practices and processes in these markets and regions.

Company performance was considered in three core areas, as indicated in Table 2. These were (a) Management Commitment & Policy, assessing the company's policy framework for managing farm animal welfare, including its policies on specific animal welfare issues. Some animal welfare issues are covered by national legislation, however the Benchmark took the approach that companies still needed to have formal policies on these issues, to cover the interests of the animals they have an impact on; (b) Governance & Management, assessing the company's systems and processes for managing farm animal welfare (responsibilities, objectives and targets, internal controls); (c) Leadership & Innovation, assessing the company's efforts to advance farm animal welfare more widely.

Table 2. Business Benchmark on Farm Animal Welfare: core areas and key elements of the scoring framework.

Core Area and Key Elements	No. of Points	Weighting (% of Total Score)
1. Management Commitment:	65	36%
• General account of why farm animal welfare is important to the business, including discussion on the risks and business opportunities. • Overarching farm animal welfare policy that sets out core principles and beliefs on farm animal welfare and that explains how these are addressed and implemented throughout the business. • Specific policy positions on key welfare concerns such as the close confinement of livestock, animals subjected to genetic engineering or cloning, routine mutilations, slaughter without stunning, and long distance live transportation.		
2. Governance & Management:	85 (75 in 2012)	47% (44% in 2012)
• Defined responsibilities for the day-to-day management of animal welfare-related issues as well as strategic oversight of how the company's policy is being implemented. • Objectives and targets including process and performance measures (with an explanation of how they are delivered and how progress is monitored). • Outcomes in terms of performance against objectives and targets, performance against company policy and animal welfare outcomes. • Internal controls such as employee training in farm animal welfare and the actions to be taken in the event of non-compliance with the farm animal welfare policy. • Policy implementation through supply chains, including formalising farm animal welfare in supplier contracts, supply chain monitoring and auditing processes, and supporting suppliers in meeting the company's standards on farm animal welfare.		
3. Leadership & Innovation:	30	17%
• Company involvement in research and development programmes to advance farm animal welfare. • Company involvement in industry or other initiatives directed at improving farm animal welfare. • Acknowledgement of farm animal welfare performance from notable award or accreditation schemes. • Company initiatives to promote higher farm animal welfare amongst customers or consumers.		
Total	180 (170 in 2012)	100%

The core areas were weighted as indicated in Table 2. These initial weightings were chosen to align with similar investor benchmarks and to ensure that the results and subsequent company rankings were not overly sensitive to weightings. They also reflected the fact that reporting on farm animal welfare was relatively immature and so it was considered premature to assign very high weightings to performance on impact. Investors, who were seen as the primary audience for the benchmark, were formally consulted on the weightings and were supportive of the approach adopted [23,28]. Within each of the core areas, each company was evaluated against a number of criteria with scores awarded according to how close the company was to best practice. The number of points awarded for specific criteria within a question corresponded to the level of detail in a company statement. In general, the more detail, the broader the scope and the higher the level of commitment, the more points were awarded (implicit in this is that a higher level of commitment has a potential higher impact on animal welfare). An example is presented in Table 3.

Table 3. Business Benchmark on Farm Animal Welfare: sample scoring of an individual question.

Does the company have a clear position on the avoidance of close confinement for livestock (i.e., no sow stalls, concentrated animal feeding operations (CAFOs), feedlots, farrowing crates, single penning, battery cages, tethering or veal crates)?	
No stated position	Score: 0 points
The company has made a partial commitment to the avoidance of confinement but the scope (in terms of geography, species, products) is not clearly defined.	Score: 1 point
The company has made a partial commitment to the avoidance of confinement and the scope of the commitment (in terms of geography, species, and products) is clearly defined.	Score: 3 points
Universal commitment to avoid confinement across all relevant species, own-brand products and geographies.	Score: 5 points
Maximum Score:	**5 points**

2.4. Changes in Methodology: 2013 vs. 2012

The criteria used in the 2012 Benchmark were broadly the same as for the 2013 Benchmark. There were two minor changes. First, one new question was added in the Governance & Management section in 2013, on internal controls (specifically on employee training and on the actions to be taken in the event of non-compliance with the farm animal welfare policy). The effect of this change was to increase the total number of points from 170 to 180 and to increase the total number of points for the Governance & Management section from 75 to 85, representing an increase in the proportion of points for this section from 44% to 47% of the total score (Table 2). Second, the interpretation of the question on whether companies had specific policies on genetically modified animals (GMO) was changed, so that in 2013 companies were required to explicitly state that they would not use genetically modified animals as opposed to more general corporate commitments on the avoidance of genetically modified organisms (e.g., in feed), as was the case in 2012.

2.5. Statistical Analysis

In order to test if the increase in overall Benchmark score from 2012 to 2013 was a significant change, a Wilcoxon Signed-Rank test (SPSS v. 22) was performed. The data used for this test were the scores for 68 companies that were assessed in both 2012 and 2013 (excluding the new demerged companies that were added in 2013). The absolute (not the relative) values of the Benchmark scores were analysed, but the scores for 2013 were corrected by deducting the points that (some) companies obtained for the new question on internal controls that was added in 2013, so that a like-for-like comparison could be made between the years. The Wilcoxon Signed-Rank test analysed 55 samples, after removing 13 tied scores (as tied scores cannot be ranked).

3. Results

Approximately 70% of the companies assessed in the two Benchmark years acknowledged farm animal welfare as a business issue (71% in 2012, 70% in 2013). Figure 1 shows a comparison of the results of the 2012 and 2013 Benchmarks and illustrates that companies are increasingly reporting that they pay attention to farm animal welfare. The overall benchmark score (the mean of all the overall company scores) increased from 23% in 2012 to 28% in 2013. The Wilcoxon Signed-Rank test showed that this increase in the overall scores was significant (Wilcoxon Signed-Rank test, one-tailed $p < 0.001$). Over this period, the proportion of companies with a published farm animal welfare policy (within the core area 'Management Commitment') increased, as did the proportion of companies with published objectives and targets for farm animal welfare.

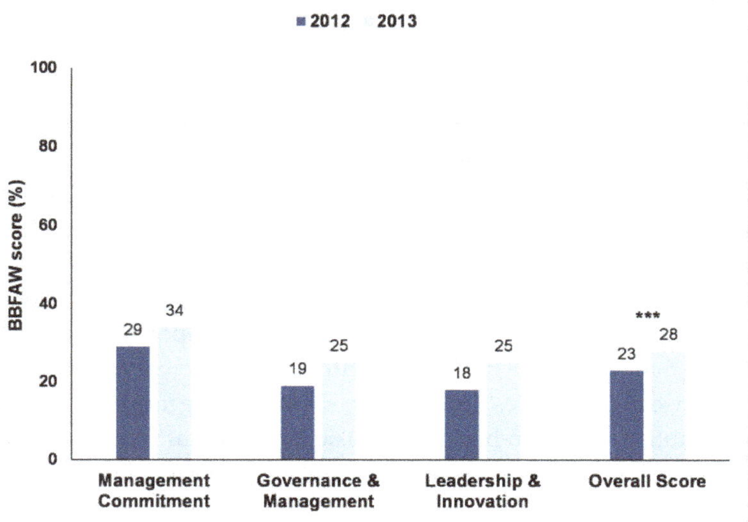

Figure 1. Business Benchmark on Farm Animal Welfare scores (as % of the total possible points) of 68 companies in 2012 and 70 companies in 2013. Average company scores for three core areas of the assessment and average overall company scores in both years. Statistical significance is indicated with *** ($p < 0.001$).

At the individual company level, companies that achieved an overall Benchmark score of less than 26% were classified in Tiers 5 and 6 (the bottom tiers). These companies provided no evidence that they recognised farm animal welfare as a business issue (Tier 6) or they provided very limited evidence (and limited information on implementation) that this subject was on the business agenda (Tier 5), let alone reporting that they were taking action to address the business risks and opportunities presented by farm animal welfare. In 2013, 53% of the companies were classified in Tier 5 and 6; this figure was higher in 2012 (62% of companies in Tier 5 or 6), showing a slight improvement in overall ranking.

At the other end of the spectrum, companies that achieved an overall Benchmark score of more than 62% were classified in Tiers 1 and 2. Tier 1 companies were those that were considered to be showing leadership through having strongly stated commitments to animal welfare and detailed reporting on how these were being implemented [23,26,28,29]. Tier 2 companies were those that were considered to have animal welfare as an integral part of their reported business strategies, with well-developed (published) management systems and processes and a clear focus on farm animal welfare outcomes. In 2013, 10% of companies were placed in Tiers 1 and 2 (seven companies in total). This was an increase in comparison to the 2012 Benchmark, when only 4% of companies (three companies) were assessed to have this highest level of performance. While there was a general increase in total scores (44 companies) and movement of companies towards higher tiers (11 companies jumped up one tier and eight companies jumped up two tiers in 2013), there were also five companies that dropped by at least one tier and 13 companies that did not change scores. In most cases a fall in tier level appeared to have been caused by changes in reporting (e.g., revamping of corporate websites and, in the process, removing relevant information that was previously published), rather than changes in published policies and practices.

The addition of the new question on internal controls to the Governance & Management section in 2013 had no effect on the tier ranking of companies. While some companies saw a modest (typically less than 1%) increase or decrease in their percentage scores, none saw their overall tier rankings increase or decrease.

3.1. Overarching Farm Animal Welfare Policies

The Benchmark differentiated between companies that were considered to have a basic farm animal welfare policy (broadly defined as having a clear published statement of commitment to farm animal welfare and/or farm animal welfare-related principles, but providing limited information on how the policy was to be implemented) and those that published a comprehensive farm animal welfare policy. In order to be considered comprehensive, a farm animal welfare policy needed to include most or all of the following elements: a clear statement of the reasons why farm animal welfare is important to the business; a commitment to compliance with relevant legislation; a clear position with regard to expected standards of farm animal welfare; a description of the processes in place to ensure that the policy is effectively implemented; and a commitment to public reporting on performance [28,29]. The number of companies that had published (any) farm policies, increased from 2012 to 2013 (Figure 2), mainly due to an increase in companies publishing comprehensive farm animal welfare policies. Of the companies that had published comprehensive or basic farm animal welfare policies, 77% and 79% (for 2012 and 2013, respectively) applied these policies to all geographies; 58% and 68% (for 2012 and 2013, respectively) applied these policies to all relevant animal species and 48% and 45% (for 2012 and 2013, respectively) applied these policies to all products produced, manufactured or sold.

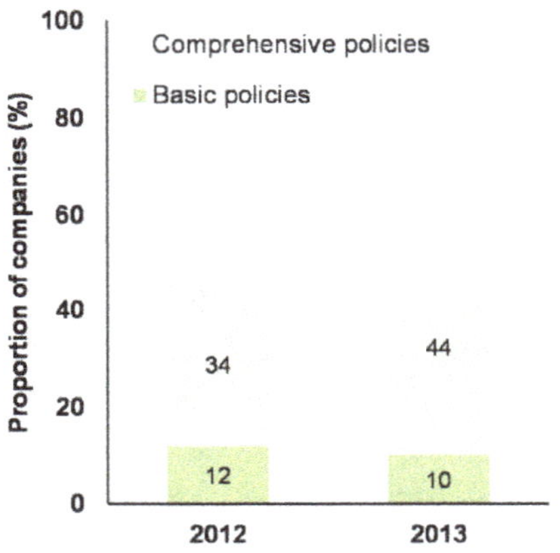

Figure 2. Business Benchmark on Farm Animal Welfare: the proportion of 68 companies in 2012 and 70 companies in 2013 that published either basic or comprehensive policies on farm animal welfare.

3.2. Policies on Specific Farm Animal Welfare Issues

Figure 3 indicates the proportion of companies that made at least partial commitments to six key farm animal welfare-related issues included in the Benchmark. Between 2012 and 2013, there was an increase in the proportion of companies with published policies on each of these issues, with the exception of long-distance transport (not changed) and Genetically Modified Organisms (GMOs), which showed a decrease. The change in the proportion of companies considered to have GMO policies related to the change in methodology, discussed in the methods section, on how this question was interpreted between 2012 and 2013.

Overall, few companies published specific policies on animal welfare issues, the exception being policies on close confinement. Relatively few companies made commitments to the complete avoidance of various welfare practices; most only made partial commitments. Policies were, generally limited to particular species, geographies or product segments. For example, for routine mutilations, of the 13% of companies with a policy on this issue in 2013, 4% were unclear about the scope of their commitment,

6% limited the scope to particular geographic regions, species or products, and just 3% had made a universal commitment to the avoidance of routine mutilations.

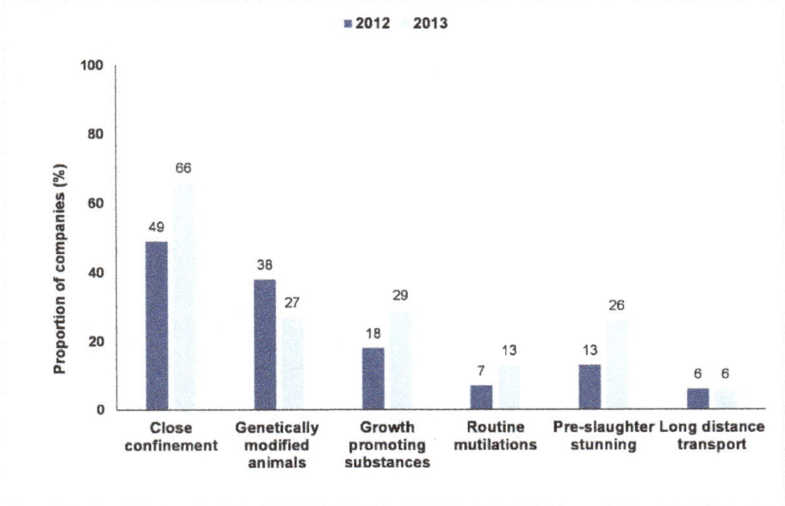

Figure 3. Business Benchmark on Farm Animal Welfare: the proportion of 68 companies in 2012 and 70 companies in 2013 with specific policies on six key farm animal welfare-related issues.

3.3. Governance and Management

More than half of the companies (59% in 2012; 54% in 2013) assessed did not provide any information on who was responsible for farm animal welfare, at either a senior management or operational level.

Around 40% of companies provided this information in 2012 and this percentage increased in 2013 (Figure 4). These companies specified who (i.e., the individual or the position) had operational responsibility for farm animal welfare, who at senior management or board level had oversight responsibility for farm animal welfare, or provided information on both operational and strategic responsibilities (Figure 4).

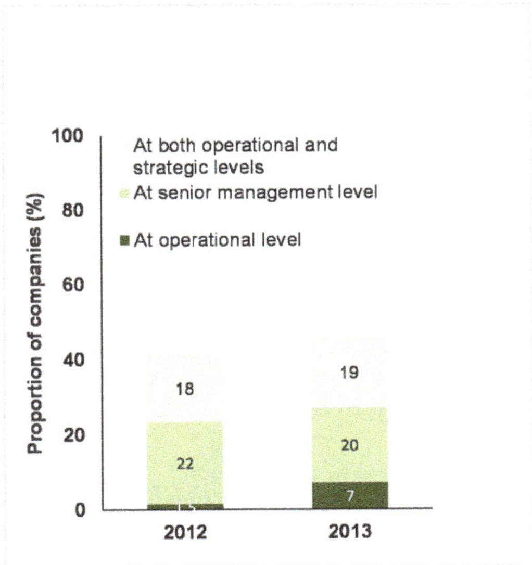

Figure 4. Business Benchmark on Farm Animal Welfare: the proportion of 68 companies in 2012 and 70 companies in 2013 that published who was responsible for managing farm animal welfare and at which level.

3.4. Objectives and Targets

In 2013 there was an increase in the number of companies that reported that they had set farm animal welfare-related objectives and targets (Figure 5). There was also an increase in the number of these companies that provided a reasonable amount of information on how the targets were to be achieved (for example, who was responsible, what resources were allocated, what were the key steps or actions towards the target) (see Figure 5).

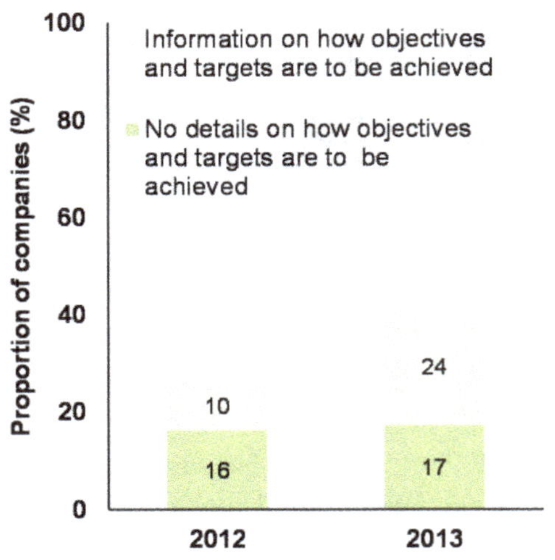

Figure 5. Business Benchmark on Farm Animal Welfare: the proportion of 68 companies in 2012 and 70 companies in 2013 that published objectives and targets for their farm animal welfare policy, with or without details on how to achieve these objectives.

3.5. Supply Chain Management

The number of companies discussing how farm animal welfare was included in supplier contract conditions increased between 2012 and 2013 (Figure 6). The majority of these companies reported that they included farm animal welfare in all relevant contracts, suggesting they have a comprehensive approach to farm animal welfare in their supply chains. The other companies reported that they included farm animal welfare in some contracts but did not specify the proportion of contracts where farm animal welfare was included.

While companies increased the amount of information they provided on their supply chain management processes, most provided limited information on the actual standards of farm animal welfare in their supply chains. For example, in 2013, only 43% of the companies (35% in 2012) described how they audited their suppliers, and only 34% described their supplier education and capacity-building initiatives (31% in 2012).

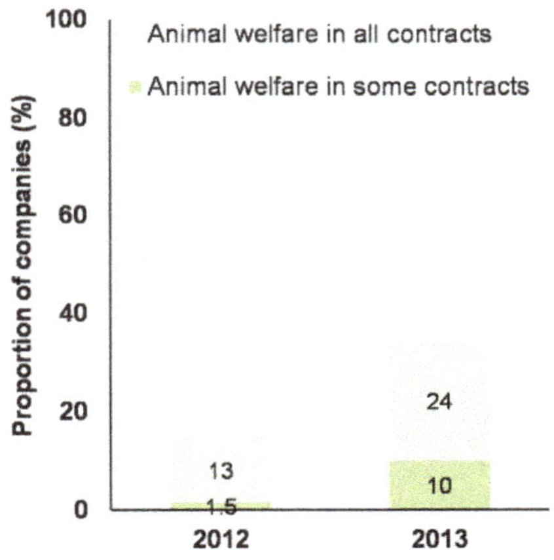

Figure 6. Business Benchmark on Farm Animal Welfare: the proportion of 68 companies in 2012 and 70 companies in 2013 had farm animal welfare in all or some of their supplier's contracts.

3.6. Reporting on Farm Animal Welfare Performance

Performance reporting by companies remains relatively underdeveloped. In the 2013 Benchmark, only 17% of the 70 companies reported on how they performed against their policy commitments, and 30% reported on their performance against their objectives and targets. However, these numbers did represent increases from the 2012 Benchmark (Figure 7).

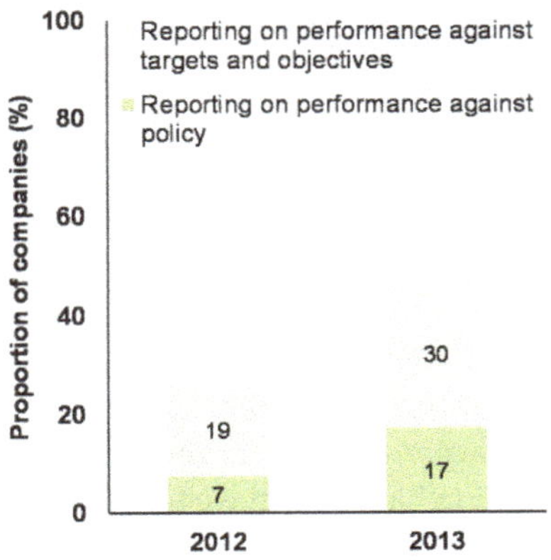

Figure 7. Business Benchmark on Farm Animal Welfare: the proportion of 68 companies in 2012 and 70 companies in 2013 that reported on their performance against their farm animal welfare policy or targets and objectives.

3.7. Assurance Schemes

About half of the companies (50% in 2012; 60% in 2013) assessed by the Benchmark provided at least some information on the assurance schemes (or standards) to which their animals were reared, transported and slaughtered (Figure 8). None of the companies, in either 2012 or 2013, had all of their products audited to higher level assurance standards (e.g., NL Beter Leven 3*). The largest proportion were companies stating that a proportion of their products or farms were audited to a basic farm assurance standard (e.g., UK Red Tractor), but they provided no information on the balance.

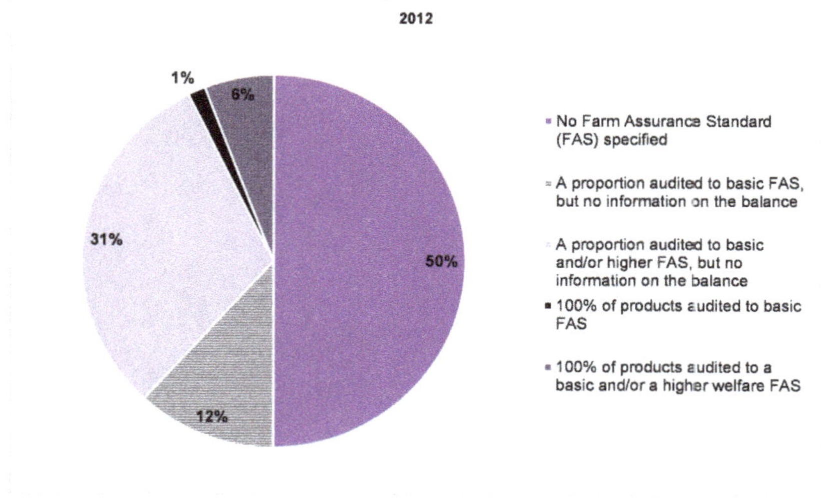

(a)

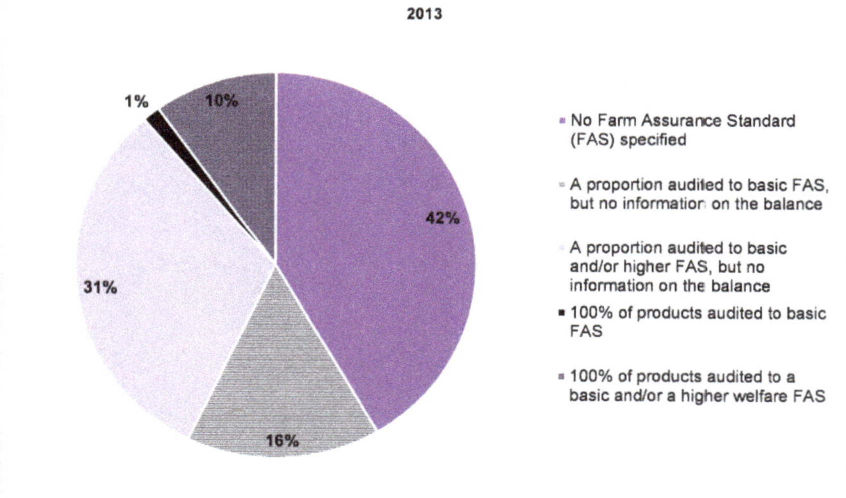

(b)

Figure 8. Business Benchmark on Farm Animal Welfare: (**a**) proportion of 68 companies in 2012 and (**b**) proportion of 70 companies in 2013 that reported on the farm assurance schemes, FAS (or standards) to which their animals were reared, transported and slaughtered. None of the companies (in either year) had 100% of their products audited to higher level welfare farm assurance standards.

3.8. Promoting Farm Animal Welfare

Figure 9 shows that there was an increase in the number of companies that provided information to their customers or consumers on farm animal welfare. The number of companies that presented multiple examples increased in 2013.

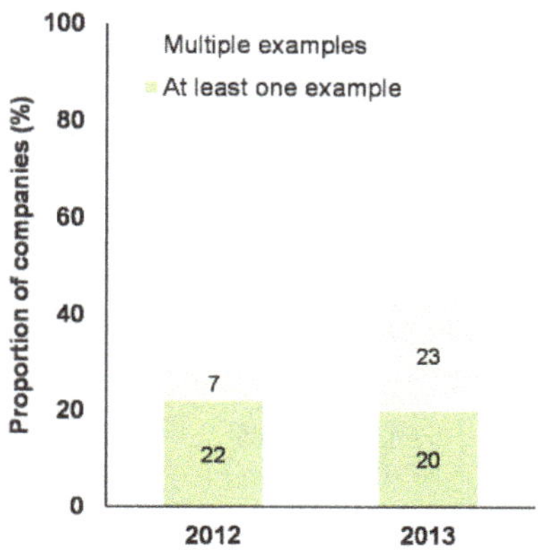

Figure 9. Business Benchmark on Farm Animal Welfare: the proportion of 68 companies in 2012 and 70 companies in 2013 that provided information (either at least one example or multiple examples) to their customers or consumers on farm animal welfare.

3.9. Sectoral Analysis

Figure 10 presents the results broken down for each sub-sector. In comparison to the 2012 Benchmark, the overall average scores for both Food Retailers & Wholesalers and for Food Producers had increased by 7%, but the average score for Restaurants & Bars had only increased by 2%.

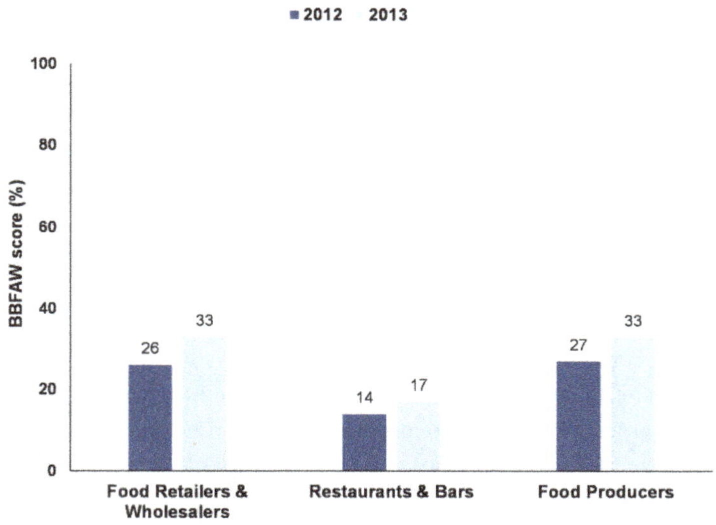

Figure 10. Business Benchmark on Farm Animal Welfare scores (as % of the total possible points) of 68 companies in 2012 and 70 companies in 2013. Average scores for the three sub-sectors (company types) in each year.

3.10. Company Feedback

Companies were given the opportunity to comment on the confidential initial results reports they received. In 2012, approximately half of the companies covered by the Benchmark provided additional information, compared to approximately one-third in the 2013 iteration. A number of individual company scores were revised based on the information provided, as per the Benchmark criteria for revising scores (see Materials and Methods).

4. Discussion

The overall score (the mean of the scores across the three core areas: Management Commitment, Governance & Management, and Leadership & Innovation) for the 2013 Benchmark had increased by 5% in comparison with 2012, which was a significant change. This was caused by 44 companies (out of 68 companies) increasing their total Benchmark score (and 19 companies actually moving up at least one tier) and shows that companies were increasingly reporting on the attention they pay to farm animal welfare.

The overall findings are similar to findings presented by Janssens and Kaptein, who analysed the websites of the 200 largest corporations in the world (selected from the 2012 Fortune Global 500 list) for statements of responsibility towards animals [19]. Their assessment of 21 companies involved in animal-based food products concluded that 76% of these companies had made statements of responsibility towards animals, a broadly similar figure to the 71% (2012) and 70% (2013) of the companies covered by the Benchmark who acknowledged farm animal welfare as a business issue.

The 2013 Benchmark shows that companies are increasingly reporting on their management infrastructure (starting with policies, then management systems and processes, and then performance reporting) to ensure that they manage farm animal welfare effectively. There could be various reasons why companies are doing this. One likely reason is pressure on companies from stakeholders to effectively manage farm animal welfare-related issues. This pressure comes from stakeholders who see ethical animal issues as part of business ethics, global consumers who believe it is important to protect the welfare of farmed animals and who are interested in buying higher welfare products [21,31,32], animal welfare NGOs and, regulation, in particular within the EU [33].

Companies have sought to reassure consumers by publishing more information on the management of their supply chains, covering issues such as monitoring, testing, supplier training and auditing. They have also sought to reassure their investors that they are effectively managing the risks related to food provenance, traceability and quality. Many companies have concluded that ignoring supply chain-related issues may create business risks, with many deciding that it is in their financial best interests to proactively prepare a comprehensive strategy for managing supply chain CSR [2].

Despite the positive trends in corporate disclosures, less than half of the companies included in the 2013 Benchmark (46%) had published comprehensive farm animal welfare policies, and another 10% had only published a basic policy statement. While this shows progress in comparison with the 2012 Benchmark where just 34% of companies had published comprehensive policies (and 12% had basic policy statements), it is important to recognise that many of these policies were quite limited in terms of their scope and there was only limited information on how the policy was to be implemented. The existence of a policy does not guarantee implementation in western multinationals (see discussion on this topic below), however, the absence of a policy is often interpreted by investors as a signal that the issue in question is not on the corporate agenda [23,25] (pp. 70–72) [34] (p. 17). This suggests that farm animal welfare is not seen as a corporate priority by many companies, and that companies may not have the management infrastructure necessary to ensure that these policies are implemented effectively. The implication, supported by the data presented in this article, is that many companies are not monitoring or managing farm animal welfare in their supply chains and this increases the risk that the welfare of farm animals is not being properly monitored and managed. It may also mean that companies will be slow to act in situations where animal welfare is being compromised or that the actions taken may be ineffective.

The slightly higher average company scores for Management Commitment compared to Governance & Management and Leadership & Innovation suggest that companies' awareness of the importance of farm animal welfare is growing and that they are starting to develop the policy frameworks needed to effectively manage these issues. Despite this, progress is variable, as many of the published overarching policies on farm animal welfare have limitations in terms of the geographic regions they apply to, the species and/or the products covered (e.g., own-brand vs. all products).

More than half of the companies assessed (2012: 59%, 2013: 54%) did not provide any information on who is responsible for farm animal welfare, at either a senior management or operational level. Furthermore, it was frequently difficult to tell how much, if any, senior management attention was actually focused explicitly on farm animal welfare. In most cases, farm animal welfare was presented as just one of a whole range of corporate responsibility-related issues that were reportedly overseen by senior management.

Within the group of companies with overarching farm animal welfare policies, only a few also reported to have established formal policies on specific farm animal welfare issues (such as on mutilations, transport duration etc.). This reflects the normal evolution of corporate responsibility practice—as has been seen in areas such as climate change [35,36]—where companies tend to start with high level policies and, over time, as they gain greater knowledge of the issue in question, formulate more detailed policies on specific issues.

A number of the companies that provided feedback on the 2012 and 2013 Benchmark results argued that the fact that some of these issues are covered by legislation removed the need for them to have a formal policy on it. A commonly cited example is the requirement for pre-slaughter stunning (with a few exemptions), that is currently part of EU law. However, one of the limitations of legislation is that it is rarely comprehensive across all species (either in terms of the issues covered or geographic scope), and its effectiveness is dependent on the level of enforcement [33]. This is particularly important when considering the complexity of supply chains where product ingredients can be sourced from a variety of jurisdictions. Therefore, the approach adopted in the Benchmark was to require companies to have public formal policies that applied to their own operations and their suppliers, and covered the interests of the animals they have an impact on.

In line with the general trends in improvements across all aspects of the Benchmark in this period, the 2013 Benchmark showed an increase in the number of companies that published formal public policy commitments on specific animal welfare issues, except for long distance transport (no change) and on the use of genetically modified animals (decrease). However, this was primarily caused by the change in the Benchmark methodology (as the focus of this question was narrowed to require companies to explicitly state that they would not use genetically modified animals as opposed to more general public corporate commitments on the avoidance of genetically modified organisms, e.g., in feed).

One specific farm animal welfare issue, close confinement, was an exception in terms of companies having a formal public policy on it. Almost 50% (2012) and 66% (2013) of companies published a specific policy on this issue. Most policies related, depending on the company, to the sourcing of cage-free eggs or to the sourcing of pig meat from sows who had not been constrained in sow (gestation) stalls and/or farrowing crates. It is likely that this trend has been driven by a range of factors, notably public pressure, NGO campaigns and changes in legislation in some parts of the world (e.g., the EU Pig Council Directive 2008/120/EC, which lays down minimum standards for the protection of pigs [37]; see also [38] (pp. 101–115)). In some cases companies have made formal, public commitments to the elimination of such practices, even in the absence of an overarching farm animal welfare policy.

With regards to farm animal welfare-related objectives and targets, the increase in the number of companies with published specific objectives and targets was encouraging. In most cases, the targets reflected the relative novelty of farm animal welfare as a management issue, with companies tending to focus on management processes (for example, to formalise their farm animal welfare management systems, to introduce audits) and/or on a single farm animal welfare-related issue (for example, to eliminate the use of gestation crates, or to move towards cage-free eggs).

The Benchmark indicates that reporting on farm animal welfare, although improving, remains relatively underdeveloped. In addition, companies regularly put their actions or results in documents of marginal importance or 'hidden' documents of a low status or with a temporary nature (e.g., company magazines or blogs) rather than in online corporate-level documents. This limits communication about such actions or outcomes [19]. There may be various reasons for this. Companies generally have multiple animal species in their supply chain that are frequently managed to different standards which makes it difficult to reduce performance to a single data point. Furthermore, companies may be concerned that reporting on performance will lead to them being criticised, especially if other companies do not provide equivalent levels of disclosure, or if disclosure is impeded due to commercial confidentiality [39].

It was reassuring to find that 60% of the companies covered by the 2013 Benchmark provided at least some information on the assurance schemes (or standards) to which their animals are reared, transported and slaughtered. Assurance schemes can play an important role in promoting and applying farm animal welfare standards [40]. Membership of assurance schemes and the associated inspections of livestock holdings can lead to a reduced risk of non-compliance with legislation and reduced unnecessary pain or distress present on farm [41]. Assurance schemes also provide many of the core process elements (e.g., on auditing, on traceability) that companies need if they are to implement effective farm animal welfare management processes in their supply chains.

The number of companies that provided information on farm animal welfare to their stakeholders (e.g., customers and investors) almost doubled in 2013 (43%) compared to 2012 (25%). This suggests that farm animal welfare has become a more integral part of customer messaging and engagement, rather than a one-off initiative. This may be a consequence of the public Benchmark assessments, putting pressure on companies to strengthen their disclosures on farm animal welfare. This should be a win-win situation, as a company that communicates what it stands for and how it performs, makes it easier for customers and other stakeholders to hold them to account and also stimulates the company to fulfil its responsibilities [42,43].

The Benchmark results for each sub-sector (i.e., Food Retailers & Wholesalers, Restaurants & Bars and Food Producers) show that performance across all three of the sectors is relatively poor. Furthermore, the Restaurants & Bar sector continues to be a noticeably poorer performer than the other two sectors and the gap with these sectors widened further in 2013, compared to 2012. The sub-sectors also show a different degree of change, the Restaurants & Bars sector only marginally increasing their overall score. The reasons for this variation in performance are unclear, but it may be because not all companies in the Restaurants & Bars sector are highly visible to the general public. For example, the sub-set of companies in the Restaurants & Bars sector that had a strong high street presence and who traded under their corporate brand name, had an average score of 27%, which was broadly similar to the average score for the other two sub-sectors (Food Retailers & Wholesalers and Food Producers).

One of the main conclusions from the 2012 and 2013 Business Benchmarks on Farm Animal Welfare is that, despite signs of progress, reporting of corporate practice on farm animal welfare lags behind practice of reporting on other corporate responsibility issues (see, for example, data on climate change in [35,36]). This may, in part, reflect the focus on published information in the Benchmark methodology, as many companies have argued that they do much more behind the scenes, but that they do not always report on these activities. Even if this were true, the lack of disclosure (and the consequent low overall Benchmark scores) suggests that companies are not well prepared to report on their performance. Businesses that refrain from publishing policy documents that are coupled to actual behaviour and operations aligned to the goals of that policy will increasingly be confronted with stakeholders who want to know why a policy is not viewed as a desirable instrument to manage ethics, integrity and social responsibility [42,44].

An important point of discussion is whether the Benchmark affects the quality of life of farm animals, as that is its ultimate aim. There may be a lack of connection between companies' disclosures and their actual performance (see for a comparable issue on CSR policy disclosure: [45]). It is possible that some companies scored well in the Benchmark, simply because they have good disclosure, instead of actually achieving good performance outcomes. That is, there may be a structural disconnect (or decoupling) between organisational policies and organisational practices [44]. Addressing this disconnect requires ethics programmes to be truly integrated into other performance management processes, such as performance monitoring systems tracking (un)ethical behaviour, training programs integrating instruction on ethical processes, and reward/punishment systems attending not only to business results, but also on how those results were achieved [44]. Another possibility is that some companies are doing an excellent job of managing farm animal welfare but score poorly because their disclosure is not sufficiently detailed or robust, or document location is inadequate (e.g., not at corporate level). One of the priorities for future iterations of the Benchmark is to progressively introduce performance criteria, with a greater focus on performance outcomes, into the Benchmark. This will include reporting on both input-based and outcome-based measures [39]. Input-based measures focus on housing and husbandry provisions, such as the type of production system (e.g., cage, barn, free-range), relevant aspects of housing (e.g., space allowance, provision of environmental enrichment), treatments and procedures, breed use, feeding and health management (e.g., the use of preventative antibiotics), and transport and slaughter practices. Outcome-based measures focus on an animal's current welfare state, while integrating long-term consequences of past husbandry [46]. These measures are specific to individual species, e.g., lameness and mastitis in dairy cows, gait score and footpad dermatitis in broilers, tail-biting and lameness in pigs, and bone breakage and feather cover in laying hens. Outcome-based measures are not confined to physical measures of wellbeing but also include aspects of mental wellbeing (e.g., reaction to humans or novelty, fear, comfort) and behaviour (e.g., time spent lying down, resting or ruminating; or being active, foraging, perching, dust-bathing or socialising). Relevant outcome- and input-based measures correlate with increased welfare [46].

The Business Benchmark on Farm Animal Welfare will be repeated on an annual basis, with the plan being to increase its coverage to 100 global food companies and to increase its focus on animal welfare performance (reporting on outcome measures). This reflects the trend in responsible investment to move beyond a process-focused approach of management systems and reporting, to focus more on the actual social and environmental performance of companies [25]. A focus on performance should allow for simultaneous tracking and influencing of corporate practices on farm animal welfare and, over time, contribute to meaningful improvements in welfare that make a real difference to farm animals.

5. Conclusions

The findings from the 2012 and 2013 Business Benchmarks on Farm Animal Welfare suggest that many of the world's largest food companies have yet to formally and publicly acknowledge farm animal welfare as a business issue. Furthermore, many companies have yet to establish robust farm animal welfare management systems and processes, and many have yet to provide a comprehensive account to their stakeholders and to wider society of their approach to farm animal welfare. Corporate practice and reporting on farm animal welfare remains relatively underdeveloped, but there are encouraging signs of progress. One of the reasons for this may be the contribution of the Business Benchmark on Farm Animal Welfare. While it is premature to offer a definitive assessment, there are signs that the Benchmark is driving change by enabling companies to benchmark themselves against their industry peers, and by providing companies with a clear set of expectations [47].

It may be too early to see the impact of the BBFAW on animals' lives, but there are clear signs of an increase in the attention being paid to animal welfare in company disclosures, as illustrated by the significant increase in scores that companies achieved in the first two years of Benchmarking discussed in this paper.

Ultimately, the views that food companies hold about the welfare of animals in their realm of influence and their management practices and processes are of critical importance in determining the welfare of billions of farm animals.

Acknowledgments: The BBFAW programme was established by two animal welfare non-governmental organisations (Compassion in World Farming and World Animal Protection). These organisations also provided the funding for the 2012 and 2013 Benchmarks that are the subject of this article.

Author Contributions: Nicky Amos and Rory Sullivan conceived and designed the Business Benchmark on Animal Welfare and managed the Benchmarking process in 2012 and 2013. Heleen van de Weerd performed the statistical analysis of the data. Rory Sullivan, Nicky Amos and Heleen van de Weerd wrote the paper.

Conflicts of Interest: The authors declare no conflicts of interest. The 2012 and 2013 iterations of the BBFAW programme were sponsored by Compassion in World Farming and World Animal Protection, but the company analysis and interpretation of results were conducted by the independent BBFAW Secretariat. Some members of staff of both Compassion in World Farming and World Animal Protection participated in the Technical Steering Group that guided the programme, and reviewed the accuracy of some of the company assessments as part of the overall quality assurance process. These members of staff were not involved in data collection for the company assessments. The founding sponsors had no role in the writing of this manuscript, or in the statistical analysis of the data.

References

1. Steinfeld, H.; Gerber, P.; Wassenaar, T.; Castel, V.; Rosales, M.; De Haan, C. *Livestock's Long Shadow: Environmental Issues and Options*; FAO: Rome, Italy, 2006. Available online: http://www.fao.org/docrep/010/a0701e/a0701e00.HTM (accessed on 15 February 2017).
2. Maloni, M.J.; Brown, M.E. Corporate Social Responsibility in the supply chain: An application in the Food Industry. *J. Bus. Ethics* **2006**, *68*, 35–52. [CrossRef]
3. Harrison, R. *Animal Machines the New Factory Farming Industry*; Vincent Stuart: London, UK, 1964.
4. Broom, D.M.; Fraser, A.F. *Domestic Animal Behaviour and Welfare*, 5th ed.; CABI: Wallingford, UK, 2015.
5. European Food Safety Authority (EFSA); Panel on Animal Health and Welfare (AHAW). Scientific opinion on Animal Health and Welfare in Fattening Pigs in Relation to Housing and Husbandry. *EFSA J.* **2007**, *564*, 1–14.
6. European Food Safety Authority (EFSA); Panel on Animal Health and Welfare (AHAW). Scientific opinion on the overall effects of farming systems on dairy cow welfare and disease. *EFSA J.* **2009**, *1143*, 1–38.
7. European Food Safety Authority (EFSA); Panel on Animal Health and Welfare (AHAW). Scientific opinion on the influence of genetic parameters on the welfare and the resistance to stress of commercial broilers. *EFSA J.* **2010**, *8*, 1666.
8. Janczak, A.M.; Riber, A.B. Review of rearing-related factors affecting the welfare of laying hens. *Poult. Sci.* **2015**, *94*, 1454–1469. [CrossRef] [PubMed]
9. Farm Animal Welfare Committee. Opinion on the Welfare of Farmed Fish, 2014. Available online: https://www.gov.uk/government/uploads/system/uploads/attachment_data/file/319323/Opinion_on_the_welfare_of_farmed_fish.pdf (accessed on 21 December 2016).
10. European Food Safety Authority (EFSA). Scientific Opinion on Food Safety, Animal Health and Welfare, and Environmental Impact of Animals Derived from Cloning by Somatic Cell Nuclear Transfer (SCNT) and their Offspring and Products Obtained from those Animals. *EFSA J.* **2008**, *767*, 1–49.
11. European Food Safety Authority (EFSA). Update on the State of Play of Animal Health and Welfare and Environmental Impact of Animals derived from SCNT Cloning and their Offspring, and Food Safety of Products Obtained from those Animals. *EFSA J.* **2012**, *10*, 2794.
12. Sutherland, M.A.; Tucker, C.B. The long and short of it: A review of tail docking in farm animals. *Appl. Anim. Behav. Sci.* **2011**, *135*, 179–191. [CrossRef]
13. Glatz, P.C. (Ed.) *Poultry Welfare Issues: Beak Trimming*; Nottingham University Press: Nottingham, UK, 2005.
14. European Food Safety Authority (EFSA); Panel on Animal Health and Welfare (AHAW). Scientific opinion concerning the Welfare of Animals during Transport. *EFSA J.* **2011**, *9*, 1966.
15. European Food Safety Authority (EFSA). Opinion of the Scientific Panel on Animal Health and Welfare on a Request from the Commission related to Welfare aspects of the main systems of stunning and killing the main commercial species of animals. *EFSA J.* **2004**, *45*, 1–29.
16. O'Neill, J. Tackling Drug-Resistant Infections Globally: Final Report and Recommendations. The Review on Antimicrobial Resistance. Available online: https://amr-review.org/sites/default/files/160525_Final%20paper_with%20cover.pdf (accessed on 20 December 2016).

17. Davis, M.F.; Price, L.; Liu, C.; Silbergeld, E.K. An ecological perspective on U.S. industrial poultry production: The role of artificial ecosystems on the emergence of drug-resistant bacteria from agricultural environments. *Curr. Opin. Microbiol.* **2011**, *14*, 244–250. [CrossRef] [PubMed]
18. Brinkmann, J. Looking at consumer behavior in a moral perspective. *J. Bus. Ethics* **2004**, *51*, 129–141. [CrossRef]
19. Janssens, M.; Kaptein, M. The ethical responsibility of companies toward animals: A study of the expressed commitment of the fortune global 200. *J. Corp. Citizensh.* **2016**, *63*, 42–72. [CrossRef]
20. Napolitano, F.; Girolami, A.; Braghieri, A. Consumer liking and willingness to pay for high welfare animal-based products. *Trends Food Sci. Technol.* **2010**, *21*, 537–543. [CrossRef]
21. Evans, A.; Miele, M. Consumers' Views about Farm Animal Welfare: Part 2, European Comparative Report based on Focus Group Research. Welfare Quality® Report No. 5. 2008. Available online: www.welfarequality.net/everyone/43215/7/0/22 (accessed on 20 June 2016).
22. Broom, D.M. Animal Welfare: An aspect of care, sustainability, and food quality required by the public. *J. Vet. Med. Educ.* **2010**, *37*, 83–88. [CrossRef] [PubMed]
23. Amos, N.; Sullivan, R. *The Business Benchmark on Farm Animal Welfare: 2012 Report*; Business Benchmark on Farm Animal Welfare: London, UK, 2013; Available online: www.bbfaw.com/media/1061/bbfaw_report_2012.pdf (accessed on 15 June 2016).
24. Krosinsky, C.; Purdom, S. *Sustainable Investing: Revolutions in Theory and Practice*; Routledge: Abingdon, UK, 2017.
25. Sullivan, R. *Valuing Corporate Responsibility: How Do Investors Really Use Corporate Responsibility Information?* Greenleaf Publishing: Sheffield, UK, 2011.
26. Amos, N.; Sullivan, R. *The Business Benchmark on Farm Animal Welfare: 2013 Report*; Business Benchmark on Farm Animal Welfare: London, UK, 2013; Available online: www.bbfaw.com/media/1058/bbfaw-report-2013.pdf (accessed on 15 June 2016).
27. Deloitte. *Global Powers of Retailing*, 2012 ed.; Deloitte: London, UK, 2012.
28. Amos, N.; Sullivan, R. *The Business Benchmark on Farm Animal Welfare: 2012 Methodology Report*; Business Benchmark on Farm Animal Welfare: London, UK, 2013; Available online: www.bbfaw.com/media/1062/bbfaw_methodology_report_2012.pdf (accessed on 15 June 2016).
29. Amos, N.; Sullivan, R. *The Business Benchmark on Farm Animal Welfare: 2013 Methodology Report*; Business Benchmark on Farm Animal Welfare: London, UK, 2014; Available online: www.bbfaw.com/media/1059/bbfaw_2013_methodology_report.pdf (accessed on 15 June 2016).
30. Frostenson, M.; Helin, S.; Sandström, J. Organising corporate responsibility communication through filtration: A study of web communication patterns in Swedish retail. *J. Bus. Ethics* **2011**, *100*, 31–43. [CrossRef]
31. Special Eurobarometer 442. Attitudes of Europeans towards Animal Welfare. Survey by TNS Opinion & Social at the Request of the European Commission, Directorate—General for Health and Food Safety. Available online: www.eurogroupforanimals.org/wp-content/uploads/2016/02/Eurobarometer-2016-Animal-Welfare.pdf (accessed on 14 December 2016).
32. Miranda-de la Lama, G.C.; Estévez-Moreno, L.X.; Sepúlveda, W.S.; Estrada-Chavero, M.C.; Rayas-Amor, A.A.; Villarroel, M.; María, G.A. Mexican consumers' perceptions and attitudes towards farm animal welfare and willingness to pay for welfare friendly meat products. *Meat Sci.* **2017**, *125*, 106–113. [CrossRef] [PubMed]
33. Rayment, M.; Asthana, P.; Van de Weerd, H.A.; Gittins, J.; Talling, J. Evaluation of the EU policy on Animal Welfare and Possible Options for the Future. GHK Consulting Ltd. and ADAS Report for DG SANCO, 2010. Available online: www.eupaw.eu (accessed on 2 September 2016).
34. Lydenberg, S. *How to Read a Corporate Social Responsibility Report: A User's Guide*; Boston College Centre for Corporate Citizenship: Boston, MA, USA, 2010.
35. Sullivan, R. The management of Greenhouse Gas Emissions in large European companies. *Corp. Soc. Responsib. Environ. Manag.* **2009**, *16*, 301–309. [CrossRef]
36. Sullivan, R. An assessment of the climate change policies and performance of large European companies. *Clim. Policy* **2010**, *10*, 38–50. [CrossRef]
37. EU Pig Directive. Council Directive 2008/120/EC of 18 December 2008 Laying Down Minimum Standards for the Protection of Pigs (Codified Version). Available online: http://eur-lex.europa.eu/LexUriServ/LexUriServ.do?uri=OJ:L:2009:047:0005:0013:EN:PDF (accessed on 20 July 2016).

38. Edwards, S.A. Current developments in pig welfare. In *The Appliance of Pig Science*; Thompson, J.E., Gill, B.P., Varley, M.A., Eds.; BSAS Publication 31; Nottingham University Press: Nottingham, UK, 2003.
39. Amos, N.; Sullivan, R. *Reporting on Performance Measures for Farm Animal Welfare*; Investor Briefing No. 14; Business Benchmark on Farm Animal Welfare: London, UK, 2014; Available online: www.bbfaw.com/media/1074/investor-briefing-14_briefing-on-performance-measures.pdf (accessed on 15 June 2016).
40. Cooper, M.D.; Wrathall, J.H.M. Assurance schemes as a tool to tackle genetic welfare problems in farm animals: Broilers. *Anim. Welf.* **2010**, *19*, 51–56.
41. KilBride, A.L.; Mason, S.A.; Honeyman, P.C.; Pritchard, D.G.; Hepple, S.; Green, L.E. Associations between membership of farm assurance and organic certification schemes and compliance with animal welfare legislation. *Vet. Rec.* **2012**, *170*, 152. [CrossRef] [PubMed]
42. Kaptein, M. Codes of multinational firms: What do they say? *J. Bus. Ethics* **2004**, *50*, 13–30. [CrossRef]
43. Reid, E.M.; Toffel, M.W. Responding to public and private politics: Corporate disclosure of climate change strategies. *Strateg. Manag. J.* **2009**, *30*, 1157–1178. [CrossRef]
44. MacLean, T.; Litzky, B.; Holderness, D.K. When organizations don't walk their talk: A cross-level examination of how decoupling formal ethics programs affects organizational members. *J. Bus. Ethics* **2015**, *128*, 351–368. [CrossRef]
45. Font, X.; Walmsley, A.; Cogotti, S.; McCombes, L.; Häusler, N. Corporate social responsibility: The disclosure-performance gap. *Tour. Manag.* **2012**, *33*, 1544–1553. [CrossRef]
46. Webster, A.J.F.; Main, D.C.J.; Whay, H.R. Welfare assessment: Indices from clinical observation. *Anim. Welf.* **2004**, *13*, S93–S98.
47. Amos, N.; Sullivan, R. *How Are Companies Using the Business Benchmark on Farm Animal Welfare?* Investor Briefing No. 16; Business Benchmark on Farm Animal Welfare: London, UK, 2014. Available online: www.bbfaw.com/media/1071/investor-briefing-no-16_how-are-companies-using-the-benchmark.pdf (accessed on 15 June 2016).

13

Dairy Cows Produce Less Milk and Modify Their Behaviour during the Transition between Tie-Stall to Free-Stall

Jan Broucek [1,*], Michal Uhrincat [1], Stefan Mihina [2], Miloslav Soch [3], Andrea Mrekajova [1] and Anton Hanus [1]

1 Research Institute of Animal Production Nitra, 951 41 Luzianky, Slovakia; uhrincat@vuzv.sk (M.U.); mrekajova@vuzv.sk (A.M.); hanus@vuzv.sk (A.H.)

2 Faculty of Engineering, Slovak Agriculture University Nitra, 949 01 Nitra, Slovakia; stefan.mihina@gmail.com

3 Faculty of Agriculture, University of South Bohemia Ceske Budejovice, 370 05 Ceske Budejovice, Czech Republic; soch@zf.jcu.cz

* Correspondence: broucek@vuzv.sk; jbroucek@hotmail.com

Academic Editor: Rachel A. Grant

Simple Summary: The purpose of this study was to evaluate the influence of moving cows from the barn with stanchion-stall housing to free-stall housing on their behaviour and production. Cows lay down up to ten hours after removing. The cows in their second lactation and open cows tended to lie sooner after removing than cows in their first lactation and pregnant cows. The times of total lying and rumination were increasing from the first day to the tenth day after removing. Cows'produced 23.3% less milk at the first day following the transfer than at the last day prior to moving (23.76 ± 7.20 kg vs. 30.97 ± 7.26 kg, $p < 0.001$). Loss of milk was gradually reduced and on the 14th day, cows achieved maximum production. The difference was found in milk losses due to the shift between cows in first and second lactation.

Abstract: Transfer of cattle to an unknown barn may result in a reduction in its welfare. Housing and management practices can result in signs of stress that include a long-term suppression of milk efficiency. The purpose of this study was to evaluate the influence of moving cows from the stanchion-stall housing to free-stall housing on their behaviour and production. The Holstein cows were moved into the new facility with free-stall housing from the old barn with stanchion-stall housing. Cows lay down up to ten hours (596.3 ± 282.7 min) after removing. The cows in their second lactation and open cows tended to lie sooner after removing than cows in their first lactation and pregnant cows. The times of total lying and rumination were increasing from the first day to the tenth day after removing (23.76 ± 7.20 kg vs. 30.97 ± 7.26 kg, $p < 0.001$). Cows produced 23.3% less milk at the first day following the transfer than at the last day prior to moving ($p < 0.001$). Loss of milk was gradually reduced and maximum production was achieved on the 14th day. The difference was found in milk losses due to the shift between cows on the first and second lactation ($p < 0.01$). The results of this study suggest that removing from the tie-stall barn with a pipeline milking system into the barn with free-stall housing and a milking parlour caused a decline in the cows' milk production. However, when the cows are moved to a better environment, they rapidly adapt to the change.

Keywords: dairy cow; milk yield; behaviour; housing; milking

1. Introduction

Many dairy buildings are relatively old and cow size has increased progressively in past decades [1]. Tethering cows restricts their freedom of movement [2]. Therefore, a number of farms are currently changing from individual housing to group housing. However, the relocation process has been implicated as one of the major aversions for received cattle [3]. The welfare impairment associated with removing and arrival at a new facility can be one of the most stressful situations an animal experiences and can cause a number of physiological and behavioural changes including altered hormones, metabolites of energy and protein metabolism, and also changes in milk production [4–6]. During relocation, cattle are subjected to noise, strange surroundings, odours and companions, overcrowding or sometimes isolation, hot or cold conditions, and a change of feed [7–9]. All of these factors contribute to stress and potential performance losses.

There is limited information describing changes in production associated with relocation of lactating dairy cows. Soch et al. [10] recorded decrease of milk yield from 19.0 kg before treatment on 10.2 kg in the first day after moving from stanchion-stall to free-stall housing. Norell and Appleman [11] declared that there was no effect on yearly milk production the first year following relocation. After changing from stanchion-stall to free-stall housing, cows produced nearly 200 kg less 3.5% fat-corrected milk than stanchion herds. However, yearly differences in milk yield did not occur between the changes of the four housing systems and control herds significantly. Brakel and Leis [12] observed that the milk yield of regrouped cows decreased by 3% on the first day following regrouping. Phillips and Rind [13] found that milk yield of mixed cows was 3% less in the first week than cows in the unmixed groups, and 1% less in the sixth week after mixing. The reduction in milk yield was similar for first- and second-lactation cows.

Dairy cows are relatively adaptable to a wide range of environments. Various criteria have been proposed to identify inappropriate management and housing conditions for them [14–16]. Some researchers have placed emphasis on criteria of well-being [17]. Cow comfort is widely recognized as an important effect on propensity to produce milk [18–20]. When a cow is not standing comfortably in the milking parlour, she can be stressed. Stress factors in the milking parlour include small stands, inconvenient hygiene, inappropriate microclimate, and insufficient floors. Cow reactions to parlour stress include not entering the parlour voluntarily, kicking off the milking cluster, defecating in the parlour or refusing milk let-down [8,18]. A modern dairy facility requires quality design, construction and ultimately management to provide a cow-centered living space. The novelty of this study is examining the impact of the manner of milking and parlour change on cows.

The purpose of this study was to evaluate the influence of the acute stress associated with the environment transition when relocating dairy cows to different housing on their milk production and maintenance behaviour. We hypothesized that treatment would affect time spent lying, standing and ruminating, and also milk yield.

2. Materials and Methods

2.1. Animals

In order to evaluate the adaptability after relocation to a new housing type, 41 Holstein cows on first and second lactation with the average age of 1244 ± 113 days (from 1080 to 1380 days) and live body weight of 558 ± 45 kg were observed. Eighteen cows were in their first lactation and 23 in their second lactation. Cows were kept in two pens (movement area 7.4 m^2 per animal, concrete alleys 2.6 m wide) with free stalls (1.15 × 2.0 m). There were 21 cows in Pen 1, and 20 cows in Pen 2. The groups were balanced according to parity. There were 9 first-parity cows and 12 second-parity cows in Pen 1 (1284 ± 107 days, 559 ± 45 kg), and 9 first-parity cows and 11 second-parity cows in Pen 2 (1200 ± 102 days, 556 ± 47 kg). Feed was available throughout the 24 h period, except during milking. The average day in milk and days pregnant for all cows were 203 ± 135 days and 52 ± 75 days. In total, there were 21 cows pregnant and 20 open. Eleven cows were pregnant and 10 open in Pen 1. Ten heads from both categories were in Pen 2 (10 pregnant and 10 open). The mean stages of gravidity were 101 ± 69 days in the first-lactation group and 37 ± 53 days in second-lactation cows.

2.2. Treatment and Housing

The cows were moved into the new facility with free-stall housing from the old barn with stanchion-stall housing. The average daily air temperature and relative humidity in the housing facility were 14.4 ± 1.52 °C and 79.1% ± 3.15%, respectively, during the last week period.

On the morning of the relocation day, farm employees led cows to a rebuilt facility with free-stall housing, a distance of 120 m. All cows were moved to the new housing together, at the same time (at once). The mean daily air temperature and relative humidity in the housing facility were 14.6 ± 1.6 °C and 81.3% ± 3.0%, respectively, during the experimental period of 25 days. These parameters were continuously recorded using data loggers.

Before relocation, cows were housed in stanchion-stall housing in the old facility (with low arch chain ties, dimensions of platform 1.125 m width, 1.65 m length). The new facility was group housing with free stalls, which consists of a solid concrete floor sloping towards a drain in the middle. Manure was removed with a scraper. Cows were kept in two pens (movement area 7.4 m^2 per animal, concrete alleys 2.6 m wide). Free stalls (1.15 × 2.0 m) were concrete-bedded with straw mattresses above the concrete floor covered lightly with sawdust. Both pens were continuously illuminated throughout the experiment.

The experimental period lasted for 25 days (during the months of November–December 2012). The study was performed in Nitra, Slovakia. The local climate is Cfb, according to the Köppen climate classification. The "C" climate is defined as one with the coldest month's average temperature below 18 °C and above −3 °C; the warmest month's average temperature is above 10 °C. The letter "f" represents a climate where no dry season occurs, the "b" the warmest month < 22 °C, but at least 4 months > 10 °C [21].

2.3. Milking

Cows in the stanchion-stall housing were milked with a pipeline milking system with the vacuum level of 50 kPa (standard high-line system), pulsation rate 57 cycles per min, and pulsation ratio 60:40. The last two individual milk yields were recorded during the evening and morning milking before shifting (05:00 and 16:00 h). Individual milk yields were weighted as the amount of milk from the milk can (pail) on an electronic scale.

After removing in the free-stall housing facility, the cows were milked in a double-five herringbone design (with vacuum level 50 kPa, pulsation rate 55 cycles per minute and pulsation ratio 60:40) and both side-opening at the same time as before removing. The parameters of both milking systems were similar.

Individual milk yields were measured electronically. Each electronic milk meter was checked the last day before starting the trial and then two times per each week in order to calculate its deviation level. This was done by comparison of the amount of milk weighed on an electronic scale. All electronic meters had a tolerance level to within 3%. The cows were not habituated to milking in a free-stall barn.

The first milking after removing was at the evening milking, and the second one the next morning. Individual milk yields (25 days) were recorded electronically at each morning and evening milking. Individual milk yield per day was calculated as the sum of the evening and morning yields. The decrease in the amount of milk in the first day (D01) and the increase in the amount of milk on the 14th day (D141) were calculated.

The cows were milked twice a day at 05:00 and 16:00 after being driven by the herdsman a short distance within the barn to a holding area, which measured 13.5 m × 4.5 m, adjacent to the milking parlour. The time spent by each cow in the holding area before milking varied from 15 to 30 min. Total times of standing during a milking session (i.e., before and after milking in the holding area, standing in the milking parlour) represented in the first, second and tenth days were 118 ± 30 min,

95 $\pm$ 22 min, and 100 $\pm$ 23 min, respectively. Cows entered the parlour individually once a milking stall was available. Upon exiting the parlour, cows remained in a separate holding area until all other cows in the group were milked. The cows then walked through an alley, passed over a set of scales, and had access to their free-stall pens immediately.

2.4. Feeding

The cows were fed a total mixed ration consisting of maize silage (6.72 kg·DM^{-1}), alfalfa haylage (6.18 kg·DM^{-1}), alfalfa hay (3.22 kg·DM^{-1}), barley straw (0.38 kg·DM^{-1}), brewer's grain (0.38 kg·DM^{-1}), sugar-beet pulp (0.57 kg·DM^{-1}), and concentrate mixture for high-yielding cows (1.73 kg·DM^{-1}) throughout the study. Feed ration contained 19.2 kg DM, 131.0 MJ NEL, 1.84 kg PDI, 2.89 kg of crude protein and 3.41 kg of crude fibre. Diet was the same from one week prior to relocation through four weeks following the move. Following relocation, all cows had daily feed prepared in troughs. Subsequently, they were fed once daily at 10:00 h. The total mixed diet was administered to troughs in the new cubicle barn by a feeding wagon. Feeding was allowed throughout the 24 h period, except during milking. Feed bunks were located centrally to the free-stall pens; raised 0.68 m above ground and 0.7 m of feeding space. Cows did not receive concentrates separately. Automatic watering troughs were located next to feed bunks and at the end of free-stall pens.

2.5. Ethological Observations

Cows were observed for three 24 h periods (first, second and tenth day) after moving into the new facility with cubicle housing, and behavioural observations were recorded at 10 min intervals (from 10:00 h). Behavioural data were obtained by video observations and electronic measurements. The barn was equipped with video cameras for continuous filming of the cows' activities. There were computer technics and software for evaluation (cameras: Samsung SCB-3000P, HDD recorder Versatile H.264 DVR), the Observer XT, Noldus (software on transmitten behavioural activities into numerical data). During the 24 h of observation, two observers were always present (for controlling); one person for each pen. They changed after 12 h; four observers were totally (they were mostly the authors of the manuscript), highly experienced in the evaluation of welfare and health. They received directives, of course, one day before the start of the experiment; and also conducted training. Data were obtained from the video, done by the author of the article.

Thus, the lying-down behaviour of cows can be an important criterion in the assessment of comfort after relocation, which we used in the research and comprehensive assessment of laterality, including lying on the left and right sides. Time spent lying (on the left and right side), standing (with or without movement, including time spent in the milking parlour), feeding, ruminating (ruminating while standing plus ruminating while lying), as well as the number of activity bouts were calculated. Based on this data, activity latencies and episodes were calculated.

The latency for each cow to the initiation of defined behaviour after relocation was determined. Latency time for total lying is first time lying down, regardless of which side the cow lay on. Latency for lying on the left or right side was calculated as first time lying down on the observed side (time between the onset of a stimulus and the initiation of the response, or recording how long it takes the cow to initiate the behaviour after a particular event occurs). Duration of behaviour episodes were summarized as a continuous series of records of the same activity lengths after relocation.

2.6. Statistical Evaluation

The data were analysed using a General Linear Model analysis of variance (ANOVA) as repeated measures by the statistical package STATISTIX [22]. The normality of data distribution was evaluated by the Wilk–Shapiro/Rankin Plot procedure. All data conformed to a normal distribution. Significant differences between groups were tested by Comparisons of Mean Ranks. Differences among means were tested by Bonferroni's method. Values are expressed as means $\pm$ SD (standard deviation).

The evaluated factors were day (of ethological observation or milk recording), lactation number (first, second), and gestation (pregnant, open). The dependent variables included daily milk yield; times spent lying, lying on the left and right sides, standing, feeding, and ruminating; latencies of lying, lying on the left and right sides, feeding, and ruminating; lengths and numbers of episodes lying, lying on the left and right sides, standing, feeding, and ruminating.

3. Results

3.1. Maintenance Behaviour

The observed cows generally lay down up to ten hours (596.3 $\pm$ 282.7 min). The cows in their second lactation and open cows lay down sooner after relocation to a new barn than cows in their first lactation and pregnant cows. The biggest differences—but statistically insignificant—were found between first and second lactation in the times of the first lying on the left side (793.3 $\pm$ 453.2 min vs. 648.7 $\pm$ 382.2 min, df 1,1,40; $p \geq 0.05$), and lying on the right side (1053.3 $\pm$ 663.7 min vs. 745.0 $\pm$ 380.0 min, df 1,1,39; $p \geq 0.05$). Latencies of first lying tended to vary also between pregnant and open cows in the lying on the left side (760.5 $\pm$ 504.9 min vs. 661.5 $\pm$ 300.5 min, df 1,1,40; $p \geq 0.05$), and lying on the right side (980.0 $\pm$ 654.8 min vs. 777.4 $\pm$ 371.0 min, df 1,1,39; $p \geq 0.05$) (Figure 1).

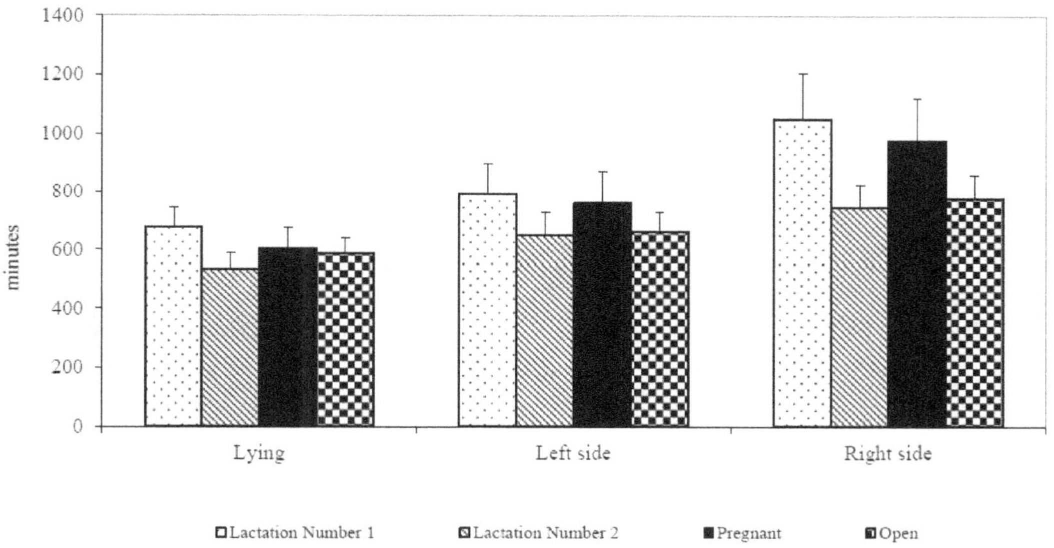

Figure 1. Latency of resting behaviour after removing (minutes).

Dissatisfaction or restlessness of cows is reflected in the frequently changing positions and behaviour activities. This study recorded a trend of non-significantly lower episode numbers of total lying (1.25 $\pm$ 0.5 vs. 2.67 $\pm$ 1.63; 1.6 $\pm$ 0.55 vs. 2.6 $\pm$ 1.95), lying on the left (1.33 $\pm$ 0.58 vs. 1.71 $\pm$ 1.11; 1.5 $\pm$ 0.58 vs.1.67 $\pm$ 1.21) and right sides (1.0 $\pm$ 0.0 vs. 1.4 $\pm$ 0.55; 1.0 $\pm$ 0.0 vs.1.5 $\pm$ 0.58) in first-lactation cows and pregnant cows in comparison to second-lactation cows and open cows.

First-lactation cows had shorter lying episodes on the left side (40.0 $\pm$ 29.4 min vs. 58.6 $\pm$ 33.9 min; df 1,1,10; $p < 0.001$), and the lying time on the right side (15.0 $\pm$ 7.1 min vs. 52.5 $\pm$ 5.0 min; df 1,1,5; $p < 0.001$) than second-lactation cows. Pregnant cows exhibited longer duration of the first lying episode on the left side (53.3 $\pm$ 30.8 min vs. 50.0 $\pm$ 37.4 min; df 1,1,10; $p < 0.01$), and a longer time of ruminating (18.5 $\pm$ 14.2 min vs. 15.5 $\pm$ 9.4 min; df 1,1,10; $p < 0.05$) than open cows following the relocation (Figure 2).

Large differences were recorded in the evaluation of the behaviour during the first, second and tenth day after the transfer. The times of total lying (336.3 $\pm$ 171.1 min; 628.0 $\pm$ 181.2 min; 756.1 $\pm$ 140.3 min; df 1,122; $p < 0.001$) were increasing from the first day to the tenth day after relocation.

The cows lay longer on their left sides during all days. A similar course—but not significant—was found at rumination time (318.0 ± 58.7 min; 325.4 ± 74.1 min; 440.5 ± 77.4 min). The total time of standing was decreasing (1103.7 ± 171.1 min; 811.9 ± 181.2 min; 683.9 ± 140.3; df 1,122; $p < 0.001$) from the first day to the tenth day.

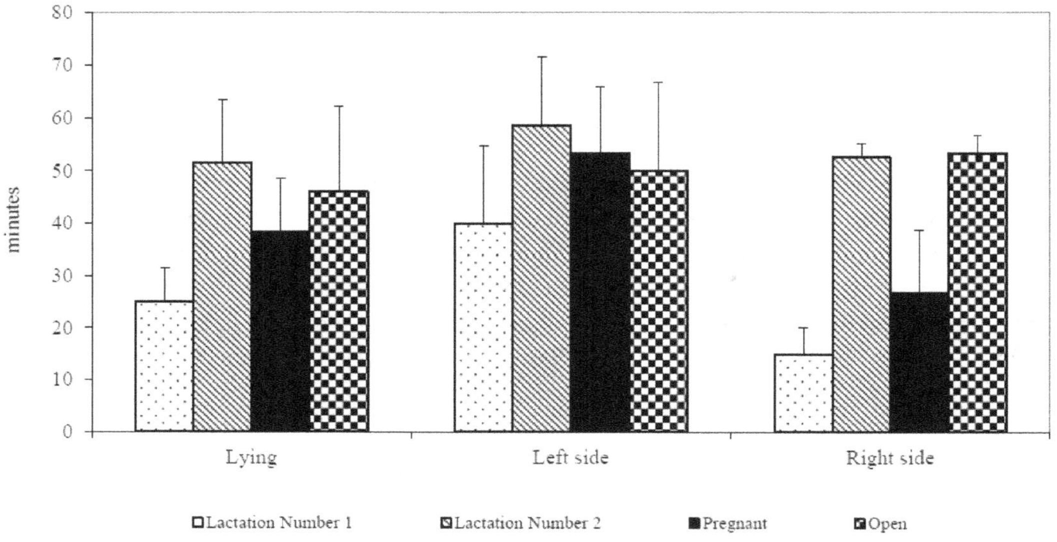

Figure 2. Lengths of resting episode after removing (minutes).

3.2. Milk Yield

Relocation had a large significant negative impact on daily milk yield. Cows produced 7.21 kg milk (23.3%) less at the first day (D1) following the transfer than at the last day prior to moving (D0) (23.76 ± 7.20 kg vs. 30.97 ± 7.26 kg, df 1,1,40; $p < 0.001$); on the second day milk production increased from D1 (27.53 ± 8.09 kg). Loss of milk was gradually reduced and, on the 14th day (D14), achieved maximum production (32.16 ± 8.87 kg), which represented an increase of 8.84 kg (28.5%) of milk yield. The difference between D1 and D14 was also significant (df 1,1,40; $p = 0.015$). A similar course was observed in all evaluated factors (lactation number, pregnancy) (Figure 3).

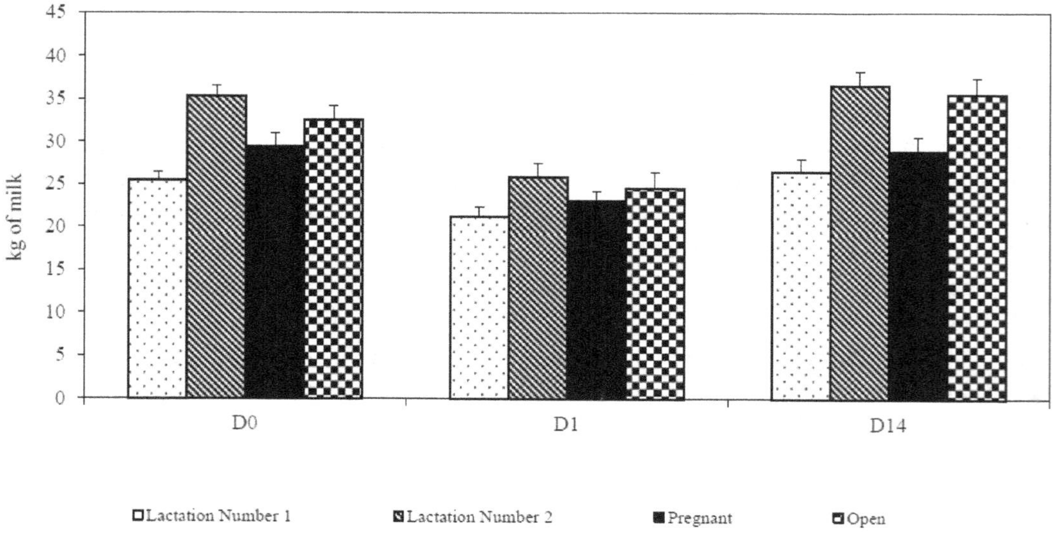

Figure 3. Milk yield after removing (D0 = last day before removing; D1 = first day after removing; D14 = fourteenth day after removing).

Open cows decreased milk yield immediately after relocating more so than pregnant cows (8.08 ± 7.37 kg vs. 6.38 ± 5.00 kg). Decrease of milk immediately after removing (30.95 ± 7.23 kg vs. 23.77 ± 7.22 kg, df 1,1,40; $p < 0.01$) was significantly lower in first-lactation cows (4.26 ± 3.74 kg vs. 9.51 ± 6.89 kg, $p < 0.01$), but the increase on the 14th day (D141) (23.76 ± 7.20 kg vs. 31.82 ± 8.96 kg, df 1,1,40; $p < 0.05$) was also lower than in second-lactation cows (5.29 ± 6.45 kg vs. 10.83 ± 6.92 kg, df 1,1,40; $p < 0.05$) (Figure 4).

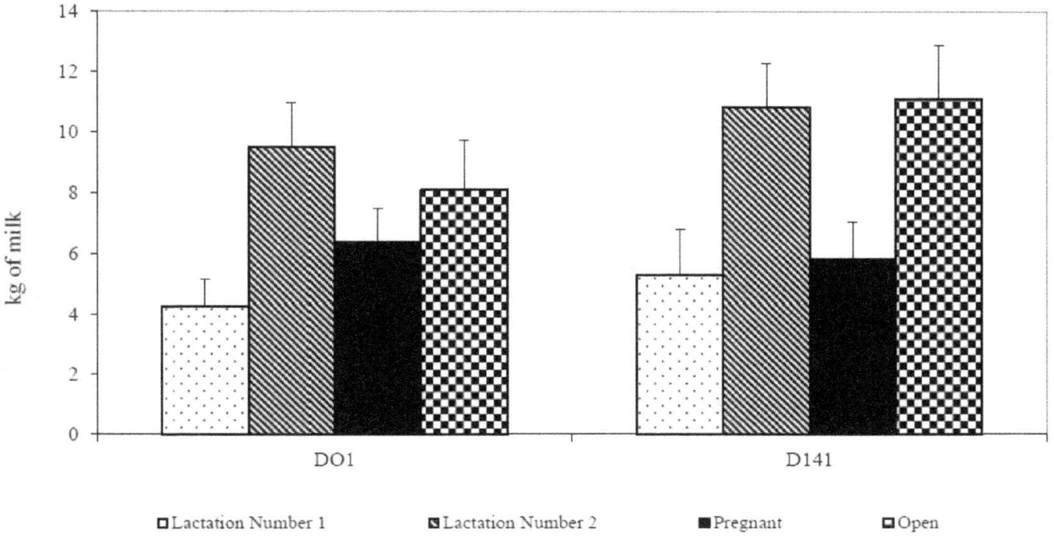

Figure 4. The decrease and increase in milk yield due to the shift of cows (D01 = decrease in the amount of milk on the first day after removing; D141 = increase in the amount of milk on the fourteenth day after removing).

4. Discussion

This study was conducted to determine that the factors of day, lactation number, and reproductive status have an effect on behaviour after dairy cows shift from a stanchion-stall barn to a free-stall barn. The central topics to be considered in this paper are the ways in which behaviour and production responses change during unknown environment. Relocation had a negative impact on daily milk yield [10,11]. Therefore, we used the following measures as latency to perform the first rest, eating and ruminating, lengths and number the first maintenance episodes of behaviours. Dissatisfaction or restlessness of cows is reflected in the frequently changing positions and behaviour activities [8,12].

Dairy cattle present few behaviour problems in general probably because farmers have selected docile animals as well as high producers. Farmers tend to cull cows from the herd that do not conform to the system. Dairy cows are docile by nature and only change their behaviour in response to environmental or management changes. The biggest problems that arise with dairy cows concern changes in their management [23]. However, there are many other factors that cause these cows to suffer when adapting. A sudden novel event can be highly stressful to an animal [24,25]. When cows that had formerly been milked and fed simultaneously in a stanchion barn are placed in free stalls and are not fed concentrates when they are milked, they may be dissatisfied and restless [23]. Dairy cows pay greater attention to moving objects than to static ones, but their perception of motion is less fluid than in humans. Therefore, sudden movements by humans may cause anxiety and even panic reactions [24]. Cattle perception ability is also used to characterize individual animal behaviours under farm conditions—especially social behaviour.

After relocation, it is important to find how long it takes dairy cows to calm down and start performing normal patterns of behaviour. After moving, cows spend more time standing. The number of standing cows after relocating depends on the suitability of the barn conditions and the experience of the operator who moves the cows [26]. When unfamiliar cows first meet, fighting often occurs to establish rank. Once hierarchical structure within a group is established, negative interactions become less common except for a limited feed, or preferred lying areas [27–30].

Cows occasionally manifested a reduced latency to first lying down. Chaplin et al. [31] reported about significant findings when cows showed more lying after deprivation of lying. Lying deprivation can cause a rest disturbance. Krohn and Munksgaard [32] and Bolinger et al. [33] wrote that cows changed their lying position depending on whether or not they were restricted from lying down before.

Second-lactation cows and open cows lay down sooner than cows in their first lactation or pregnant cows. It could be due to multiple causes. According to our opinion, second-lactation cows needed less time to lie down due to a faster adaptation phase; they are older and more experienced. After a housing change, cows could have more interrupted attempts at lying and explore the lying area more prior to lying down. Yet why open cows lie down earlier than pregnant cows is difficult to explain. There is a lack of sources; nobody has probably dealt with this problem except for us. One reason could be that these cows were placed in the social order at a higher level. Evaluating the adaptation of dairy cows to new housing is a significant indicator length of the first episode of specific behaviour activity. First-lactation cows had shorter lying episodes on the left side and the lying time on the right side than second-lactation cows. Pregnant cows exhibited longer duration of the first lying episode on the left side, and a longer time of ruminating than open cows following the relocation. However, neither lactation order nor pregnancy status statistically affected latency to lie down, ruminating or feeding. Similarly, lengths of maintenance activities episodes after removing were not significantly different among treatment groups following relocation. Another possible explanation is that the open cows are higher yielding than those cows early after parturition or pregnant.

We found that after relocation and a housing change, the times of lying increased gradually from the first to the third observation in the tenth day. A similar course was found at rumination time. Relocation and mixing of unfamiliar cows resulted in modification of behaviour immediately following the change. The lying time was reduced; an increase in time spent standing was recorded. However, these modifications were clearly evident only during the first day after moving and a change of housing type. The expected decreased lying time on the day of relocation may have been due to some cows being much less willing to displace others to gain access to a preferred free stall [34], especially primiparous cows, which when first exposed to free stalls in a competitive environment at feeding, may have decreased lying time [35]. However, in this work, the groups were not equally balanced; obtained results can be questionable.

The lying-down behaviour of cows is an important criterion in comfort; therefore, we used comprehensive assessment of laterality, including lying on the left and right sides. Laterality of lying can be an interesting welfare indicator; particularly in assessing changes in health status in dairy cattle [36,37]. Laterality is not random, but is motivated by the amount of rumen fill, slope of the floor, stage of gestation, occupancy of an adjacent stall [38,39].

In this study, the cows lay longer on their left sides during all days. First-lactation cows had shorter lying episodes on the left side, and the lying time on the right side, than second-lactation cows. Pregnant cows exhibited longer duration of the first lying episode on the left side, and a longer time of ruminating than open cows following the relocation. It is very difficult to explain this phenomenon. However, after analyzing the cows according to pregnancy stage, we discovered that mean stages of gravidity were 101 days in the first-lactation group and 37 days in second-lactation cows. Our findings are comparable with the pattern within the literature. Previous studies [38,39] have suggested that cows in later stages of pregnancy tend to lie down more on the left side because the fetus is located mainly on the right side of the body.

Relocation to a new facility did not involve long transport periods or constantly changing surroundings. Therefore, we expected that there would be an increase in behavioural and physiological indicators of stress, but these responses would be short-lived as the cows became acclimated to their new habitat. The welfare of dairy cattle can be maintained and their needs met under a variety of management systems [40]. The coping of cattle with different environment processes certainly involves a change in neuronal and endocrine functions [41–43]. Therefore, the relocation process is generally recognized as a stressful event. We have few objective criteria with which to judge the well-being in dairy farms yet, and many of the fundamental questions remain unanswered and often unconsidered [8,17,44]. For another animal that has had no previous experience with being restrained, it may react violently and physiological stress indicators will be high. The stress response is affected by an interaction between the animal's previous experiences and temperament and other inherited behavioural traits. On the other hand, according to [25], attempts to acclimate older cows to handling have been less successful. This may be due to previous experiences with aversive handling. The findings of Adamczyk et al. [30] indicate that the behaviour of loose-housed cows depends on whether their behavioural needs have been met. Once the animals were provided with a sufficient number of feeding places and ad libitum access to good quality feed, no social behaviours associated with temperament or dominance/submissiveness relationships at the feeding table were observed. Regrouping may also affect the stocking density within the pen, which also affects competitive encounters among cows. Increasing stocking density can increase competition over feed and decrease the time cows spend feeding and lying down.

Cows produced significantly less milk at the first day following the relocation than at the last day before. This important result can be explained on the basis of many aspects. Animals in group housing are often regrouped with unfamiliar herd-mates [34,45,46], and this regrouping can induce stress-related behavioural and physiological reactions. Generally, after relocating into the unknown barn, cows stand only, do not want to eat and lie down, and relax. Rest and rapid food intake are guidelines for preventing a decline in milk following a change of environment [7,13]. Dairy herds require a high level of management to ensure that the health and welfare standards for the cows are met. While removing is part of the relocation process, there are multiple factors such as pathogen exposure, commingling, feed ration changes, handling, and acclimation to a novel environment that contribute to the shipping stress complex generally associated with newly received cattle [15]. It is common to regroup cows during lactation and it is said that this usually does not adversely affect production. However, a reduction in milk production after regrouping has been reported by Hasegawa et al. [47] and this can reach 4% during the first five days [48]. When cows were overstocked, there was increased competition for free stalls. They spent less time lying in the free stalls and more time standing in the alley. Also, rumination time is decreased [26,49]. The relocation process is generally regarded as stressful to cattle and includes both physical and psychological stimuli that can cause detrimental physiological and endocrine changes. These physiological and endocrine changes can often potentiate or alter other physiological, immunological, or endocrine responses. Consequently, the release of cortisol associated with the relocation process can cause cattle to be more susceptible to disease through immunosuppression [15,39]. Sevi et al. [50] studied the stressfulness in sheep. They found that regrouping and relocation induced short-term effects on production performance. Moving and mixing were found to cause increased cortisol secretion and to have a slight and short-term effect on the productive traits of lactating cows. Former trials have shown that regrouping has no or very slight impact on milk yield in dairy cattle [12,39], except when this practice involves animals that are presumably more susceptible to stress, such as primiparous cows of low social rank [47].

As stated by Grandin [3] and Broom [17], animal housing and management systems are the most important causes of poor welfare, which impairs the production. These responses would vary with the environmental terms but are also considerably affected by the metabolic pressures on the individual [6,15]. Environmental conditions that elicit physiological coping responses in animals cause deterioration of well-being and slow adaptation of cattle [38]. Changes in the milking parlour can also affect cow behaviour [51]. Being milked in an unfamiliar environment can cause the inhibition of milk ejection.

In the present study, relocation caused prolonged stress, but cows quickly adapted to the new facility. Milk yield was gradually increased and, on the 14th day, reached maximum production.

Why did it not last longer? We can explain this through better conditions in the new barn and better well-being in free-stall housing. The freedom of movement in tie stalls is restricted in a way that unnecessary suffering is caused [2]. Also, Holroyd et al. [52] found that growth decrease of beef cattle was recovered after five days. Phillips and Rind [13] reviewed more papers and highlight the various milk reductions after cows mixing. The social relationship between cows is usually established within a few days after grouping [13,52,53]. During these first days, aggressive encounters are common. When the dominance relationships are established, conflicts are often resolved through threats and avoidance. It seems that the consequences of relocation have no long-lasting effect in dairy cows.

In this study, cows on second lactation yielded more than first-lactation cows, not only prior to removing, but also following the move. However, second-lactation cows had higher declines in milk yield than first-lactation cows. This can be explained by their yielding. The response of high-yielding cows is usually greater than low-yielding cows. However, the drop did not last too long. Cows in free-stall housing have greater freedom for movement and exercise, and have more opportunity to improve production. Also, open cows decreased milk yield immediately relocating more than pregnant cows. Open cows are much more likely to be sensitive. They may not yet be adapted after a difficult delivery, but, generally, it depends on the stage of lactation or pregnancy [54,55]. The impact of pregnancy on milk yield depends on lactation stage. The effect is higher in mid-lactation than in late-lactation [54]. A significant effect of pregnancy on milk yield is usually observed from the fifth month of gestation onwards [54–56].

It should be noted again that the first-lactation cows and pregnant cows could be included in the lower level of the social hierarchy. The behavioural studies by more authors [57–59] showed that the low-ranked cows had a more efficient eating pattern with less time spent in an eating area and fewer visits made to the feed troughs. However, dominance becomes important only when there is a very limited amount of food for which to compete [38,48].

We studied the available literature, but the effect of pregnancy on the social behaviour or hierarchy of dairy cows was dedicated to a minimum number of authors. Pregnant cows are likely to be problematic; therefore, the study of their behaviour is absent. However, they are always present in the herd, so we wanted to include this in our observations. According to our empirical experience, we have compiled a design experiment.

The mixing of new cows after relocation can create social tension [12]. The previous research has considered only the grouping of similar cattle, for example, as they change between feeding groups. Yet the cows usually end up in approximately the same relative position in the dominance hierarchy as previously, with no loss of position as a result of the move [12,13]. However, there is little evidence to support the conclusion that pregnant animals suffer in terms of well-being when forced to compete with open cows. According to the majority of authors, the most important factors in determining social position are age and live body weight. However, it is almost impossible to control for other factors, as age is often associated with seniority in the group, weight and experience [38,48].

With advancing gestation, dairy cows become slower. For most pregnant cows, sexual activity is not clear but some accompany open cows in heat and can even jump on them. They do not try to improve their social order, avoid conflicts and tend to seek contact in the form of social contacts, such as sniffing and licking. Other cows respect their condition. However, we cannot say that they are strictly submissive. Rather, they only protect the fetus and therefore do not enter into conflict. Pregnant cows are, therefore, often considered to suffer the most in new social environments [38,60].

Moreover, the effects of combining stressors may cause aggravated welfare, especially the adaptation to a distinct type of housing [34,48,51]. Cows in the present study decreased their lying times directly after removing on the first day but partially returned to usual levels the following day. Overcrowding could contribute to a high variation in individual lying bouts [34,59]. A better understanding of the components associated with the relocation process is needed to identify the major stressors and at what point multiple stressors begin to impact endocrine, immunological and metabolic functions in a manner that jeopardizes the health and well-being of cattle [15,43].

5. Conclusions

The observed cows generally lay down up to ten hours after moving, milk production decreased significantly immediately after relocation and return to baseline occurred after 14 days.

The results of this study suggest that removing cows from the tie-stall barn with a pipeline milking system into the barn with free-stall housing and a milking parlour caused a decline in cows' milk production. However, when the cows are moved to a better environment, they rapidly adapt to change.

Acknowledgments: This article was written based on data projects APVV-0632-10 and APVV-15-0060 of the Slovak Research and Development Agency, Bratislava, Slovakia. This study was made possible also through projects NAZV QJ1210144, Czech Republic, and CEGEZ 26220120073, supported by the Operational Programme Research and Development funded from the European Regional Development Fund.

Author Contributions: Jan Broucek, Michal Uhrincat and Stefan Mihina conceived and designed the experiments; Michal Uhrincat, Anton Hanus, and Andrea Mrekajova performed the experiments; Jan Broucek and Miloslav Soch analysed the data; Jan Broucek wrote the manuscript with contributions from Stefan Mihina and Miloslav Soch.

Conflicts of Interest: The authors declare no conflict of interest.

References

1. Bewley, J.; Palmer, R.W.; Jackson-Smith, D.B. A Comparison of Free-Stall Barns Used by Modernized Wisconsin Dairies. *J. Dairy Sci.* **2001**, *84*, 528–541. [CrossRef]
2. Norring, M.; Valros, A.; Munksgaard, L. Milk yield affects time budget of dairy cows in tie-stalls. *J. Dairy Sci.* **2012**, *95*, 102–108. [CrossRef] [PubMed]
3. Grandin, T. Handling methods and facilities to reduce stress in cattle. *Vet. Clinics N. Am. Food Anim. Pract.* **1998**, *14*, 325–341. [CrossRef]
4. Loerch, S.C.; Fluharty, F.L. Physiological changes and digestive capabilities of newly received feedlot cattle. *J. Anim. Sci.* **1999**, *77*, 1113–1119. [CrossRef]
5. Van Reenen, C.G.; O'Connaell, N.E.; van der Werf, J.T.; Korte, S.M.; Hopster, H.; Jones, R.B.; Blokhuis, H.J. Responses of calves to acute stress: Individual consistency and relations between behavioural and physiological measures. *Physiol. Behav.* **2005**, *85*, 557–570. [CrossRef] [PubMed]
6. Koolhaas, J.M.; De Boer, S.F.; Coppens, C.M.; Buwald, B. Neuroendocrinology of coping styles: Towards understanding the biology of individual variation. *Front. Neuroendocrinol.* **2010**, *31*, 307–321. [CrossRef] [PubMed]
7. Johansson, B.; Redbo, I.; Svennersten-Sjaunja, K. Effect of feeding before, during and after milking on dairy cow behaviour and the hormone cortisol. *Anim. Sci.* **1999**, *68*, 597–604. [CrossRef]
8. Keeling, L.; Jensen, P. Behavioural Disturbances, Stress and Welfare. In *The Ethology of Domestic Animals: An Introductory Text*, 1st ed.; Jensen, P., Ed.; CAB International: Wallingford, UK, 2002; pp. 79–98.
9. Grandin, T. Transferring results of behavioral research to industry to improve animal welfare on the farm, ranch and the slaughter plant. *Appl. Anim. Behav. Sci.* **2003**, *81*, 215–228. [CrossRef]
10. Soch, M.; Kolarova, P.; Rehout, V.; Kosvanec, K.; Hajic, F.; Citek, J. Effect of dairy cows moving from tie-stall to loose housing system on their production and behaviour. *Sbornik ZF JU Ceske Budejovice-Zootechnicka Rada* **1997**, *14*, 77–86.
11. Norell, R.J.; Appleman, R.D. Change of milk production with housing system and herd expansion. *J. Dairy Sci.* **1981**, *64*, 1749–1755. [CrossRef]
12. Brakel, W.J.; Leis, R.A. Impact of social disorganization on behavior, milk yield, and body weight of dairy cows. *J. Dairy Sci.* **1976**, *59*, 716–721. [CrossRef]
13. Phillips, C.J.C.; Rind, M.I. The Effects on Production and Behavior of Mixing Uniparous and Multiparous Cows. *J. Dairy Sci.* **2001**, *84*, 2424–2429. [CrossRef]
14. Grandin, T. Assessment of stress during handling and transport. *J. Anim. Sci.* **1997**, *75*, 249–257. [CrossRef] [PubMed]
15. Falkenberg, S.M.; Carroll, J.A.; Keisler, D.H.; Sartin, J.L.; Elsasser, T.H.; Buntyn, J.O.; Broadway, P.R.; Schmidt, T.B. Evaluation of the endocrine response of cattle during the relocation process. *Livest. Sci.* **2013**, *151*, 203–212. [CrossRef]

16. Adamczyk, K.; Pokorska, J.; Makulska, J.; Earley, B.; Mazurek, M. Genetic analysis and evaluation of behavioural traits in cattle. *Livest. Sci.* **2013**, *154*, 1–12. [CrossRef]
17. Broom, D.M. Behaviour and welfare in relation to pathology. *Appl. Anim. Behav. Sci.* **2006**, *97*, 73–83. [CrossRef]
18. Mihina, S.; Kazimirova, V.; Copland, T.A. *Technology for Farm Animal Husbandry*; Slovak Agricultural University: Nitra, Slovakia, 2012.
19. Wilkes, C.O.; Pence, K.J.; Hurt, A.M.; Becvar, O.; Knowlton, K.F.; McGilliard, M.L.; Gwazdauskas, F.C. Effect of relocation on locomotion and cleanliness in dairy cows. *J. Dairy Res.* **2008**, *75*, 19–23. [CrossRef] [PubMed]
20. Schirmann, N.; Chapinal, D.M.; Weary, D.M.; Heuwieser, W.; von Keyserlingk, M.A.G. Short-term effects of regrouping on behavior of prepartum dairy cows. *J. Dairy Sci.* **2011**, *94*, 2312–2319. [CrossRef] [PubMed]
21. Melo, M.; Lapin, M.; Damborska, I. Methods for the design of climate change scenario in Slovakia for the 21st century. *Bull. Geogr. Phys. Geogr. Ser.* **2009**, *1*, 77–90.
22. Anonymous. *Statistix for Windows, Version 9.0, User's Manual*; Analytical Software: Tallahassee, FL, USA, 2009; pp. 1–454.
23. Houpt, K.A. *Domestic Animal Behavior for Veterinarians and Animal Scientists*; Iowa State University Press: Ames, IA, USA, 1991.
24. Adamczyk, K.; Górecka-Bruzda, A.; Nowicki, J.; Gumułka, M.; Molik, E.; Schwarz, T.; Earley, B.; Klocek, C. Perception of environment in farm animals—A review. *Ann. Anim. Sci.* **2015**, *3*, 565–589. [CrossRef]
25. Grandin, T.; Shivley, C. How Farm Animals React and Perceive Stressful Situations Such As Handling, Restraint, and Transport. *Animals* **2015**, *5*, 1233–1251. [CrossRef] [PubMed]
26. Fregonesi, J.A.; Tucker, C.B.; Weary, D.M. Overstocking Reduces Lying Time in Dairy Cows. *J. Dairy Sci.* **2007**, *90*, 3349–3354. [CrossRef] [PubMed]
27. Huzzey, J.M.; DeVries, T.J.; Valois, P.; von Keyserlingk, M.A.G. Stocking density and Feed Barrier Design Affect the Feeding and Social Behaviour of Dairy Cattle. *J. Dairy Sci.* **2006**, *89*, 126–133. [CrossRef]
28. Broucek, J.; Uhrincat, M.; Soch, M.; Kisac, P. Genetics of behaviour in cattle. *Slovak J. Anim. Sci.* **2008**, *41*, 166–172.
29. Adamczyk, K.; Gil, Z.; Felenczak, A.; Skrzyński, G.; Zapletal, P.; Choroszy, Z. Relationship between milk yield of cows and their 24-hour walking activity. *Anim. Sci. Pap. Rep.* **2011**, *29*, 185–195.
30. Adamczyk, K.; Slania, A.; Gil, Z.; Felenczak, A.; Bulla, J. Relationships between milk performance and behaviour of cows under loose housing conditions. *Ann. Anim. Sci.* **2011**, *11*, 283–293.
31. Chaplin, S.J.; Tierney, G.; Stockwell, C.; Logue, D.N.; Kelly, M. An evaluation of mattress and mats in two dairy units. *Appl. Anim. Behav. Sci.* **2000**, *66*, 263–272. [CrossRef]
32. Krohn, C.C.; Munksgaard, L. Behaviour of dairy cows kept in extensive (loose housing/pasture) or intensive (tie stall) environments. II. Lying and lying-down behaviour. *Appl. Anim. Behav. Sci.* **1993**, *37*, 1–16. [CrossRef]
33. Bolinger, D.J.; Albright, J.L.; Morrow-Tesch, J.; Kenyon, S.J.; Cunningham, M.D. Restraint using self-locking stanchions. *J. Dairy Sci.* **1997**, *80*, 2411–2417. [CrossRef]
34. Von Keyserlingk, M.A.G.; Olenick, D.; Weary, D.M. Acute Behavioral Effects of Regrouping Dairy Cows. *J. Dairy Sci.* **2008**, *91*, 1011–1016. [CrossRef] [PubMed]
35. Chaplin, S.J.; Munksgaard, L. Effects of stage of lactation and parity on the lying behaviour of dairy cows in tie–stalls. In Proceedings of the 33rd International Congress of the International Society for Applied Ethology (ISAE'99), Lillehammer, Norway, 17–21 August 1999; p. 195.
36. Tucker, C.B.; Cox, N.R.; Weary, D.M.; Spinka, M. Laterality of lying behaviour in dairy cattle. *Appl. Anim. Behav. Sci.* **2009**, *120*, 125–131. [CrossRef]
37. Ledgerwood, D.N.; Winckler, C.; Tucker, C.B. Evaluation of data loggers, sampling intervals, and editing techniques for measuring the lying behavior of dairy cattle. *J. Dairy Sci.* **2010**, *93*, 5129–5139. [CrossRef] [PubMed]
38. Albright, J.L.; Arave, C.W. *The Behaviour of Cattle*, 1st ed.; CAB International: Wallingford, UK, 1997; p. 320.
39. Phillips, C.J.C.; Llewellyn, S.; Claudia, A. Laterality in bovine behavior in an extensive partially-suckled herd and an intensive dairy herd. *J. Dairy Sci.* **2003**, *86*, 3167–3173. [CrossRef]
40. Broucek, J.; Uhrincat, M.; Hanus, A. Maintenance and competitive behaviour study in dairy calves. *Slovak J. Anim. Sci.* **2011**, *44*, 28–33.
41. Friend, T. Recognizing behavioral needs. *Appl. Anim. Behav. Sci.* **1989**, *22*, 151–158. [CrossRef]

42. Huzzey, J.M.; Nydam, D.V.; Grant, R.J.; Overton, T.R. The effects of overstocking Holstein dairy cattle during the dry period on cortisol secretion and energy metabolism. *J. Dairy Sci.* **2012**, *95*, 4421–4433. [CrossRef] [PubMed]
43. Kovács, L.; Tözsér, J.; Bakony, M.; Jurkovich, V. Changes in heart rate variability of dairy cows during conventional milking with nonvoluntary exit. *J. Dairy Sci.* **2013**, *96*, 7743–7747. [CrossRef] [PubMed]
44. Watters, A.M.E.; Meijer, K.M.A.; Barkema, H.W.; Leslie, K.E.; Keyserlingk, M.A.G.; DeVries, T.J. Associations of herd- and cow-level factors, cow lying behavior, and risk of elevated somatic cell count in free-stall housed lactating dairy cows. *Prev. Vet. Med.* **2013**, *111*, 245–255. [CrossRef] [PubMed]
45. Schirmann, K.; Chapinal, N.; Weary, D.M.; Heuwieser, W.; Von Keyserlingk, M.A.G. Rumination and its relationship to feeding and lying behavior in Holstein dairy cows. *J. Dairy Sci.* **2012**, *95*, 3212–3217. [CrossRef] [PubMed]
46. Silva, P.R.B.; Moraes, J.G.N.; Mendonça, L.G.D.; Scanavez, A.A.; Nakagawa, G.; Ballou, M.A.; Walcheck, B.; Haines, D.; Endres, M.I.; Chebel, R.C. Effects of weekly regrouping of prepartum dairy cows on innate immune response and antibody concentration. *J. Dairy Sci.* **2013**, *96*, 7649–7657. [CrossRef] [PubMed]
47. Hasegawa, N.; Nishiwaki, A.; Sugawara, K.; Iwao, I. The effects of social exchange between two groups of lactating primiparous heifers on milk production, dominance order, behavior and adrenocortical response. *Appl. Anim. Behav. Sci.* **1997**, *51*, 15–27. [CrossRef]
48. Bouissou, M.F.; Boissy, A.; Le Neindre, P.; Veissier, I. The social behaviour of cattle. In *Social Behaviour in Farm Animals*, 1st ed.; Keeling, L.J., Gonyou, H.W., Eds.; CAB International: Wallingford, UK, 2001; pp. 113–145.
49. Smith, J.F.; Harner, J.P., III; Brouk, M.J.; Armstrong, D.V.; Gamroth, M.J.; Meyer, M.J.; Boomer, G.; Bethard, G.; Putnam, D. Relocation and Expansion Planning for Dairy Producers. In *Special Publication MF2424*; Kansas State University Agricultural Experiment Station and Cooperative Extension Service: Manhattan, KS, USA, 2000; pp. 1–20.
50. Sevi, A.; Taibi, L.; Albenzio, M.; Muscio, A.; Dell'Aquila, S.; Napolitano, F. Behavioral, adrenal, immune, and productive responses of lactating ewes to regrouping and relocation. *J. Anim. Sci.* **2001**, *79*, 1457–1465. [CrossRef] [PubMed]
51. Hillerton, J.E.; Ohnstad, I.; Baines, J.R.; Leach, K.A. Performance differences and cow responses in new milking parlours. *J. Dairy Res.* **2001**, *69*, 75–80. [CrossRef]
52. Holroyd, R.G.; Doogan, V.J.; Jeffery, M.R.; Lindsay, J.A.; Venus, B.K.; Bortolussi, G. Relocation does not have a significant effect on the growth rate of Bos indicus cross steers. *Aust. J. Exp. Agric.* **2008**, *48*, 608–614. [CrossRef]
53. Kondo, S.; Hurnik, J.F. Stabilization of social hierarchy in dairy cows. *Appl. Anim. Behav. Sci.* **1990**, *27*, 287–297. [CrossRef]
54. Olori, V.E.; Brotherstone, S.; Hill, W.G.; McGuirk, B.J. Effect of gestation stage on milk yield and composition in Holstein Friesian dairy cattle. *Livest. Prod. Sci.* **1997**, *52*, 167–176. [CrossRef]
55. Roche, J.R. Effect of pregnancy on milk production and body weight from identical twin study. *J. Dairy Sci.* **2003**, *86*, 777–783. [CrossRef]
56. Haile-Mariam, M.; Bowman, P.J.; Goddard, M.E. Genetic and environmental relationship among calving interval, survival, persistency of milk yield and somatic cell count in dairy cattle. *Livest. Prod. Sci.* **2003**, *80*, 189–200. [CrossRef]
57. Arave, C.W.; Albright, J.L.; Armstrong, D.V. Effects of isolation of calves on growth, behavior, and first lactation milk yield of Holstein cows. *J. Dairy Sci.* **1992**, *75*, 3408–3415. [CrossRef]
58. Huzzey, J.M.; Grant, R.J.; Overton, T.R. Relationship between competitive success during displacements at an overstocked feed bunk and measures of physiology and behavior in Holstein dairy cattle. *J. Dairy Sci.* **2012**, *95*, 4434–4441. [CrossRef] [PubMed]
59. Von Keyserlingk, M.; Weary, D. Improving the Welfare of Dairy Cattle: Implications of Freestall Housing on Behavior and Health. In Proceedings of the Western Dairy Management Conference, Reno, NV, USA, 11–13 March 2009; pp. 43–52.
60. Cook, N.B.; Nordlund, K.V. Behavioral needs of the transition cow and considerations for special needs facility design. *Vet. Clin. Food. Anim.* **2004**, *20*, 495–520. [CrossRef] [PubMed]

PERMISSIONS

All chapters in this book were first published in ANIMALS, by MDPI; hereby published with permission under the Creative Commons Attribution License or equivalent. Every chapter published in this book has been scrutinized by our experts. Their significance has been extensively debated. The topics covered herein carry significant findings which will fuel the growth of the discipline. They may even be implemented as practical applications or may be referred to as a beginning point for another development.

The contributors of this book come from diverse backgrounds, making this book a truly international effort. This book will bring forth new frontiers with its revolutionizing research information and detailed analysis of the nascent developments around the world.

We would like to thank all the contributing authors for lending their expertise to make the book truly unique. They have played a crucial role in the development of this book. Without their invaluable contributions this book wouldn't have been possible. They have made vital efforts to compile up to date information on the varied aspects of this subject to make this book a valuable addition to the collection of many professionals and students.

This book was conceptualized with the vision of imparting up-to-date information and advanced data in this field. To ensure the same, a matchless editorial board was set up. Every individual on the board went through rigorous rounds of assessment to prove their worth. After which they invested a large part of their time researching and compiling the most relevant data for our readers.

The editorial board has been involved in producing this book since its inception. They have spent rigorous hours researching and exploring the diverse topics which have resulted in the successful publishing of this book. They have passed on their knowledge of decades through this book. To expedite this challenging task, the publisher supported the team at every step. A small team of assistant editors was also appointed to further simplify the editing procedure and attain best results for the readers.

Apart from the editorial board, the designing team has also invested a significant amount of their time in understanding the subject and creating the most relevant covers. They scrutinized every image to scout for the most suitable representation of the subject and create an appropriate cover for the book.

The publishing team has been an ardent support to the editorial, designing and production team. Their endless efforts to recruit the best for this project, has resulted in the accomplishment of this book. They are a veteran in the field of academics and their pool of knowledge is as vast as their experience in printing. Their expertise and guidance has proved useful at every step. Their uncompromising quality standards have made this book an exceptional effort. Their encouragement from time to time has been an inspiration for everyone.

The publisher and the editorial board hope that this book will prove to be a valuable piece of knowledge for researchers, students, practitioners and scholars across the globe.

LIST OF CONTRIBUTORS

Jennifer Hood, Carolyn McDonald, Bethany Wilson and Paul McGreevy
Faculty of Veterinary Science, University of Sydney, Room 206, R.M.C. Gunn Building, Sydney 2006, New South Wales, Australia

Phil McManus
School of Geosciences, University of Sydney, Room 435, F09, Madsen Building, Sydney 2006, New South Wales, Australia

Christa A. Rice, Nicole L. Eberhart and Peter D. Krawczel
Department of Animal Science, University of Tennessee, Knoxville, 2506 River Dr. 258 Brehm Animal Science Knoxville, Knoxville, TN 37996, USA

Sara Shields and Andrew Rowan
Humane Society International, 1255 23rd Street, Northwest, Suite 450,Washington, DC 20037, USA

Paul Shapiro
Humane Society of the United States, 700 Professional Drive Gaithersburg, MD 20879, USA

Jan Hassink and Marjolein Elings
Wageningen University and Research, P.O. Box 16, 6700 AA Wageningen, The Netherlands

Simone R. De Bruin
National Institute for Public Health and the Environment, Centre for Nutrition, Prevention and Health Services, P.O. Box 1, 3720 BA Bilthoven, The Netherlands

Bente Berget
Agderforskning, Gimlemoen, P.O.Box 422, 4604 Kristiansand, Norway

Terry Spencer and Nancy Hardt
College of Medicine, University of Florida, Gainesville, FL 32611, USA

Linda Behar-Horenstein
Colleges of Dentistry, Education, Veterinary Medicine, & Pharmacy, University of Florida, Gainesville, FL 32611, USA

Joe Aufmuth
George A. Smathers Libraries, University of Florida, Gainesville, FL 32611, USA

Jennifer W. Applebaum
College of Liberal Arts and Sciences, University of Florida, Gainesville, FL 32611, USA

Amber Emanuel
College of Health & Human Performance, University of Florida, Gainesville, FL 32611, USA

Natalie Isaza
College of Veterinary Medicine, University of Florida, Gainesville, FL 32611, USA

Emily Weiss, Emily D. Dolan, Heather Mohan-Gibbons, Shannon Gramann and Margaret R. Slater
Research and Development, Community Outreach, American Society for the Prevention of Cruelty to Animals (ASPCA ®), New York, NY 10128, USA

Peta S. Taylor, Paul H. Hemsworth, and Jean-Loup Rault
AnimalWelfare Science Centre, Faculty of Veterinary and Agricultural Sciences, University of Melbourne, Parkville, VIC 3010, Australia

Peter J. Groves
Poultry Research Foundation, School of Veterinary Science, Faculty of Science, The University of Sydney, Camden, NSW 2570, Australia

Sabine G. Gebhardt-Henrich
Research Centre for Proper Housing: Poultry and Rabbits (ZTHZ), Division of Animal Welfare, University of Bern, CH-3052 Zollikofen, Switzerland

Daniela Luna
Programa Doctorado en Ciencias Silvoagropecuarias y Veterinarias, Universidad de Chile, Santa Rosa 11315, La Pintana, Santiago 8820000, Chile

Rodrigo A. Vásquez
Instituto de Ecología y Biodiversidad, Departamento de Ciencias Ecológicas, Facultad de Ciencias, Universidad de Chile, Las Palmeras 3425, Ñuñoa, Santiago 7800003, Chile

Manuel Rojas
Departamento de Ingenieria Industrial, Facultad de Ciencias Físicas y Matemáticas, Universidad de Chile, Beauchef 851, Santiago 8370456, Chile

Tamara A. Tadich
Departamento de Fomento de la Producción Animal, Facultad de Ciencias Veterinarias y Pecuarias, Universidad de Chile, Santa Rosa 11735, La Pintana, Santiago 8820000, Chile

Gabriella Gronqvist, Chris Rogers, Erica Gee and Charlotte Bolwell
Massey Equine, Institute of Veterinary, Animal and Biomedical Sciences, Massey University, Private Bag 11-222, Palmerston North 4442, New Zealand

Audrey Martinez
École Nationale Vétérinaire de Toulouse 23 Chemin des Capelles, BP 87614, 31076 Toulouse CEDEX 3 France

Ihab Erian and Clive J. C. Phillips
Centre for AnimalWelfare and Ethics, School of Veterinary Science, University of Queensland, Gatton, Queensland 4343, Australia

Hiroyuki Yamauchi and Nobuyo Ohtani
Department of Animal Science and Biotechnology, Azabu University Graduate School of Veterinary Science, 1-17-71 Fuchinobe, Chuo-ku, Sagamihara, Kanagawa 252-5201, Japan

Masashi Hayakawa and Tomokazu Asano
Hayakawa Institute of Seismo Electromagnetics Co. Ltd., UEC (University of Electro-Communications) Incubation Center, 1-5-1 Chofugaoka, Chofu, Tokyo 182-8585, Japan

Mitsuaki Ohta
Department of Human and Animal-Plant Relationships, Tokyo University of Agriculture, 1737 Funako, Atsugi, Kanagawa 243-0034, Japan

Rory Sullivan
Centre for Climate Change Economics and Policy, School of Earth and Environment, University of Leeds, Leeds LS2 9JT, UK

Nicky Amos
Nicky Amos CSR Services Ltd., Old Broyle Road, Chichester, West Sussex PO19 3PR, UK

Heleen A. van de Weerd
Cerebrus Associates Ltd., The White House, 2 Meadrow, Godalming, Surrey GU7 3HN, UK

Jan Broucek, Michal Uhrincat, Andrea Mrekajova and Anton Hanus
Research Institute of Animal Production Nitra, 951 41 Luzianky, Slovakia

Stefan Mihina
Faculty of Engineering, Slovak Agriculture University Nitra, 949 01 Nitra, Slovakia

Miloslav Soch
Faculty of Agriculture, University of South Bohemia Ceske Budejovice, 370 05 Ceske Budejovice, Czech Republic

Index